Basic Electronics Theory with Projects and Experiments—3rd Edition

Basic Electronics Theory
with Projects and
Experiments—3rd Edition

Delton T. Horn

TAB BOOKS
Blue Ridge Summit, PA

THIRD EDITION
FIFTH PRINTING

© 1989 by **TAB Books**.
Earlier editions © 1981 and 1985 by TAB Books.
TAB Books is a division of McGraw-Hill, Inc.

Library of Congress Cataloging-in-Publication Data

Horn, Delton T.
    Basic electronics theory / by Delton T. Horn.—3rd ed.
        p.   cm.
    Includes index.
    ISBN 0-8306-0395-6   ISBN 0-8306-3195-X (pbk.)
    1. Electronics. 2. Electronics—Experiments. I. Title.
TK7816.H69   1989
621.381—dc19   88-32312
                                            CIP

TAB Books offers software for sale. For information and a catalog, please contact
TAB Software Department, Blue Ridge Summit, PA 17294-0850.

Technical Editor: David Gauthier
Cover Design: Lori E. Schlosser

Contents

Introduction

Electronics is a very rapidly growing field. To try to keep up with new developments, and to expand the coverage of certain subjects, *Basic Electronics Theory—3rd Edition* is being released. Very little has been deleted from the first two editions of this book. However, quite a bit of new material has been added. As with the earlier editions, the goal of this volume is to serve as both an introductory text and as a reference volume.

The new material in this third volume includes information on superconductivity, expanded coverage of photosensitive devices and other sensors, motors, microprocessors, circuit diagrams, and Kirchhoff's Laws, among other topics.

Most of the chapters in this volume feature tests to help you master the material. The answers to these self-help tests appear in the Appendix.

I sincerely hope you find this third edition helpful and informative. I also hope you have fun learning about the exciting, multi-faceted field known as electronics.

1

What Is Electronics?

Today, in our increasingly high-tech world, almost anywhere you turn you're likely to encounter electronics products. Many items we take for granted would have been unthinkable outside the realm of science fiction 50 years ago. Some products that are fairly common would have been nearly impossible ten years ago.

A few examples are digital watches, pocket TVs, lasers, MIDI synthesizers, personal computers, CD players, and VCRs. This list is scarcely comprehensive or complete, there are hundreds of other things that could be included.

There is no denying that electronics is having an increasingly important impact on our lives and society. Since you are reading this book, it's reasonable to assume you are interested in this field, whether as a profession or as a hobby.

Just what is included in the field of electronics? As you might have guessed, it is a pretty large area, with many sub-categories. But all of these various sub-categories have much in common. In the simplest terms, electronics is the study, design, and use of electrical circuits using a variety of components to manipulate electrical signals in some way

ELECTRONIC COMPONENTS

While there are countless variations, there are just a few fundamental components. Most sophisticated, specialized devices are modified versions of some common component type. The basic component categories include:

resistors
capacitors
coils
semiconductor junctions (diodes)
crystals
switches
digital gates (actually these are made up of simpler components from the above list).

Each of these component types and their more common variations will be discussed in this book.

BASIC CIRCUITS

There may seem to be an infinite variety of electronics circuits. In a way, there are. But all complex circuits are really made up of smaller sub-circuits. Once again, there are just a few basic circuit types (with countless variations and modifications) that can be combined in a wide variety of ways. Basic circuit types include:

amplifiers
power supplies
switching circuits
oscillators
filters
timers
digital gates (these can be considered simple sub-circuits, or components, depending on the context. This will be explained later, when digital circuitry is discussed).

DON'T PANIC

While electronics can seem like a terribly complex field, the most complex devices, circuits, and concepts can usually be broken down into simpler sub-units. If a particular chapter seems hard to understand, just take it slow. Sometimes, coming back to a tough chapter after reading a later chapter will make some concept a little clearer.

To work with electronics theory, there is no getting around the need for mathematics; this is intimidating for many people. Don't panic, just take it one step at a time. I've made every attempt to make the math and theory in this book as painless as possible. Sometimes things are hard because we expect them to be. In high school I assumed that calculus was beyond me, and didn't take that class. Meanwhile, I was getting involved with electronics as a hobbyist. Some years later, when I started writing in the field, I realized with some surprise that I had been using calculus for some time. Without that intimidating name, it wasn't really so bad after all.

Some of the details of the electronics theory aren't too critical for practical work. If some details here and there seem fuzzy to you, don't worry about it, just try to get a general feel for what is going on. You could discover those troublesome concepts will make a lot more sense to you later, once you are more comfortable with the field.

The most important thing to remember is that you do *not* have to be a genius or a scientific whiz to understand electronics theory. I believe anyone can learn the basics if they just try and don't get scared off by the seeming complexity, and the unfamiliar terms and concepts.

Interestingly, many of the people who start out most intimidated by the field end up being the most enthusiastic.

Don't expect to learn everything all at once. You will probably have to go through this book two or three times before you have a firm grasp on electronics theory.

I can't over-emphasize the importance of doing the experiments described in the text. Hands-on experience is the best way of learning. It will make a lot more sense and stick with you a lot better, if you actually see the principles in operation than to just read about them.

2

Electrons and Electricity

To understand electricity and electronics, it is necessary to have at least a basic grasp of the theoretical structure of atoms.

ATOMS AND THEIR STRUCTURE

All substances are made up of tiny particles called *atoms*. There are approximately one hundred different kinds of atoms (ninety-two occur in nature, others are man-made). A substance that is made up entirely of just one type of atom is called an *element*. Copper, hydrogen, carbon, gold, and oxygen are a few familiar elements.

On the other hand, two or more elements can be chemically combined into a more complex *compound* substance. For example, water is made up of hydrogen and oxygen atoms. The smallest unit of matter that is recognizable as a compound, rather than as its component elements, is a *molecule*. If we tried to break a single molecule of water into smaller particles, we'd have two hydrogen atoms and one oxygen atom. Billions of different substances can be formed by various combinations of the basic elements, just as the twenty-six letters of the alphabet can be arranged into millions of different words.

While an atom is the smallest particle recognizable as a specific element, atoms themselves are not indivisible. They are made up of still smaller particles called *protons, neutrons,* and *electrons*. Recent discoveries have indicated the presence of a large number of additional sub-atomic particles, but for our purposes, they can be ignored. These three kinds of particles are all roughly of equal size, but protons and neutrons are considerably more massive than electrons.

If 250 million hydrogen atoms were laid end to end, they'd span only about one inch. It would take one hundred thousand electrons (or protons or neutrons) laid side by side to span the width of a single hydrogen atom. Atoms don't contain anywhere near that many particles. The hydrogen atom typically consists of only a single proton and a single electron. Most of the space of an atom is empty. The protons and neutrons are clumped together in the center, forming a structure called the *nucleus*. The electrons revolve around the nucleus. The basic structure of a typical atom (carbon) is usually drawn as in Fig. 2-1.

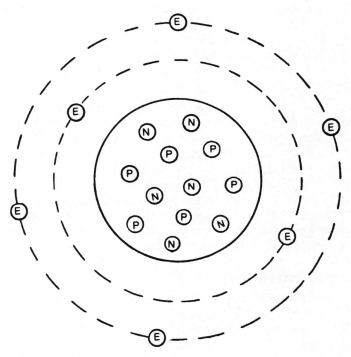

Fig. 2-1. Structure of an atom.

You'll notice that this arrangement is roughly similar to the solar system. The various planets revolve around the sun, like the electrons around the nucleus.

It is important to realize that a lead proton is exactly the same as, say, a gold proton. What differentiates between the elements is simply the number of these particles contained within the atom.

Electrical Charge

When an electron is isolated from an atom, it exhibits a tiny *electric charge*. The basic unit for measuring electric charge is the *coulomb*. The combined charge of 6,250,000,000,000,000,000 (6.25×10^{18}) electrons equals a charge of one coulomb.

There are actually two types of electrical charge. The type exhibited by an electron is arbitrarily called a *negative charge*.

A proton has the same amount of electrical charge as an electron, but it is the opposite type. In this case, it is referred to as a *positive charge*.

Two similarly charged particles (that is, two electrons, or two protons) will tend to repel each other. On the other hand, two oppositely charged particles (an electron and a proton) will tend to attract each other. This is one of the factors that keeps the electrons in their orbit around the oppositely charged protons in the nucleus.

Ordinarily, an atom has an equal number of electrons and protons, and therefore, the atom as a whole has no electrical charge. That is, it is electrically *neutral*. But in most cases an extra electron can be added to an atom, giving the atom, as a unit, a negative charge. Conversely, an electron can be deleted, leaving the atom, as a unit with a positive electrical charge.

Neutrons, which are contained within the nucleus, along with the protons, have no electrical charge. As their name implies, they are neutral.

Isotopes

The number of particles in an atom determines what kind of atom it is. For instance, an ordinary hydrogen atom (hydrogen is the simplest element) consists of a single proton, a single electron, and no neutrons. Sometimes, however, a hydrogen nucleus does contain one or even two neutrons (heavy hydrogen). In this case, it is still a hydrogen atom, but it will have a few different properties (the specifics of the differences are not relevant to our purposes here). This kind of atomic variation is called an *isotope*.

Just because an atom contains neutrons doesn't necessarily mean it is an isotope, of course. Many elements contain a number of neutrons in their basic form. For example, the nucleus of an ordinary lead atom contains eighty-two protons and one hundred and twenty-five neutrons. Its isotopes contain still more neutrons.

Atomic Number and Atomic Weight

Elements are often identified by their *atomic number*, which is simply the number of protons they contain. Each element has a unique atomic number. For example, hydrogen has an atomic number of one. Helium's atomic number is two, carbon is six and lead is eighty-two.

Atomic weight, on the other hand, is the total number of both the protons and the neutrons.

Ordinary hydrogen's atomic number and atomic weight are identical—one (this is the only element where this is true). However, a hydrogen isotope can have an atomic weight of two or three, but the atomic number is always one. Ordinary lead's atomic number is eighty-two, but its atomic weight is two hundred and seven. It is the atomic number that determines what kind of element the atom is.

Electrons and most other sub-atomic particles can easily be ignored in determining atomic weight, because they have little mass in comparison with protons and neutrons. They do not significantly affect the weight of the atom. They do have some mass, however, because they are matter, and all matter has mass. Remember, an electron is a physical unit, not a unit of energy. Electrons possess potential energy in the form of their negative charge, but they are physical objects. Often people get confused on this point.

Electron Rings

Look again at the diagram of the atom in Fig. 2-1. Notice that the electrons circle the atom in a number of fixed, concentric rings. These rings have a definite pattern in the maximum number of electrons each ring can contain.

The first ring, the one closest to the nucleus, can only hold one or two electrons. If the atom has three electrons, two will be in this innermost ring, and the third will be in the second ring, farther out. This second ring can hold up to eight electrons. The third ring can hold eighteen, the fourth thirty-two, the fifth fifty, and the sixth can contain seventy-two. No known atom has more than six rings. Usually the inner rings are completely filled before the outer rings are started, but this isn't invariable.

Obviously, the easiest electrons to remove, giving the atom a positive charge, are those in the outermost ring. They are the easiest to strike with an external force, and they are the farthest from the attraction of the positively charged protons that try to hold the electrons in place.

Conductors and Insulators

Certain substances will give up an electron (or accept an extra electron) more readily than others. Such substances (typically metals) are called *conductors*, because they can conduct electricity. That is, they allow an electric current to pass through them. This concept will be explained in greater depth in the next section of this chapter.

A substance which has strong internal attraction, and is thus resistant to releasing or accepting electrons is an *insulator*. Electric current can pass through an insulator, but it takes a far greater amount of force than for a conductor.

Any atom can be made to give up an electron. Conductors are simply those substances that will give up electrons without a great deal of external force.

ELECTRIC CURRENT

When an electron is knocked free from an atom, it drifts through space until it collides with a second atom which accepts it, and throws off one of its own original electrons. This electron then strikes a third atom, and so forth. Each individual electron doesn't travel very far, but the energy of electron movement can be transmitted along the length of the conductor.

A simplified model of this kind of process is shown in Fig. 2-2. A cardboard tube is filled with three ping pong balls. When an extra ball "X" is pushed into the tube, it displaces ball 1, which displaces ball 2, which shoves ball 3 out of the far end of the

Fig. 2-2. Model of current flow.

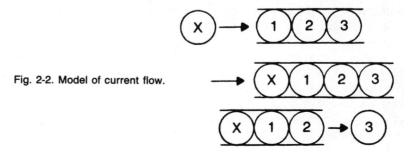

tube. This all takes place almost instantaneously—when "X" is pushed in, "3" is shoved out. Each ball moved very little, and fairly slowly, but the energy was quickly transmitted through the tube.

When this process takes place with electrons in a conductor it is called *electricity*, or *electron current*.

Another way to put it would be to consider current the effective flow of electrons (usually simplified as *electron flow*). If a coulomb (6.25×10^{18} electrons) flows past a given point within one second we say the current equals one *ampere*.

The ampere is the basic unit of measurement for electric current, but in many practical cases, it is too large a unit to be convenient. In these instances, it is simpler to use *milliamperes*, which equal one thousandth of an ampere, or *microamperes*, which equal one millionth of an ampere (one thousandth of a milliampere). For example, 50 milliamperes equals 0.05 amperes, or 50,000 microamperes. Generally we can use whichever term that gives the most manageable figures.

The word ampere is often abbreviated as *amp*, or simply the letter *A*. Similarly, milliampere can be written as *milliamp*, or *mA*. Microampere is usually abbreviated as *μA*.

In electrical equations current is usually represented by the letter *I*. I's value is generally assumed to be in amperes unless otherwise stated.

VOLTAGE

Since current specifies the number of electrons moving past a given point within a given time period, it can be considered the speed of the electron flow.

If we visualize electron flow as a toy car, we can draw a clear analogy to the basic elements of electricity. If we set the toy car on a perfectly flat table top, the speed (current) will be zero—there is nothing to make it go. But if we tilt up one end of the table, the car will start to roll down the hill. The steeper the slope, the faster it will go.

What is making the car move is the difference between the highest point and the lowest point of the slope. In electricity the "highest point" is a point with a surplus of electrons (negative charge), and the "lowest point" is an electron shortage (positive charge). Since like charges repel, and opposite charges attract, a stream of electrons will flow from the most negative point to the most positive point.

Remember, each individual electron doesn't move very far, but the disturbance travels through the entire *circuit*. A circuit is a complete path for current flow. If the path is broken at any point, no current will flow.

How strongly the current flows depends on the difference in charge between the most negative point and the most positive point of the circuit. This *difference of electrical potential* is also called *voltage* or *electromotive force*.

Voltage is measured in units called *volts*. One volt will push one ampere of current through one *ohm* of *resistance*. Resistance will be discussed in the next chapter.

If the volt is too large a unit, voltage can be measured in *millivolts* (mV). One thousand millivolts equals one volt. In other conditions, the volt is too small a unit. In these cases the *kilovolt* is used as the unit of measurement. One kilovolt equals 1000 volts. For example, 25 volts equals 25,000 millivolts, or 0.025 kilovolts. Similarly, 2,300 volts equals 2.3 kilovolts, or 2,300,000 millivolts.

In electrical equations, voltage is usually represented by the letter *E*. E is usually given in volts, unless otherwise stated.

POWER

If we want to determine how much work the circuit is doing, we need to consider both the voltage and the current. The total energy consumed is called *power*, and is measured in *watts*. One watt of power is consumed when one volt pushes one ampere through a circuit.

The relationship between power, voltage and current is stated in the following formula:

$$P = EI \qquad \text{Equation 2-1}$$

P is power, E is voltage, and I is current. So power in watts equals voltage in volts times current in amperes.

This formula can be rearranged if you know the wattage and the voltage and need to find the current. In this case:

$$I = P/E \qquad \text{Equation 2-2}$$

The meaning of the variables is the same as in Equation 2-1.

The final possibility is if you know the wattage and the current, but not the voltage. Here you can write the equation as:

$$E = P/I \qquad \text{Equation 2-3}$$

Again, the variables are the same as above. P is usually given in units of watts. Kilowatts or milliwatts can be used if they are more convenient. The conversion is the same as with volts to kilovolts or to millivolts.

BATTERIES

Now, how can we generate this electrical force to push a current through a circuit? We can convert another form of energy (such as mechanical, chemical, heat, or even light) into electrical energy. One of the most common, and simplest of these methods is chemically. Other methods of producing electricity will be discussed in later chapters.

Wet Cells

If a little hydrochloric acid is poured into a jar of water, the compound will start to break down chemically, producing negative and positive *ions*. An ion is simply an electrically charged particle. This process is called *ionization* and the acid/water mixture is called an *electrolyte*.

Now, if we put a strip of zinc and a strip of copper into the acid/water solution, and connect the two metal strips with some wire, an electrical current will start to flow between them. See Fig. 2-3.

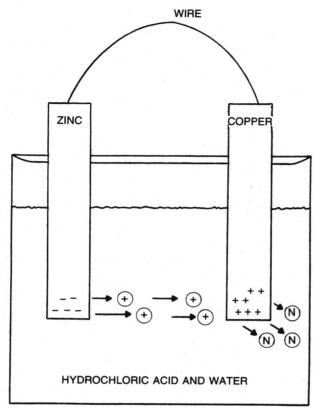

Fig. 2-3. A wet cell.

When the zinc strip is placed into such an acid/water solution, it starts to dissolve, emitting positive ions. But every positive ion that leaves the zinc strip, leaves behind two electrons, so the strip itself soon has a surplus of electrons. That is, it has a negative charge.

Meanwhile, as the positive ions float through the electrolyte, some of them collide with the copper strip. Copper gives up electrons quite easily (it's an excellent conductor), so the positive ions take electrons away from the copper strip and are thus neutralized. Since the copper strip now has a deficiency of electrons, it takes on a positive electrical charge.

The difference between the negative charge on the zinc strip and the positive charge on the copper strip creates an electrical potential, or voltage between the two strips, or *electrodes*. If the electrodes are connected with a conductive material (such as a piece of wire), an electric current will flow between them from negative to positive.

This device is known as an *electric cell*. Since the electric cell we have been describing has an electrolyte in liquid form, it is called a *wet cell*.

This type of cell will generate about one volt. You'll notice that no mention has been made about the size of the cell. This is because the size has no effect on the voltage produced. A gallon of electrolyte will produce no more voltage than a pint. However, a larger cell will allow more current to be drawn from it before it is used up. That is, it lasts longer, and can handle a heavier work load.

Different materials can be used for the electrolyte and electrodes. The voltage produced will depend on the specific materials used.

Wet cells have a number of obvious disadvantages. They have to be rather large to produce a useful amount of current. And, since the electrolyte is a liquid it can easily be spilled.

Also, bubbles can form a sheath around the positive electrode, preventing any further ions from striking it and being neutralized. This, of course, will stop some of the chemical action, and the voltage will quickly drop off to zero. This process is called *polarization*. There are methods of preventing this effect (or, at least, limiting it) but, naturally, they increase the cost and complexity of the cell.

Dry Cells

These problems can be largely avoided by using an electrolyte that is in a paste-like form. Such a cell is called a *dry cell*.

A typical dry cell uses a carbon rod for the positive electrode, and a pasty electrolyte consisting primarily of ammonia and chlorine. The cell is contained in a zinc can that also serves as the negative electrode. Obviously, such a cell is called a *carbon-zinc cell*.

The voltage put out by such a cell is about 1.45 to 1.55 volts when fresh—the voltage drops as the cell ages. It is said to have a nominal voltage of 1.5 volts. This is the figure usually used in circuit calculations for such cells.

Once again, the size of the cell affects the amount of current the cell can handle, but the voltage remains constant.

Table 2-1 shows typical current handling capabilities of the four most common cell sizes. Notice that current handling capability increases with the size of the cell.

The amount of current that is actually drawn from the cell is determined by the circuit that the battery is connected to (this will be explained in later chapters). These figures are maximum ratings. Exceeding the current ratings given in Table 2-1 could result in damage or premature failure of the cell.

The primary disadvantage of carbon-zinc cells is their relatively low current handling capabilities. If they are used to operate even a moderately heavy circuit, they will have a short operating life.

In addition, these cells have a limited *shelf life*. That is, they can go bad even if they aren't used.

Current is drawn from a battery only when an external circuit, or *load* is connected between its electrodes, or *terminals*. Obviously, if the terminals are not connected, there

**Table 2-1. Current Handling Capabilities
of Standard Dry Cell Sizes—Carbon-zinc.**

AAA	20 mA	1.5 volt
AA	25 mA	1.5 volt
C	80 mA	1.5 volt
D	150 mA	1.5 volt

is virtually no path for the current to flow between them (air is a very poor conductor). However, a certain amount of internal current leakage within the cell itself is inevitable, and this discharges (uses up) the cell at a slow, but definite rate, even if no external current is being drawn from it. Also, low-level chemical reactions are occurring constantly between the various substances within the cell, and this too can cause eventual deterioration.

Finally, while carbon-zinc cells are carefully sealed, no seal is 100 percent perfect. The moisture in the electrolyte can evaporate, causing the chemical activity of the cell to cease.

These processes are somewhat temperature related, so the way in which a cell is stored will affect its shelf life. High temperatures tend to speed up chemical reactions, but lower temperatures slow them down. Plainly, it's a good idea to store dry cells in a cool, dry place. However, they should *not* be frozen. If a carbon-zinc battery is frozen, it will have an extremely short life when thawed.

The best temperature for storing these cells is about 40 to 50° F. (or 4 to 10° C.). At these temperatures a carbon-zinc cell can have a shelf life of up to two or three years, as opposed to about six months when stored at room temperature. If these cells are kept refrigerated, it's usually a good idea to let them slowly warm up to room temperature before use. Some people believe that refrigerating an old dry cell will rejuvenate it, but this is not true.

The way these cells generate a voltage is by a chemical that gradually eats away the negative electrode. Clearly, if the electrode is completely destroyed, or even badly damaged, the cell is useless, because there is no way the electrode can be replaced.

However, if a small current is applied to a cell before it is completely dead with *reverse polarity* (that is, negative to positive and positive to negative), then some (not all) of the chemical action within the cell can be partially reversed. This procedure can extend the life of a dry cell somewhat. It won't be as good as new, because much of the chemical activity is irreversible, and the negative electrode will only be restored unevenly. Also, the process will work only a limited number of times for any individual cell.

This process is called *charging*, or, more correctly, *recharging* the cell. The recharging current must be kept very small, otherwise, the reverse polarity will destroy the cell. Ordinarily you must always avoid interconnecting two voltage sources with their polarities reversed.

The low recharging current limitation means the recharging process must take about twelve to sixteen hours. It is also important to remove the recharging current when the cell is fully charged.

For the recharging process to be at all effective, the cell's voltage mustn't be allowed to drop below one volt (two thirds of the nominal value), and it must be recharged immediately after it is taken out of service. Once the cell is recharged it should be used as soon as possible because the shelf life of a recharged cell is quite short.

There are a number of commercially manufactured recharger units available today. They can represent a substantial savings to the user if the inherent limitations are kept in mind.

Alkaline Cells

By using an alkaline material instead of an acid as the electrolyte, a greater current handling capability can be achieved. Table 2-2 shows the current handling capability for the four most popular cell sizes of *alkaline cells*.

Alkaline cells come in exactly the same sizes as regular carbon-zinc cells, and also have a nominal voltage (when fresh) of 1.5 volts. Alkaline cells are directly interchangeable with their carbon-zinc counterparts.

Alkaline cells are usually more expensive than carbon-zinc cells, but their greater current handling capability means that they'll last longer in high current drain applications. A single alkaline cell may last as long as five or six carbon-zinc cells, so they often come out less expensive in the long run.

However, in very low current drain applications alkaline cells may offer no real advantage, because of a relatively short shelf life.

One major disadvantage of alkaline cells is that they *cannot* be recharged. Applying a reverse polarity current to an alkaline cell can damage the cell, or even cause it to explode.

Nickel-Cadmium Cells

Another type of electric dry cell that is becoming increasingly more popular these days is the *nickel-cadmium cell* (often shortened to *Ni-Cad*). These cells are specifically designed to handle moderately large current drains, and to be fully rechargeable. A typical Ni-Cad can generally be recharged five-hundred to a thousand times before it finally fails.

Usually, when a Ni-Cad does fail it is because of an internal short between the electrodes. Often the cause of this is allowing the cell voltage to drop down too low before recharging. A Ni-Cad's voltage should never be allowed to drop below 1.05 volts.

When a short does develop in a Ni-Cad cell, it can sometimes be "blown" away by *briefly* applying a moderately large, reverse polarity current. However, this doesn't always work.

The recharging requirements for Ni-Cads made by different manufacturers vary widely. Some are designed to be recharged in just three or four hours. Most require about fourteen hours for a full charge.

Most Ni-Cads are designed so that no damage is done if the charging current is continually applied even after the cell is fully charged. These cells, unlike carbon-zinc cells, can be left in the recharging unit for extended periods of time. Another advantage of Ni-Cads is that they are not as temperature sensitive as other types of dry cells.

Nickel-cadmium cells cost four to eight times as much as most carbon-zinc cells, but since they can be reliably recharged and reused so many times, they are considerably less expensive over time.

**Table 2-2. Current Handling
Capabilities of Standard Dry Cell Sizes—Alkaline.**

AAA	200 mA	1.5 volt
AA	300 mA	1.5 volt
C	500 mA	1.5 volt
D	500 mA	1.5 volt

One disadvantage of this type cell is that they only put out 1.25 volts when fully charged, as opposed to the 1.5 volts produced by a fresh carbon-zinc or alkaline cell. In many applications this won't matter, but there are many applications where that missing quarter volt can make a significant difference. Fortunately, this can be overcome by using multiple cells. This problem and its solution are the subject of the next section.

Series and Parallel Batteries

The cells discussed on the last few pages are often incorrectly called *batteries*. A true battery consists of two or more interconnected cells. These cells can be wired together either in *series* or in *parallel*.

If we need more voltage than is generated by a single dry cell, we can connect two or more cells in series. That is, one after another. This is illustrated in Fig. 2-4. The negative electrode, or *terminal* of cell A is connected to the positive terminal of cell B, and the negative terminal of cell B is connected to the positive terminal of battery C. The load circuit is connected between the positive terminal of cell A and the negative terminal of cell C. The current must therefore pass through each of this series of cells sequentially. Their voltages will add. In our examples three 1.5 volt cells are connected in a series battery, so the total voltage will be 3 × 1.5 or 4.5 volts.

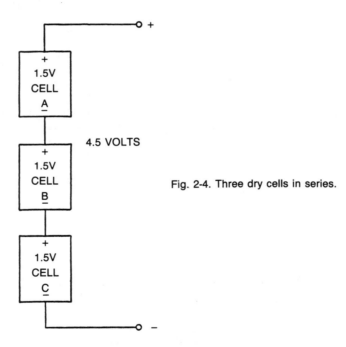

Fig. 2-4. Three dry cells in series.

Similarly, two 1.5 volt cells would produce three volts, eight cells would generate a total of twelve volts, and so forth. The current handling capability of the entire battery is the same as that for a single cell.

Here is where the quarter volt difference of nickel-cadmium cells can become significant. Eight regular carbon-zinc cells would add up to twelve volts, but eight Ni-

Cads would generate only ten volts. It would take ten Ni-Cads to produce a voltage of 12.5 volts, which would usually be close enough.

In most applications more nickel-cadmium cells are needed than carbon-zinc or alkaline cells. In some pieces of equipment there is no space for the extra cells, which usually means nickel-cadmium cells can't be used. This is the case, for example, with most portable cassette tape recorders.

A lot of modern electronic equipment makes allowances for this difference, however. The battery compartment is made large enough to hold the required number of nickel-cadmium cells, and a spacer is provided to fill the extra space if higher voltage carbon-zinc or alkaline batteries are used.

Now, let's look at another possible situation. Suppose your dry cell has enough voltage, but won't handle the required current. Remember that the amount of current drawn from a cell (or battery) is determined by the demands of the load circuit it is powering, but there are definite limits as to how much current a cell can supply. Too high a current drain can prematurely age or destroy a cell. Therefore, it is sometimes necessary to form a battery that increases the current handling capability, rather than the voltage.

Such a battery, shown in Fig. 2-5, is called a *parallel battery*. The positive terminals are all connected together, as are all the negative terminals. The cells can thus work together, each cell providing only part of the current drawn by the load circuit. In a three cell battery, such as the one shown, each cell provides only one third of the total current.

The total voltage of a parallel battery is the same as for a single cell, but the current handling capability is multiplied by the number of cells in the battery. For example, three size C batteries in parallel can handle up to 240 milliamps if they are the carbon-zinc type, or 1,500 milliamps (1.5 amps) if alkaline cells are used. Cell types should not be mixed.

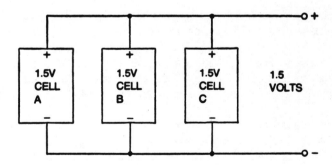

Fig. 2-5. Three dry cells connected in parallel.

In actual practice the parallel battery is rarely used. Most battery operated equipment is designed for low current drain. If a high current drain is necessary, some other type of power source will be used. Other power sources will be discussed in later chapters. If both higher voltage and higher current are needed, both series and parallel techniques can be used simultaneously.

SCHEMATIC SYMBOLS

In electronic diagrams it is generally unnecessary and inefficient to actually draw each of the components, especially in complex circuits. Instead, a form of visual shorthand is used. Drawings using this system are called *schematic diagrams*.

Each kind of electronic component is represented by a specific symbol. In some cases different symbols are used by different technicians, but usually the differences are minor. In this book, as each type of component is discussed, its schematic symbol (or symbols) will also be introduced. Schematic diagrams will be discussed in detail in Chapter 14.

The standard schematic symbol for an electrical cell is shown in Fig. 2-6A, and the symbol for a battery is shown in Fig. 2-6B. Notice that the symbol for a battery is just two (sometimes three) cells together. Usually only two or three cells are shown in a battery on the schematic, regardless of how many cells are actually employed in the circuit. This convention is simply one of convenience. It would be quite awkward to show each individual cell in a 22.5 volt or 45 volt battery.

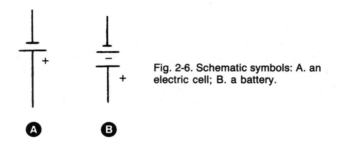

Fig. 2-6. Schematic symbols: A. an electric cell; B. a battery.

It should also be pointed out that the symbol in Fig. 2-6B is generally used only for series batteries. If parallel batteries are used the individual cells are indicated.

In some schematics the "+" sign is not shown at the positive terminal, because it isn't really necessary. The longer line always represents the positive terminal, whether it is marked as such or not. The "+" sign simply makes the polarity easier to see at a glance. Similarly a "−" sign may or may not be included at the negative terminal (short line).

Interconnecting wires are generally shown in schematic diagrams as solid lines. These lines never curve, but always bend at sharp angles (usually 90°).

Some confusion can arise when two lines on a schematic diagram cross each other. Sometimes it means they are electrically connected, other times, the wires simply pass by each other with no electrical contact. Figure 2-7 shows three common systems for showing crossed wires in schematic diagrams. Notice that the no connection symbol in Fig. 2-7A is the same as the connection symbol in Fig. 2-7B. To avoid ambiguity, the symbols in Fig. 2-7C are preferred. Always make sure you know which convention is being used whenever you look at a schematic diagram.

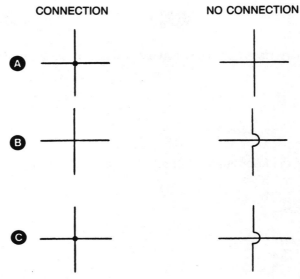

Fig. 2-7. Schematic symbols for crossed wires.

Self-Test

1. Which of the following is *not* a basic part of an atom?

A *Coulomb*
B *Electron*
C *Proton*
D *Neutron*

2. Electricity consists of which of the following flowing through a conductor?

A *Protons*
B *Isotopes*
C *Molecules*
D *Electrons*

3. The basic unit for measuring current flow is which of the following?

A *Atomic Weight*
B *Coulomb*
C *Volt*
D *Ampere*

4. The basic unit for measuring electrical power is which of the following?

A *Watt*
B *Volt*

C *Isotope*
D *Electrolyte*

5. An ordinary size D flashlight battery is which of the following?

A *Load*
B *Dry Cell*
C *Storage Cell*
D *Wet Cell*

6. Which of the following pairs of metals would make good electrodes in a wet cell battery?

A *Zinc and Copper*
B *Zinc and Iron*
C *Iron and Copper*
D *Two Strips of Zinc*

7. Connecting battery cells in series does which of the following?

A *Increases Current*
B *Decreases Current*
C *Decreases Voltage*
D *Increases Voltage*

8. Connecting battery cells in parallel does which of the following?

A *Increases Voltage*
B *Decreases Voltage*
C *Increases Current*
D *Decreases Current*

9. If the voltage applied to a circuit is 12 volts and the current flow equals 3 amperes, what is the wattage consumed by the circuit?

A *36 Watts*
B *4 Watts*
C *0.25 Watts*
D *0.4 Watts*

10. If 15 volts is applied to a circuit which consumes 75 watts, what is the current flow through the circuit?

A *0.2 Amperes*
B *11.25 Amperes*
C *6 Amperes*
D *5 Amperes*

3

Resistance and Ohm's Law

What determines how much current a load circuit will draw from a battery, or other voltage source? Let's return to our analogy of the toy car. It might seem that if we knew the slope of the table, we'd immediately know how fast the toy car would roll down the incline, but other factors also affect the speed. The most important of these factors is friction. The car will roll down a smoothly polished board much faster than it will roll down a coarse, unsanded one. Friction will slow down the car—that is, it will impede or resist the car's movement.

The electrical equivalent to friction is *resistance* (represented by the letter R). Resistance impedes, or works against the flow of current. You might think that resistance is something that should always be avoided as much as possible, but it's actually quite a useful factor in practical circuits.

Remember that the higher the current drawn from a battery or cell, the faster the battery or cell will be discharged. Resistance limits the amount of current drawn. It can also, as we'll see later, reduce the voltage in certain portions of a circuit.

OHM'S LAW

The basic unit of resistance is the *ohm*, which is sometimes written as Ω (the Greek letter omega). One volt can push one ampere of current through one ohm of resistance. The relationship between these three factors is perhaps the most important concept in electronics. This relationship is defined by a principle called *Ohm's law*. According to this law, voltage equals current times resistance, or;

$$E = IR \qquad \qquad \textbf{Equation 3-1}$$

E is the voltage in volts, I is the current in amperes, and R is the resistance in ohms. E will also be in volts if the current is in milliamperes and the resistance in *kilohms* (see below).

With a little simple algebraic manipulation, we can rearrange the equation to solve for current when voltage and resistance are known;

$$I = E/R \qquad \qquad \textbf{Equation 3-2}$$

Or, solving for resistance with known voltage and current:

$$R = E/I \qquad \qquad \textbf{Equation 3-3}$$

You'll recall that in the last chapter, we showed that power in watts equals voltage times current (P = EI). We can combine this equation with Ohm's law to find power consumed when only the resistance and the current are known:

$$
\begin{aligned}
&\phantom{\text{and so}} && P = EI \\
&\text{and} && E = IR \\
&\text{so} && P = (IR)I \text{ or} \\
&\phantom{\text{and so}} && P = I^2R
\end{aligned}
$$

$$\textbf{Equation 3-4}$$

Similarly, if we know the resistance and the voltage:

$$
\begin{aligned}
&\phantom{\text{and so}} && P = EI \\
&\text{and} && I = E/R \\
&\text{so} && P = E(E/R) \text{ or} \\
&\phantom{\text{and so}} && P = E^2/R
\end{aligned}
$$

$$\textbf{Equation 3-5}$$

These equations are quite versatile, and they are absolutely essential to your understanding of electronics.

In most practical circuits the ohm is too small a unit, so the *kilohm* (one thousand ohms) and the *megohm* (one million ohms) are often used. Kilohms is often shortened to the letter *k*, and megohms can be abbreviated as *meg*, or simply *M*.

RESISTORS

A *resistor* is an electronic component that is designed to introduce a specific amount of resistance into a circuit. Resistors are probably the most commonly used class of components in electronics.

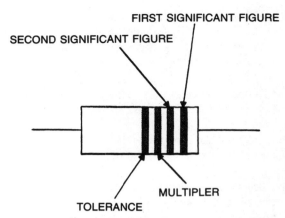

Fig. 3-1. Color coding bands on a typical resistor.

A typical resistor is shown in Fig. 3-1. Different sized resistors can handle different amounts of power, or wattage. The action of a resistor causes it to heat up. That is, it converts electrical energy into thermal energy (heat). If a resistor gets too hot it can change value, or become damaged. To prevent this problem, always use a resistor that will handle (or dissipate) the required amount of power. If in doubt, use a larger resistor. A 2 watt resistor, for example, will work fine in a ½ watt circuit.

Unfortunately, large wattage resistors tend to be more expensive and take up quite a bit of space. So generally, the smallest resistor that will comfortably handle the required wattage will be used. For most electronic circuits ½ watt resistors are more than sufficient. In circuits built around *integrated circuits* (discussed in a later chapter) ¼ watt units are commonly used.

Color Coding Resistors

Look again at Fig. 3-1. Notice that there are four bands around the body of each resistor. These bands are color coded, and are used to identify the value of the resistor. The standard resistor color code is given in Table 3-1. Most people who work with

Table 3-1. The Standard Resistor Color Code.

Color	BAND 1	BAND 2	BAND 3	BAND 4
Black	0	0	1	--
Brown	1	1	10	1%
Red	2	2	100	2%
Orange	3	3	1000	3%
Yellow	4	4	10000	4%
Green	5	5	100,000	--
Blue	6	6	1,000,000	--
Violet	7	7	10,000,000	--
Gray	8	8	100,000,000	--
White	9	9	--	--
Gold	--	--	0.1	5%
Silver	--	--	0.01	10%
No Color	--	--	--	20%

electronics use this table so frequently they automatically memorize it. It might look a little complicated at first, but it's really quite simple, once you get used to it.

Suppose you have a resistor with the following markings. The band closest to the end of the resistor (band 1) is red, the next band (band 2) is violet, band 3 is orange and the last band (band 4) is silver. What is the value of the resistor?

Since the first band is red, the chart tells us the first *significant digit* is 2. Similarly, the second band, being violet, indicates the second significant digit is 7. So we now have a value of 27.

The next band (band 3) is the *multiplier*. It is orange on our sample resistor, so we multiply the significant figures by 1,000. So the resistor has a total value of 27,000 ohms, or 27 kilohms.

The fourth band tells us the tolerance of the resistor. It is very difficult to manufacture a resistor to an exact value, and it usually isn't necessary. The actual value of a resistor may be somewhat above or below the nominal value indicated by the first three color bands.

In our example, the tolerance band is silver, which means the actual value of the resistor is within plus or minus 10% of the marked value (i.e., 27,000 ohms). That is, this particular resistor may actually be anywhere between 24,300 ohms and 29,700 ohms. For most applications this is accurate enough.

In fact, sometimes only 20% accuracy is needed. In this case, there is no fourth band at all. Assuming the resistor had a nominal value of 27,000 ohms, it could actually be anywhere from 21,600 to 32,400 ohms.

In other applications, greater accuracy is sometimes required. If the fourth band of a resistor is gold, that means the device has a 5% tolerance. Again, if the nominal value is 27,000 ohms, the actual value could range from 25,650 to 28,350 ohms.

In some very specialized applications, great precision is sometimes needed. In these cases 1% (often called *precision*) resistors are used. That is, a nominal value of 1% would require an actual value between 26,730 and 27,270 ohms. The nominal value of a precision resistor is usually printed right on the body of the device, and the color code is not used.

Most ½ watt resistors have a tolerance of 10%, although recently, the trend has been swinging towards more and more 5% units. Not surprisingly, the better the accuracy, the more the resistor is going to cost to manufacture, so if great precision is not required, it generally makes sense to use a wide tolerance resistor.

Remember, that even a 20% tolerance resistor could be exactly its nominal value, but the manufacturer only guarantees that it will be somewhere between plus or minus 20% of the indicated value.

Here are some additional examples of using the color code.

If the resistor's bands are brown, black, green, gold, the value of 1 (brown) 0 (black) × 100,000 (green), plus or minus 5% (gold), or a nominal value of 1,000,000 ohms (one megohm).

Another resistor might be marked blue, grey, red, with no fourth band. This translates to 6 (blue) 8 (grey) × 100 (red) or 6800 ohms (6.8 kilohms). The tolerance of this resistor is 20%.

As you grow accustomed to the color code, it will become second nature to you, and you'll be able to read a resistor's value directly without even thinking about it.

Here's a couple more examples for you to work on your own.

If a resistor is marked yellow, violet, yellow, silver, and its actual value is 450,000 ohms, is it within tolerance?

If a resistor is marked red, red, red, gold, and its actual value is 2000 ohms, is it within tolerance?

Types of Fixed Resistors

If only a few ohms of resistance is needed in a circuit, the resistor can simply be a piece of *nichrome* wire of suitable width and length. Nichrome wire has a much greater resistance than standard copper wire (all conductors have some resistance) so small *wirewound resistors* can be made without the length being too unreasonable. This nichrome wire (sometimes called resistance wire) is usually wound around a ceramic core, and covered with some insulating material.

Usually the resistances needed in practical electronic circuits are too large for reasonable wirewound resistors, so *composition resistors* are more commonly used. These resistors are usually made of a thin coating of carbon on a ceramic tube. Carbon is only a fairly poor conductor, so a fairly large resistance can be achieved in a relatively small space.

The upper limit for such a *carbon resistor* is generally around 10 megohms. Of course, the resistor itself is covered with an insulating body. The color coding bands are painted on the outside of the insulation.

Another common type of resistor uses a thin metallic film instead of carbon. *Metal-film resistors* can usually be made to more precise values than the carbon composition type. Metal film resistors are also less sensitive to temperature fluctuations (carbon resistors can sometimes change value at temperature extremes) and produce less internal noise (random and undesirable voltages and power fluctuations).

All of these devices are called *fixed resistors* because, unless they are in some way damaged, their value is more or less constant (all resistors change value somewhat in response to temperature fluctuations). The schematic symbol for a fixed resistor is shown in Fig. 3-2.

Fig. 3-2. Schematic symbol for a fixed resistor.

Variable Resistors

It is often necessary to be able to alter the amount of resistance in a circuit. In these cases a *variable resistor* is used. Variable resistors are usually called *rheostats* or *potentiometers* (often shortened to *pots*). These terms are more or less interchangeable, but generally rheostat is used to identify a device that is suitable for heavy-duty ac circuits (see Chapter 5), while potentiometers are generally used in relatively low power circuits.

Also, potentiometers usually have three terminals. The two outside terminals act like a simple fixed resistance—the resistance between these two terminals does not change. The center terminal, however, is attached to a slider, which is controlled by a knob. The slider is moved along the resistance element, which is either wound wire, or a strip of carbon. Depending on the slider's position, the resistance between it and either of the outside terminals will vary. See Fig. 3-3. Notice that the total of resistance

24

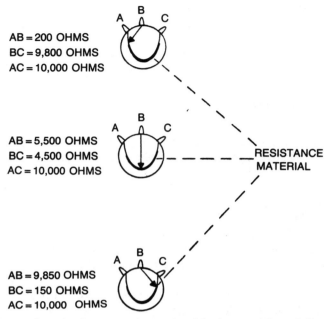

AB = 200 OHMS
BC = 9,800 OHMS
AC = 10,000 OHMS

AB = 5,500 OHMS
BC = 4,500 OHMS
AC = 10,000 OHMS

RESISTANCE
MATERIAL

AB = 9,850 OHMS
BC = 150 OHMS
AC = 10,000 OHMS

Fig. 3-3. Resistance of a potentiometer varies as the slider is moved through its range.

AB plus resistance BC always equals the constant resistance AC. As resistance AB increases, resistance BC decreases, and vice versa.

A variation on the standard potentiometer is the *slide pot*. It works in exactly the same way as a regular potentiometer, except the slider moves in a straight line rather than in a circular motion. The only advantage of using a slide pot is that in certain applications, it is easier to see where the slider is positioned.

The schematic symbol for a potentiometer is shown in Fig. 3-4. The symbol is the same whether a slide pot, or a standard round pot is used.

Fig. 3-4. Schematic symbol for a
three terminal variable resistor.

A rheostat, on the other hand, is often a two terminal device. That is, there is one fixed terminal and the moveable terminal (slider). The second fixed terminal is simply left off. Figure 3-5 shows the most common schematic symbol for a two terminal variable resistor. Alternatively, the standard symbol for a potentiometer can be used with one of the outer (fixed) terminals left disconnected.

Fig. 3-5. Schematic symbol for a two
terminal variable resistor.

Potentiometers and rheostats are generally panel controls. The user rotates a knob (which is connected to the shaft of the variable resistor) that in some way alters the operation of the circuit. For example, in an audio amplifier, potentiometers might be used to adjust the volume and the tone of the sound.

Sometimes, however, a variable resistor is needed to fine tune a circuit, perhaps to compensate for component tolerances that could throw the precise operation of the circuit off. In cases like this you want a variable resistor you can set and forget. *Trimpots* or *trimmers* are used for this kind of function.

Trimpots are simply miniature potentiometers with sliders that are positioned with a screwdriver, rather than with an external knob. Sometimes, when the correct setting is found, a tiny drop of paint or glue is placed on the screwdriver slot to prevent the slider from being moved out of position accidentally. In other applications the trimmer may need to be readjusted periodically (perhaps to compensate for changing values as components age), but not often enough to warrant a more expensive and space consuming front panel control. Also, there are often critical adjustments that should only be made with special test equipment, so they are kept inaccessible to the casual user to prevent misadjustment.

The *taper* of a potentiometer refers to the way in which the resistance changes in relation to the position of its slider. The two most popular tapers are shown graphically in Fig. 3-6.

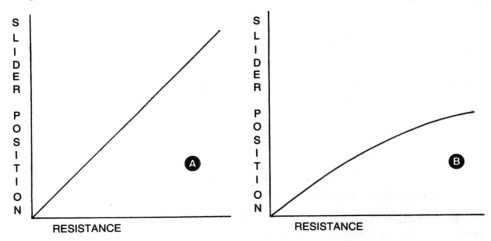

Fig. 3-6. Graphs of typical potentiometer tapers: A. linear taper; B. logarithmic taper.

The *linear taper* in Fig. 3-6A varies the resistance in direct proportion to the position of the slider. You can see that the graph is simply a straight line.

The *logarithmic taper* (Fig. 3-6B), on the other hand, has a more complex relationship between its resistance and its slider position. This relationship is based on the mathematical function of logarithms.

Which type of taper should be used depends on the nature of the specific circuit. For example, volume controls usually have a logarithmic taper because the ear hears in a logarithmic manner. If a linear taper potentiometer is used, most of the apparent range of loudness will be crowded into a relatively small portion of the knob's path of rotation. With a logarithmic taper potentiometer, the position of the knob will relate more closely to the perceived volume level.

COMBINATIONS OF RESISTORS

Practical electronic circuits consist of just a single resistance, so we need a way of determining the total value of multiple resistances in various combinations.

Series Resistances

Figure 3-7 shows a simple circuit with two resistors in series. Let's assume the battery generates three volts. R1 is 100 ohms, and R2 is 200 ohms. How do we find the current? Since the total current has to flow through both resistors, it must have the same value for each resistor. The current has to flow through 100 ohms, then it has to flow through an additional 200 ohms more. As you might have guessed, this appears to the current as a single 300 ohm resistor. Resistances in series add. Stated algebraically, the formula for resistance in series is:

$$R_T = R_1 + R_2 \ldots + R_n \qquad \textbf{Equation 3-6}$$

where R_T is the total resistance. The letter n represents the total number of resistances in the circuit.

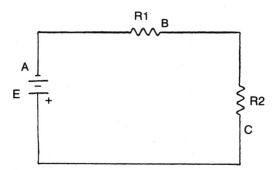

Fig. 3-7. A simple series resistance circuit.

In our example we have just two resistances. $R_T = R_1 + R_2 = 100 + 200 = 300$ ohms. Now that we know the total effective resistance in the circuit, we can use Ohm's law to find the current.

We know that I = E/R (current equals voltage divided by resistance), so the current in this circuit equals 3 volts/300 ohms, or 0.01 ampere (10 mA).

Now, what is the voltage dropped by each resistor?

Since E = IR (voltage equals current times resistance), the voltage through R1 must equal 0.01 A × 100 ohms, or one volt. Similarly, the voltage through R2 equals 0.01 A × 200 ohms, or two volts.

Notice that adding the voltages dropped across each of the resistors will give you the original source voltage. We can say that all of the voltage is used up by the resistances. This will be true of *all* circuits.

As the current passes through each resistor, the resistance causes the voltage to drop. At point A we have the full source voltage, or 3 volts. At point B, R1 has dropped one volt, so there is 3 − 1, or 2 volts. R2 drops 2 volts, so at point C the voltage is 0. The source voltage is used up.

Finally, we can calculate the total power consumed by the circuit. You'll recall that the formula is $P = EI$ (power or wattage equals voltage times current). In our example we have 3 volts × 0.01 ampere, or 0.03 watts (30 milliwatts).

Let's try another example, and use a slightly different method of solving it. We'll still be using the circuit shown in Fig. 3-7, but this time the battery generates 12 volts, R1 is 1000 ohms, and R2 is 150 ohms. The total resistance in the circuit is 1000 + 150, or 1,150 ohms.

Since we know that the total power consumed by a circuit can be calculated with the formula $P = E^2/R$, we can insert our known values. $P = (12)^2/1150 = 144/1150$ = approximately 0.125 watts (125 milliwatts).

Solving for current we can rearrange $P = EI$ to $I = P/E$, or 0.125/12 equals just over 0.01 ampere (about 10 mA).

The voltage drop across R1 is found by Ohm's law. $E = IR = 0.01$ ampere × 1000 ohms = 10 volts.

The voltage drop across R2 equals 0.01 ampere × 150 ohms, or about 1.5 volts.

You'll notice that the calculated voltage drop is 10 volts + 1.5 volts, or 11.5 volts, rather than the source voltage of 12 volts. What happened to that extra half volt? Actually, nothing. We lost in our calculations because of rounding off. The wattage consumed is actually 0.1252174 watts, but rounding this figure off to 0.125 watts made the rest of the calculations simpler. There is nothing wrong with rounding off the results of these equations, and usually the calculated values will be close enough. But when a discrepancy does show up, you may find it necessary to go back and work with the exact values.

Here's one last example that you can try solving for yourself. E = 30 volts, R1 = 250,000 ohms, and R2 = 50,000 ohms. You can use whichever method you prefer to find the current, the wattage, and the voltage drop across each resistor.

Parallel Resistances

Now, what if we have a circuit like the one shown in Fig. 3-8? In this case, the electron flow drawn from the battery (the current) is split up between the two resistors. There are two *parallel* paths for the current to follow. Some of it will flow through R1, and some will flow through R2.

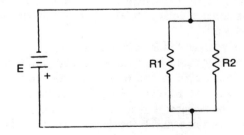

Fig. 3-8. A simple parallel resistance circuit.

Naturally, more current will flow through the path with less resistance. If both resistors are of equal value, equal currents will flow through them. As far as the voltage is concerned, a parallel circuit looks like two separate circuits, as in Fig. 3-9. The full source voltage is dropped across each resistor.

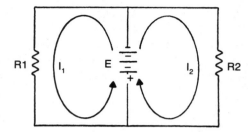

Fig. 3-9. Parallel resistance as it appears to the voltage source.

Suppose the circuit is powered by a six volt battery. R1 is 1000 ohms, and R2 is 3000 ohms. We already know that the full source voltage (6 volts) will be dropped across each resistor, so we can use Ohm's law and solve for the currents separately.

For R1, I = E/R = 6 volts/1000 ohms = 0.006 amperes (6 mA).

The current through R2 is 6 volts/3000 ohms, or 0.002 ampere (2 mA). R1 draws 6 mA from the battery, and R2 draws an additional 2 mA, so the total current drawn by the parallel circuit is 8 mA.

Solving for the equivalent resistance of the entire circuit, we can use the formula: R = E/I.6 volts/0.008 amperes = 750 ohms. Notice that this total equivalent resistance is less than the value of the smallest resistor. This will always be true in a parallel circuit. If the two resistors are equal, the equivalent resistance will be exactly one half their individual value.

Another formula for determining the equivalent resistance of n resistors in a parallel circuit is;

$$\frac{1}{R1} + \frac{1}{R2} \cdots + \frac{1}{Rn} = \frac{1}{R_T}$$

Equation 3-7

R_T is the total effective resistance of the parallel circuit.

Using the example above, we find 1/1000 + 1/3000 = 0.001 + 0.00033333 = 0.00133333. Taking the reciprocal (1/0.00133333) we get 750 ohms. We get the same answer, no matter which method is used.

The power consumed by the circuit is solved in the usual way. That is, P = EI. In this circuit 6 volts × 0.008 amperes = 0.048 watts. This can be rounded off to about 50 milliwatts.

As you can tell from Equation 3-7, any number of resistances can be combined in parallel. For example, imagine a circuit with four resistors in parallel. Their values are 1000 ohms, 2200 ohms, 6800 ohms and 10,000 ohms. So we have 1/1000 + 1/2200 + 1/6800 + 1/10,000 = 0.001 + 0.0004545 + 0.0001471 + 0.0001 = 0.0017016. Taking the reciprocal to find the total effective resistance, we get just under 590 ohms. Notice that this equivalent value is lower than the individual value of any of the separate resistors.

Here is another example for you to work on your own. Assume the circuit in Fig. 3-8 is powered by a 12 volt battery. R1 is 470 ohms, and R2 is 100 ohms. Find the equivalent resistance, the current drawn by each resistor, and the total power consumed by the circuit.

Series-Parallel Combinations

In actual practice, you'll rarely come across a circuit with just series resistances or just parallel resistances. Usually you'll find a combination of the two forms.

Take a look at Fig. 3-10. Here we have a circuit with four resistors both in series and in parallel. It might look complicated to solve for such a combination circuit, but it's easy enough if you go one step at a time.

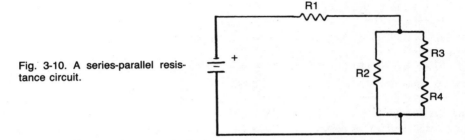

Fig. 3-10. A series-parallel resistance circuit.

Let's say our source voltage is 15 volts. R1 equals 1500 ohms, R2 equals 2200 ohms, R3 equals 470 ohms and R4 is 1000 ohms.

First you should find the equivalent resistance of the R3-R4 combination. Since these two resistances are in series, their values simply add. That is, 470 + 1000 ohms = 1,470. For simplicity, we can consider this combination as a single resistor, RA. See Fig. 3-11.

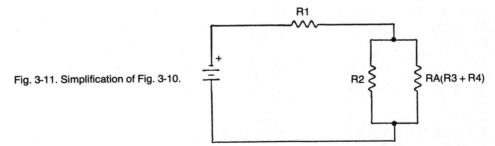

Fig. 3-11. Simplification of Fig. 3-10.

Now solve for the R2 - RA parallel combination. 1/2200 + 1/1470 = 0.0004545 + 0.0006803 = 0.0011348. So the equivalent resistance is just over 880 ohms. We'll call this value RB. See Fig. 3-12.

Now, we just have R1 in series with RB. R1 + RB = 1500 + 880 = 2380 ohms. This, then is the total equivalent resistance for the entire circuit. Solving for the total circuit current, we find that since I = E/R, the current equals 15 volts/2380 ohms, or about 0.0063 amperes (6.3 mA).

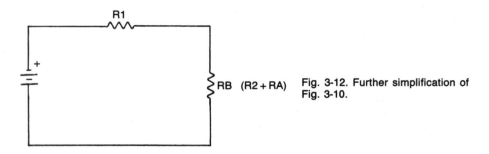

Fig. 3-12. Further simplification of Fig. 3-10.

The full current flows through R1, since there is no other path for it to take to bypass this resistor. We can solve for the voltage drop across R1. $E = IR = 0.0063$ amperes $\times$ 1500 ohms = 9.45 volts.

Looking at the RB combination as a single resistor, we find the total voltage drop is about 0.0063×880, or 5.55 volts (9.45 volts across R1, and 5.55 volts across RB equals the source voltage—15 volts).

Since RB consists of R2 and RA in parallel, we know the voltage dropped across each of these resistances is equal. Specifically, 5.55 volts. The current through R2 equals 5.55 volts/2200 ohms, or approximately 0.0025 ampere (2.5 mA). The current through the RA combination is 5.55 volt/1470 ohms, or about 0.0038 ampere (3.8 mA). RA is actually R3 and R4 in series. Since these two resistors are in series, they pass the same current—3.8 mA. The voltage drop through R3 is 0.0038 ampere $\times$ 470 ohms, or 1.77 volts. Across R4 it is 0.0038 ampere $\times$ 1000 ohms, or 3.78 volts. Notice that 3.78 volts + 1.77 volts equals 5.55 volts.

Further, the current through R2 (2.5 mA) plus the current through RA (3.8 mA) equals 6.3 mA—the same value we got for the entire circuit. You can see how all these equations are interconnected.

Finally, the power through the entire circuit is 15 volts times 0.0063 ampere, or 0.0945 watts. We can round this off to a value of about 95 milliwatts.

Now solve the various circuit values for the same circuit if the battery generates 12 volts, R1 is 4700 ohms, R2 is 100,000 ohms, R3 is 33,000 ohms, and R4 is 27,000 ohms.

Self-Test

1. The basic unit of resistance is which of the following?

A *Ohm*
B *Coulomb*
C *Mho*
D *Pot*

2. If the bands on a resistor are yellow, violet, orange, and gold, what is its value?

A *470 ohms 5%*
B *470 ohms 10%*
C *47000 ohms 5%*

D *4700 ohms 5%*
E *None of the above*

3. If the bands on a resistor are red, red, red, and silver, what is its value?

A *222 ohms 10%*
B *220 ohms 5%*
C *2200 ohms 10%*
D *2200 ohms 20%*

4. Which of the following is *not* a valid expression of Ohm's law?

A $\quad R = \dfrac{E^2}{I}$

B $\quad E = IR$

C $\quad I = \dfrac{E}{R}$

D $\quad R = \dfrac{E}{I}$

E *None of the above*

5. If a 3300 ohm resistor and a 22000 ohm resistor are connected in series, what is the total resistance?

A *18700 ohms*
B *2870 ohms*
C *25300 ohms*
D *5500 ohms*
E *None of the above*

6. If three resistors, each with a value of 560 ohms, are connected in parallel, what will be the total resistance of the combination?

A *187 ohms*
B *1867 ohms*
C *560 ohms*
D *1680 ohms*
E *None of the above*

7. Assume 12 volts is applied to a series circuit consisting of a 1200 ohm resistor and a 3900 ohm resistor. What is the current flowing through the circuit?

A *0.0425 amps*
B *0.002 amps*

C *2.35 amps*
D *0.013 amps*
E *None of the above*

8. A 9-volt battery is connected to a circuit consisting of three resistors in parallel. Their markings are as follows:

> brown, black, brown, gold
> red, violet, red, silver
> orange, orange, brown, gold

What is the current flow through this circuit?

A *0.15 amps*
B *75 amps*
C *0.75 amps*
D *0.12 amps*
E *None of the above*

9. When resistors are connected in series, what happens?

A *The total resistance is increased*
B *The total resistance is decreased*
C *The current flow is increased*
D *The tolerance is decreased*
E *None of the above*

10. When resistors are connected in parallel, what happens?

A *Nothing*
B *The current flow is increased*
C *The total resistance is increased*
D *The total resistance is decreased*
E *None of the above*

4

Kirchhoff's Laws

Kirchhoff's Laws are a handy set of tools for analyzing what's going on within an electrical circuit. You have a choice of whether to use Kirchhoff's Voltage Law or Kirchhoff's Current Law. They are just two different paths to the same ends. Ultimately, they give the same results.

Kirchhoff's Laws are especially useful in analyzing circuits that cannot be broken down into simple series and/or parallel combinations of resistances. Such a circuit is illustrated in Fig. 4-1.

For those of you who hate math, I can only say I'm sorry, but there is no way to get around using at least some math in analyzing electronic circuits. Fortunately, the math required for Kirchhoff's Laws is not too advanced or complicated. I will try to make it all as painless as possible. Once you get used to working with these equations, you probably won't have any trouble with them at all, even though they do seem intimidating at first.

If you ever took a course in algebra you're more than halfway home already. If you've never had any algebra (or if you've forgotten what you were taught), just go through each of the steps described in the text slowly and carefully and it should start making sense to you. Just don't let mathaphobia get the best of you. Don't panic. It's really not as hard as it looks. I promise.

KIRCHHOFF'S VOLTAGE LAW

According to Kirchhoff's Voltage Law, ''the algebraic sum of the voltage sources in any loop is equal to the algebraic sum of the voltage drops around the loop.'' If you

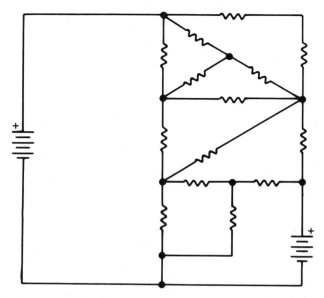

Fig. 4-1. Some circuits cannot be broken down into simple series/parallel combinations.

don't understand, don't worry about it. In somewhat simpler terms, the total amount of voltage put into the loop must be exactly cancelled out to the voltage used up (dropped) within the loop. This will become clearer as we go on. To understand Kirchhoff's Voltage Law, we first need to know what is meant by a "loop."

Loops

In the Kirchhoff system, a loop is any closed conducting path within the circuit. Loops can be made up of any combination of conductors, resistances, reactances, and/or voltage sources (but not current sources). (Reactances will be explained in later chapters, especially Chapter 6 and Chapter 8. For now all you really need to know is that a reactance is essentially an ac resistance.)

A fairly simple circuit is illustrated in Fig. 4-2. This circuit is broken up into its component loops in Fig. 4-3. Notice that the loops are redundant. While the circuit is

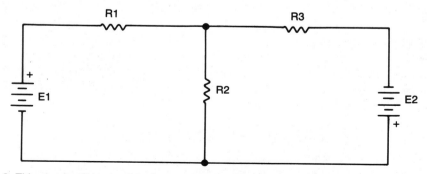

Fig. 4-2. This circuit will be used to demonstrate Kirchhoff's Laws.

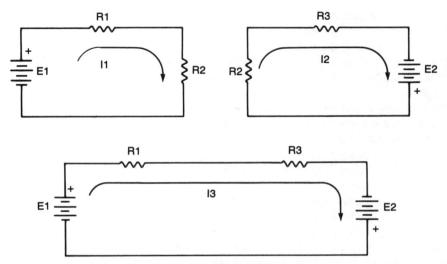

Fig. 4-3. The circuit of Fig. 4-2 broken up into component loops.

made up of three loops, any two of the loops include all of the circuit elements. The third loop is not needed. More complex circuits will break down into more than three loops, of course, but the principle is the same.

In working with Kirchhoff's Voltage Law, we will use the minimum number of loops that will include all of the circuit elements. To use any more loops than this would merely be redundant, and make unnecessary extra work for ourselves.

Loop Currents

The next term we need to define is "loop current." This definition is pretty obvious. A loop current is simply the current assumed to flow through a given loop. The loop currents are indicated in Fig. 4-3.

It is important to realize that loop currents are a theoretical concept. They exist only mathematically. If you actually measured the current at a given point in the circuit, you may or may not get a value close to the assumed loop current. There is a perfectly good reason for this. The tested point can be (and probably is) part of more than one loop. There may well be multiple loop currents flowing through that circuit point. For example, in Fig. 4-3, loop currents I1 and I2 both flow through resistance element R1.

For purposes of Kirchhoff circuit analysis, we simply artificially separate these simultaneous currents into independent loop currents. This is a somewhat confusing concept, so it might be worthwhile to go back and reread this brief discussion on loop currents before going any further.

Sign Conventions

Kirchhoff's Voltage Law deals with algebraic sums. This means that we will be dealing with positive (plus) and negative (minus) quantities. Various values will add or subtract, depending on how they are signed. Obviously we need a set of conventions or rules to determine whether a given value should be given a plus or minus sign. The procedure for determining the proper signs is as follows; pick a direction, either clockwise or counter-clockwise. It doesn't really matter which one you pick, as long as your choice remains consistent throughout the analysis process. A negative sign simply indicates that the current is moving in the opposite direction from the one we selected.

All of the loop currents must be assumed to flow in the same direction (all clockwise, or all counter-clockwise). It is important to remember that the use of a plus or minus sign here does *not* refer to ordinary electrical polarity. It is just a mathematical convention for our convenience in analyzing the circuit.

If a current flows through a resistor in the same direction as the loop current for that loop, the voltage drop across the resistance is positive. Of course, if the current flowing through the resistor is in the opposite direction, we will consider the voltage drop across the resistance to be negative.

There are two kinds of current flowing through the loop. There is the loop current for the loop presently under consideration, and there are loop currents from other loops in the circuit. As a rule of thumb, voltage drops caused by the present loop's current will almost always be positive. Voltage drops due to currents from other loops can be either positive or negative.

Loops can contain voltage sources. If a loop current passes through a voltage source from the negative terminal to the positive terminal, that current has a positive value. If it flows in the opposite direction (from the voltage source's positive terminal to the negative terminal), that current is negative. By following these rules, we can determine the proper sign for any current or voltage drop in a loop.

Kirchhoff's Voltage Law In Action

I realize that you may be scratching your head at this point. You might be starting to believe this is all beyond you, don't worry. Kirchhoff's Voltage Law is a little difficult to explain abstractly, but once we go step by step through a practical example, you should have a better grasp of the concepts described.

We will analyze the simple circuit shown in Fig. 4-2. Of course, we can't do that without knowing the voltages and resistance values in the circuit. We will assume the following values;

E1	12 volts
E2	6 volts
R1	100 ohms
R2	47 ohms
R3	220 ohms

In analyzing the circuit we can use any two of the loops illustrated in Fig. 4-3. We will use loop A and loop B. These loops are shown in Fig. 4-4. The loop currents are assumed to flow in a clockwise direction.

Loop A consists of the following elements;

E1, R1, R2

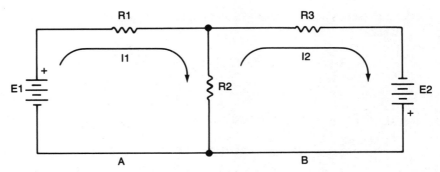

Fig. 4-4. We will use loop A and loop B from the circuit of Fig. 4-2.

The loop current for loop A is labelled I1.

Loop B contains the following elements;

$$E2, R2, R3$$

The loop current for loop B is labelled I2. Notice that resistance R2 is part of both loops. Therefore, both the I1 and I2 currents flow through this component.

Kirchhoff's Voltage Law states that the algebraic sum of all voltage sources in the loop equals the algebraic sum of all voltage drops in the circuit. In other words, the total amount of voltage put into the loop must be exactly cancelled out to the voltage used up (dropped) within the loop.

The only voltage source in loop A is E1. There is just one voltage drop across resistor R1, because only I1 flows through it. There are two voltage drops across resistor R2, however. One is due to loop current I1, and the other is due to the external current (from loop B) I2. These three voltage drops must equal the voltage supplied by E1. In algebraic terms;

$$E1 = (I1 \times R1) + (I1 \times R2) - (I2 \times R2)$$

Each voltage drop is calculated simply by Ohm's law;

$$E = I \times R$$

The (I2 × R2) voltage drop is negative because current I2 is flowing through R2 in the opposite direction as the loop current I1. The same approach is used to find the algebraic formula for loop B. Its only voltage source is E2. Only loop current I2 flows through resistor R3, so it contributes a single voltage drop. Both I1 and I2 flow through R2, so it has two voltage drops. I1 (from loop A) flows in the opposite direction of I2, so it causes a negative voltage drop. In algebraic terms we have;

$$E2 = (I2 \times R3) + (I2 \times R2) - (I1 \times R1)$$

Plugging in our component values for the circuit as listed above, the equation for loop A becomes;

$$E1 = (I1 \times R1) + (I1 \times R2) - (I2 \times R2)$$

$$12 = 100(I1) + 47(I1) - 47(I2)$$

Combining the like terms (I1), this equation can be simplified to;

$$12 = 147(I1) - 47(I2)$$

Similarly, plugging the component values into the equation for loop B, we get;

$$E2 = (I2 \times R2) + (I2 \times R3) - (I1 \times R2)$$

$$6 = 47(I2) + 220(I2) - 47(I1).$$

Once again we combine the like terms (I2, in this case), and simplify the equation to;

$$6 = 267(I2) - 47(I1)$$

I know a lot of people hate math, but there's no getting around the need to use simple algebra to solve Kirchhoff's Laws. I'll try to show each step of the process.

We need to solve for the currents (I1 and I2). Since we have two variables, we will have to define one in terms of the other. We can rearrange the equation for loop A to solve for current I1;

$$12 = 147(I1) - 47(I2)$$

$$147(I1) - 47(I2) = 12$$

(First we just reverse the order of the two halves of the equation. This is just for convenience in writing the steps in the changing equation down. No values are changed here.)

$$147(I1) - 47(I2) + 47(I2) = 12 + 47(I2)$$

(Any changes we make on one side of the equation, must be duplicated on the other side of the equation. Adding 47(I2) to the left side of the equation, cancels this factor out, but it must be added to the right side to keep the equation consistent.)

$$147(I1) = 12 + 47(I2)$$

(The 47(I2) − 47(I2) in the last step cancels out, leaving a value of zero, which can be dropped from the equation.)

$$147(I1)/147 = (12 + 47(I2))/147$$

(We divide both halves of the equation by 147 to remove this element from the left side of the equation. This moves the 147 over to the right half of the equation.)

$$I1 = (12 + 47(I2))/147$$

(The two 147s in the left hand of the equation cancel each other out and are dropped. Notice that we haven't really changed the equation. The values remain the same, we've just moved the various elements around at our convenience.

We now can define current I1 in terms of I2. Moving on to the equation for loop B, we can substitute the right half of the last equation anywhere the term "I1" appears in the second equation. That is;

$$6 = 267(I2) - 47(I1)$$

$$6 = 267(I2) - 47((12 + 47(I2))/147$$

We now have just one variable to solve for in this equation (I2). We can find the value of I2 through simple rearrangement of the terms;

$$6 = 267(I2) - ((47 \times 12) + (47 \times 47)(I2))/147))$$

$$6 = 267(I2) - ((47 \times 12)/147) + ((47 \times 47(I2))/147)$$

$$6 = 267(I2) - (564/147) - (2209(I2)/147)$$

$$6 = 267(I2) - 3.8 - 15(I2)$$

(Note — values have been rounded off in the division. This will cause a slight error in the final results, but the errors due to rounding off will be negligible.)

$$6 = 267(I2) - 15(I2) - 3.8$$

$$6 = 252(I2) - 3.8$$

$$6 + 3.8 = 252(I2) - 3.8 + 3.8$$

(3.8 is added to BOTH sides of the equation to maintain the proper balance.)

$$9.8 = 252(I2)$$

$$9.8/252 = (252(I2)/252)$$

(252 is divided into *both* halves of the equation.)

$$0.04 = I2$$

(Note we have rounded off the division again.)

Current I2 has a value of approximately 0.04 amp (or 40 mA). We defined I1, in terms of I2;

$$I1 = (12 + 47(I2))/147$$

Since we now know the value of I2 (0.04), we can simply plug in this value and solve for I1;

$$I1 = (12 + (47 \times 0.04))/147$$

$$= (12 + 1.8)/147$$

$$= 13.8/147$$

$$I1 = 0.09$$

A slight error has been introduced due to rounding off the values during the math. Usually this won't matter, but you should be aware of round-off errors, or you'll go nuts trying to figure out why the final mathematical results don't match up perfectly.

Please remember that we are working with loop currents here (I1 and I2). Loop currents are mathematical fictions. If you insert an ammeter into the circuit you will not be able to measure 0.04 amp or 0.09 amp at any point. The loop currents are assumed to exist only for the purposes of Kirchhoff's Voltage Law.

Knowing the currents, we can now go back and find the voltage drops for each of the resistances in the circuit. Remember that there are four voltage drops in this circuit, because both I1 and I2 flow through R2, while only I1 flows through R1, and only I2 flows through R3. The four voltage drops (stated via Ohm's law), are as follows;

$$I1 \times R1, \ I1 \times R2, \ I2 \times R2, \ I2 \times R3$$

We know the resistance values from our original parts list, and we have just solved for the current values. Now, to find the voltage drops, we just have to plug in the appropriate values;

(a) $I1 \times R1 = 0.09 \times 100 = 9$ volts

(b) $I1 \times R2 = 0.09 \times 47 = 4.2$ volts

(c) $I2 \times R2 = 0.04 \times 47 = 1.9$ volts

(d) $I2 \times R3 = 0.04 \times 220 = 8.8$ volts

To make things a little more convenient, I have assigned each voltage drop an identifying letter (from (a) to (d)).

Loop A includes three of the voltage drops;

(a), (b), (c)

$$E1 = (I1 \times R1) + (I1 \times R2) - (I2 \times R2)$$

$$E1 = (a) + (b) - (c)$$

Plugging in the values we now know;

$$12 = 9 + 4.2 - 1.9$$

$$12 = 11.3$$

Don't be thrown by the apparent 0.7 volt error. It is simply the cumulative result of the rounding off errors throughout the process. If you go back and use a calculator to determine the various values exactly, the error should disappear (or shrink significantly—the calculator may do some minor rounding off of its own).

Allowing for rounding off errors, the total voltage sources in the loop equal the total voltage drops in the loop, as Kirchhoff's Voltage Law predicts.

Similarly, loop B also includes three of these voltage drops;

(b), (c), (d),

$$E2 = (I2 \times R2) + (I2 \times R3) - (I1 \times R2)$$

$$E2 = (c) + (d) - (b)$$

Plugging in the known values;

$$6 = 8.8 + 1.9 - 4.2$$

$$6 = 6.5$$

Once again, we have a slight imbalance due to cumulative rounding off errors, but for all intents and purposes, the total voltage sources in the loop again equal the total voltage drops in the loop. If we recalculate all the values exactly (without any rounding off), we will find that Kirchhoff's Voltage Law accurately predicts what is happening here. The results of our analysis of this circuit using Kirchhoff's Voltage Law are illustrated in Fig. 4-5.

KIRCHHOFF'S CURRENT LAW

There is an alternate to Kirchhoff's Voltage Law, this is Kirchhoff's Current Law. This second Kirchhoff Law permits us to deal with actual currents rather than the mathematical fictions (loop currents) of Kirchhoff's Voltage Law.

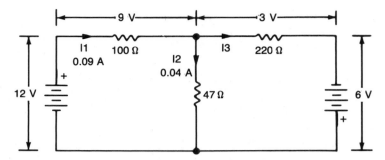

Fig. 4-5. The results of the Kirchhoff's Voltage Law analysis of Fig. 4-2.

According to Kirchhoff's Current Law, "the amount of current flowing into a node always exactly equals the current flowing out of that node." That certainly makes sense when you think about it. What you put into a node is what you get out of it. In more mathematical terms, the algebraic sum of all currents flowing through a node is zero. Obviously, before we go any further, we have to define just what we mean by the term *node*.

Nodes

A node is simply the connection point between two or more conductors. Our simple example circuit is shown in Fig. 4-6, with the nodes indicated. This particular circuit has just two nodes (A and B).

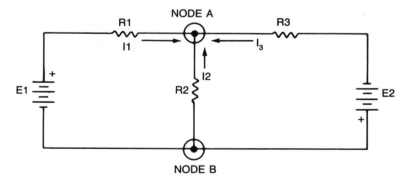

Fig. 4-6. The circuit of Fig. 4-2 with the current nodes indicated.

Current flowing into any node is always assumed to be positive. Current flowing out of the node is assumed to be negative. For voltage drops across any resistance elements, the terminal where the current enters is assumed to be at a higher potential (more positive) than the terminal where the current exits. This is illustrated in Fig. 4-7.

When using Kirchhoff's Current Law, the first step is to count the number of nodes in the circuit to be analyzed. If the circuit has N nodes, we will need to examine N-1 nodes to completely analyze the circuit. The required number of node equations is always one less than the total number of nodes in the circuit.

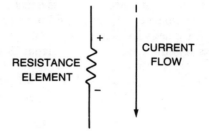

Fig. 4-7. The terminal of a resistance element where the current enters is assumed to be more positive than the terminal where the current exits.

RESISTANCE ELEMENT

CURRENT FLOW

Kirchhoff's Current Law in Action

In our sample circuit of Fig. 4-6, we have two nodes, so we only need to look at one (2 −1) to analyze the circuit.

We will use node A. We could just as easily use node B. The choice of which one to leave out is purely arbitrary.

There are three current paths into node A. These are marked in Fig. 4-6 as follows:

$$I1, \; I2, \; I3$$

According to Kirchhoff's Current Law, the algebraic sum of these three currents must be equal to zero. That is;

$$I1 + I2 + I3 = 0$$

Note that at this point we don't know which of these currents are positive and which are negative. This basic equation can't do us much good until we relate the currents to the voltages and resistances within the circuit.

Current I1 flows through resistor R1. Thanks to Ohm's law $(I = E/R)$, we know that current I1 must be equal to the voltage drop across R1, divided by the value of the resistance.

The voltage drop across R1 must be equal to the voltage going into the resistance element at the positive terminal (which is E1, in this case) minus the voltage at the negative terminal of the resistance element. We will call the voltage at the negative terminal Ea. The current direction of I2 tells us that node A is less positive (more negative) than node B, so voltage Ea takes on a negative sign. Putting all of this together, we can create a simple Ohm's law equation for current I1;

$$I1 = (E1 - (-Ea))/R1$$

The two negative signs in front of Ea cancel each other out. This leaves us with;

$$I1 = (E1 + Ea)/R1$$

Current I2 is defined by the voltage drop across R2. This is simply equal to Ea, so;

$$I2 = Ea/R2$$

Finally, current I3 is determined by the voltage drop across resistor R3. The input voltage for this resistance element is equal to the voltage put out by voltage source E2. R3's output voltage is again Ea. E2 is negative because of the polarity of the voltage source. Ea is negative because of the direction of the I2 current flow. Putting all these elements together, we find that current I3 works out to;

$$
\begin{aligned}
I3 &= (-E2 - (-Ea))/R3 \\
&= (-E2 + Ea)/R3 \\
&= (Ea - E2)/R3
\end{aligned}
$$

The next step is to substitute these individual current formulas into the original node equation;

$$I1 + I2 + I3 = 0$$

$$((E1 + Ea)/R1) + (Ea/R2) + ((Ea - E2)/R3) = 0$$

We can simplify and rearrange the equation as follows;

$$(E1/R1) + (Ea/R1) + (Ea/R2) + (Ea/R3) - (E2/R2) = 0$$

$$(Ea/R1) + (Ea/R2) + (Ea/R3) + (E1/R1) - (E2/R3) = 0$$

(We've just rearranged the order of the items in the equation to make the following steps clearer.)

$$(Ea/R1) + (Ea/R2) + (Ea/R3) + (E1/R1) - (E2/R3) - ((E1/R1) - (E2/R3)) = 0 - ((E1/R1) - (E2/R3))$$

(The same factor is subtracted from *both* sides of the equation to maintain equality.)

$$(Ea/R1) + (Ea/R2) + (Ea/R3) = -((E1/R1) - (E2/R3))$$

$$(Ea/R1) + (Ea/R2) + (Ea/R3) = -(E1/R1) + (E2/R3)$$

$$(Ea/R1) + (Ea/R2) + (Ea/R3) = (E2/R3) - (E1/R1)$$

(We've just rewritten the equation in slightly simpler form, without really changing anything.)

$$Ea \times ((1/R1) + (1/R2) + (1/R3)) = (E2/R3) - (E1/R1)$$

(Here we have just factored the left side of the equation. What the equation expresses has not been changed in any way, we've just modified the way we are expressing it.)

Before we can go any further, we will need some specific component values to work with. We will assign the following component values to the circuit;

E1	9 volts
E2	12 volts
R1	10 ohms
R2	20 ohms
R3	50 ohms

Notice that all of the elements in the equation except Ea are defined by the parts list. In a more complex circuit, there can be more than one unknown voltage element. Multiple unknowns can be solved by combining multiple node equations. In this simple circuit, we have only one node and one unknown voltage value to worry about.

Plugging the known values from the parts list into the last form of the equation, we get;

$$Ea \times ((1/R1) + (1/R2) + (1/R3)) = (E2/R3) - (E1/R1)$$

$$Ea \times ((1/10) + (1/50) + (1/20)) = (12/50) - (9/10)$$

Now, we just start solving and simplifying the values;

$$Ea \times (0.1 + 0.02 + 0.05) = 0.24 - 0.9$$

$$Ea \times 0.17 = -0.66$$

(We can divide both sides of the equation by 0.17 to remove this factor from the left side and solve for Ea.)

$$(Ea \times 0.17)/0.17 = -0.66/0.17$$

$$Ea = -0.66/0.17$$

$$Ea = -3.88 \text{ volts}$$

The negative sign here simply means that the actual polarity is the opposite of the one we assumed. As you can see, you really can't get the polarity "wrong," as long as you're consistent. The equations will work out with the proper signs.

We now have enough information to go back and solve for each of the currents;

$$
\begin{aligned}
I1 &= (E1 + Ea)/R1 \\
&= (9 + (-3.88))/10 \\
&= (9 - 3.88)/10 \\
&= 5.12/10 \\
&= 0.512 \text{ amp} \\
&= 512 \text{ mA} = I1
\end{aligned}
$$

$$I2 = Ea/R2$$
$$= -3.88/20$$
$$= -0.194 \text{ amp}$$
$$= -194 \text{ mA} = I2$$

$$I3 = (Ea - E2)/R3$$
$$= (-3.88 - 12)/50$$
$$= -15.88/50$$
$$= -0.318 \text{ amp}$$
$$= -318 \text{ mA} = I3$$

The negative values for I2 and I3 simply mean that the actual direction of current flow is the opposite of that shown in Fig. 4-6. These currents flow out of, rather than into node A. These are the actual currents in the circuit, not mathematical fictions like the loop currents used in Kirchhoff's Voltage Law.

We can double-check our work, by plugging these derived current values back into the original node equation;

$$I1 + I2 + I3 = 0$$

$$0.512 + (-0.194) + (-0.318) = 0$$

$$0.512 - 0.194 - 0.318 = 0$$

Yes, it works. Kirchhoff's Current Law gave an accurate prediction of how the currents in the circuit interact at the examined node.

Other circuits will end up with slightly different equations. Naturally, the more nodes there are in the circuit, the more equations you will have to work with.

Self-Test

1. Which of the following is *not* one of Kirchhoff's laws?

A *Kirchhoff's Voltage Law*
B *Kirchhoff's Resistance Law*
C *Kirchhoff's Current Law*

2. For Kirchhoff's Voltage Law, circuits are divided into which of the following?

A *Nodes*
B *Ohms*
C *Series/parallel circuits*
D *Loops*

3. What is a node?

A *A mathematical fiction*
B *A terminal point for a loop current*
C *A connection point between two or more conductors*
D *A high voltage point in a circuit*

4. Which of the following correctly describes Kirchhoff's Voltage Law?

A *The algebraic sum of the voltage sources in a loop equals the algebraic sum of voltage drops*
B *The algebraic sum of currents entering a node equal to the algebraic sum of currents exiting the node*
C *A voltage drop equals the current multiplied by the resistance*
D *The algebraic sum of all loop currents equals the algebraic sum of all voltage drops*

5. *Which of the following cannot be included in a loop for Kirchhoff's Voltage Law?*
A *Voltage sources*
B *Current sources*
C *Resistances*
D *Reactances*

6. How many nodes are needed to completely analyze a circuit?

A *One*
B *Two*
C *All nodes in the circuit*
D *One less than the total number of nodes in the circuit*

7. Loop currents should be assumed to flow in which direction?

A *Clockwise*
B *Counter-clockwise*
C *Either A or B can be arbitrarily selected*

8. According to Kirchhoff's Current Law, what is the algebraic sum of all currents entering and exiting a node?

A *A positive value*
B *A negative value*
C *The algebraic sum of the loop currents*
D *Zero*

9. If a resistance element is part of two loops, how many voltage drops must be calculated for that component?

A *None*
B *One*
C *Two*
D *Three*

10. In Kirchhoff's Current Law, which terminal of a resistance element is assumed to be at a higher potential (more positive) than the other?

A *The terminal where the current enters the resistance element*
B *The terminal where the current exits the resistance element*
C *The terminal closest to the node being analyzed*
D *Either A or B can be arbitrarily selected*

5

Alternating Current

So far we have been working with circuits where the current flows in only one direction. This form of electricity is called *direct current*, or *dc*.

Many other circuits, however, operate on a voltage and current that continuously vary in value, according to a repetitive, periodic pattern. Electricity in this form is called *alternating current*, or *ac*.

VARYING VOLTAGE AND CURRENT

The polarity of a voltage source determines the direction of current flow. The current will flow from the negative terminal of the voltage source to the positive terminal. If the polarity of the voltage source is reversed, the current will flow in the opposite direction.

Take a look at the circuit in Fig. 5-1. Here we have a circuit with two voltage sources, each a three volt battery. Since these batteries are connected with opposing polarities, if they were allowed to have an equal effect on the circuit, they would simply cancel each other out. No current would flow through the circuit, and the power consumed would be zero.

The two pots labeled R1 and R2 control the relative effect of the two batteries in the main circuit. The dotted line between these two schematic symbols indicates that they are mechanically tied together. That is, one knob controls both pots simultaneously. Such a multiple component is said to be *ganged*. Potentiometers can be dual, triple, or even quadruple ganged. Dual ganged pots are not uncommon, but larger combinations tend to be fairly rare.

48

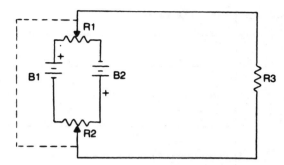

Fig. 5-1. A circuit powered by a voltage with reversible polarity.

For convenience in this discussion, we will refer to R1 and R2 as if they were a single potentiometer, since they always work in unison.

If the pot's slider is in the exact center of its path of rotation, equal resistances will be seen by each of the source voltages. So both batteries will present an equal, but opposite voltage to the main circuit. This results in the voltages cancelling each other out, and no current flows through the circuit.

If, however, the slider is moved all the way towards battery 1, that battery will see a minimum resistance, and battery 2 will see a maximum resistance. In other words, most of battery 2's voltage is dropped across the resistance of the potentiometer. But most of the voltage from battery 1 makes its way through to the external circuit. Therefore, as far as the load circuit is concerned, battery 2 doesn't exist. Battery 1 provides the power to operate the circuit.

At the other extreme of the slider's path, the situation is reversed—battery 2 is dominant, and battery 1 is ignored. At intermediate positions of the slider, the two voltages will interact in a subtractive manner. The battery closest to the slider will have the greater effect, but the other battery will cancel out some of the voltage.

For example, if the potentiometer is set so that 2.5 volts are passed from battery 1, while battery 2 is allowed to put out only 0.5 volts, the load circuit will see a voltage source of 2 volts (battery 1 - battery 2). This voltage will have the same polarity as the larger of the opposing voltages. In this case, battery 1 determines the polarity.

If we draw a graph of the effective voltage seen by the circuit, as the potentiometer's slider is rotated through its entire range, it would look like Fig. 5-2.

Now, suppose we start with the slider in its center position, then smoothly rotate the knob back and forth. A graph of the effective voltage under these circumstances would resemble the one in Fig. 5-3.

Notice that there is a repeating pattern in this graph. If you happen to be familiar with trigonometry, you might recognize this graph as a series of representations of the function called the sine of an angle. For this reason, this *waveshape* is called a *sinusoidal*, or *sine wave*. Each complete pattern, without repetition (as from point A to Point B in Fig. 5-3) is called a *cycle*, or a *wave*.

If you try a few Ohm's law equations, you'll find that the current drawn by the circuit varies in step with the fluctuations of the applied voltage. When the voltage goes up, the current goes up, and vice versa. The current is said to be *in phase* with the voltage. A purely resistive circuit does not alter the phase relationship. When the voltage and current in a circuit fluctuate in this manner, we have an *alternating current*.

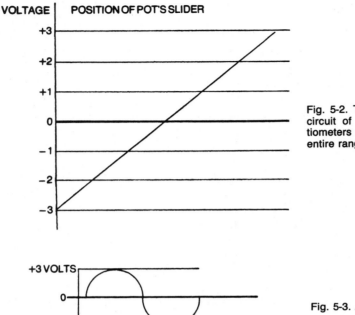

VOLTAGE | POSITION OF POT'S SLIDER

Fig. 5-2. The voltage through the circuit of Fig. 5-1 as the potentiometers are moved through their entire range.

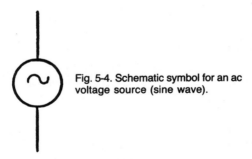

Fig. 5-3. A sine wave.

AC VOLTAGE SOURCES

Figure 5-4 shows the schematic symbol for any ac voltage source that generates a sine wave (other wave shapes will be discussed in later chapters). Notice, that since the current is constantly reversing itself, there is no fixed polarity for such a voltage source.

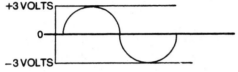

Fig. 5-4. Schematic symbol for an ac voltage source (sine wave).

An ac source changes its polarity many times each second. By counting the number of complete cycles in a second, we get the *frequency* of the wave. Frequency is measured in *cycles per second* (cps). Another name for a cycle per second is *hertz* (Hz), named after a pioneer in the field. One thousand hertz is a *kilohertz* (kHz), or a *kilocycle per second* (kcps). Similarly, a million cps is one *megahertz*, or *megacycle*.

Typically, ac power sources operate at a fairly low frequency. In the United States, house current alternates at a 60 Hz rate. Other countries use a 50 Hz standard.

It might seem that using ac would just complicate matters, but actually, the reverse is true. While for low power applications it is easier to make dc devices (i.e., batteries) than ac devices, in heavy wattage installations, ac is much more practical to generate. It is also easier to transmit ac over long lines. This is why the electric power companies operate on ac. Also, as we shall see in later chapters, many circuits will operate quite differently depending on whether the electron flow through them is ac or dc.

The most readily available source of ac for powering electronic circuits, is the house current provided by the electric company through ordinary wall sockets (see Fig. 5-5). This is a power source of about 110 to 120 volts ac (the level fluctuates somewhat) with a nominal frequency of 60 Hz. This frequency also fluctuates slightly, but the average is usually a very precise 60 Hz.

Fig. 5-5. Wall socket for ac voltage.

Many electronic circuits are designed to operate directly from an ac wall socket, so the schematic includes a line cord. The symbol is shown in Fig. 5-6A. In other cases, the power is first passed through a *transformer* (see Chapter 9) to change the voltage, or through a power supply (see Chapter 32) to convert the ac to a dc voltage.

Many devices that operate off standard ac, also include an extra socket to carry the power on to other equipment. The schematic symbol for an ac socket is illustrated in Fig. 5-6B.

Notice that an ac line cord contains two wires. One is a reference line, and the other carries the actual current. With many circuits it doesn't matter which is which, but some

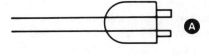

Fig. 5-6. Schematic symbols: A. ac line cord; B. ac socket.

circuits require that the power must come through the wires in a specific way. To ensure that the cord is plugged into the wall socket properly, such equipment uses a *polarized plug*. Such a plug can be plugged into the wall socket in only one way. If its prongs are reversed, it will not fit into the socket. A standard *non-polarized plug*, on the other hand can be reversed.

Sometimes, especially with equipment that draws heavy current levels (high wattage), extra protection is necessary—both for the circuit itself, and to prevent dangerous shock hazards to the user. In such equipment a third wire is needed to *ground* the circuit.

The word "ground" actually has two meanings in electronics. An *earth ground* offers protection by returning the voltage to the earth itself, via a metal post of some kind that makes a good contact with the ground. See Fig. 5-7. In this manner, any potentially dangerous voltage is shunted into the earth, rather than into the components of the circuit, or into the body of the operator. A cold water pipe is often a good place to make an earth ground connection.

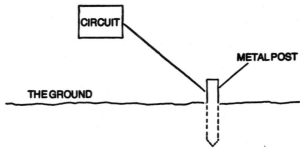

Fig. 5-7. Earth ground.

For standard ac power, the screw in the center of the socket plate is connected to earth ground. In a three prong socket, the third prong is also connected to earth ground. When a device with a three prong plug must be used with a two prong socket, an *adaptor*, is used. The wire *must* be connected to the center screw of the socket plate. Leaving this wire unconnected defeats the purpose of the third prong, and could be extremely dangerous!

The second kind of ground found in electronics work is also called a *common ground*, or simply *common*. This, as the name implies, is a point common to all signals in a circuit, so it is used as a reference point to measure voltages in the circuit.

A common, by definition, is at 0 volts, so all other voltages can be measured between the appropriate point in the circuit and the common. In dc circuits the common point is usually either the negative terminal (somewhat more frequently) or the positive terminal of the voltage source. Either can be used without affecting circuit operation. It is just a reference point for measurements.

The type of common used will affect the polarity of the voltages measured. For example, in a circuit powered by a three volt battery with a negative ground, the source voltage will be +3 volts. If a positive ground is employed, the source voltage is −3 volts. This will not affect any circuit calculations as long as you are consistent throughout the circuit. If all voltages are of the same polarity, their signs can be ignored for calculations. If the voltage is negative, the current must be negative too. This means resistance will always work out to be positive (R = E/I).

In quite a few circuits, the common ground is attached to the metal chassis, or case of the circuit. When the common point covers a large area, it is called a *ground plane*. Some circuits (especially *digital* circuits—discussed later) require a ground plane to avoid erratic operation.

Figure 5-8 shows the three most common schematic symbols for a ground point. The symbol at Fig. 5-8A is generally only used to indicate an earth ground, while the other two can represent either earth ground or common ground (sometimes also called *chassis ground*). In most cases when both types of ground are used, they are identical, i.e., the same point, but this is not always true.

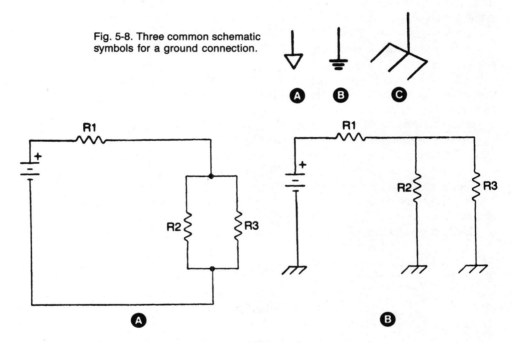

Fig. 5-8. Three common schematic symbols for a ground connection.

Fig. 5-9. Identical circuits, showing how the ground symbol is used.

All points in a circuit with a ground symbol are electrically connected with all other similarly marked points. That is, the circuits in Figs. 5-9A and 5-9B are actually identical.

AC VALUES

Since the levels in an ac circuit are continuously changing, determining voltage and current values isn't the straightforward matter it is with dc circuits. As an example, let's assume we have a sine wave that reaches a maximum, or *peak* value of 10 volts (it would also reach a negative peak of −10 volts). To say we have 10 volts ac is rather misleading, because the voltage actually reaches the full 10 volts for a brief instant during each cycle. The rest of the time, the voltage is lower.

54

Since the voltage varies between a 10 volt positive peak and a 10 volt negative peak, there is a 20 volt difference between the two peaks. That is, the voltage varies 20 volts *peak-to-peak*. While it is frequently useful to know this value, it would obviously be quite misleading to say that we had 20 volts ac.

It might seem that we could just take an average of the various instantaneous values passed through during the complete cycle, and come up with an average voltage. Unfortunately, since the positive portion of the cycle is a mirror image of the negative portion, *all* of the values are cancelled out, and we are left with an average voltage of zero.

A reasonable solution would be to take only half of the cycle—either just the positive portion, or just the negative portion—and take an average of that. This could be rather tedious to work out, but it has been mathematically proven that the average of half a cycle of a sine wave is always equal to 0.636 times the peak value. So, in our example, if the peak voltage is 10 volts, then the *average voltage* is 6.36 volts ac.

Conversely, if we know the average value, and need to find the peak voltage, we can multiply the average voltage times 1.572327. This can usually be rounded off to 1.57, or even 1.6.

While the average value can give us a fair idea of how much voltage is being passed through a circuit, a major disadvantage of using this kind of value is that the relationships stated in Ohm's law no longer hold true. What we need is a way to express ac voltage in terms that can be directly compared to an equivalent dc voltage.

Such an equivalent value can be found by taking the *root-mean-square* of the sine wave. The mathematics are fairly complex, but it works out to 0.707 times the peak value.

In our example we have 10 × 0.707, or 7.07 volts *rms*. This is the value most commonly used for ac measurements.

By using rms values, Ohm's law can be used in the same way it is used with dc circuits.

Here is a summary of the basic ac equations.

Rms	=	0.707 × Peak	**Equation 5-1**
Rms	=	1.11 × Average	**Equation 5-2**
Average	=	0.9 × Rms	**Equation 5-3**
Average	=	0.636 × Peak	**Equation 5-4**
Peak	=	1.41 × Rms	**Equation 5-5**
Peak	=	1.57 × Average	**Equation 5-6**
Peak-to-Peak	=	2 × Peak	**Equation 5-7**

Equations 5-1, 5-5 and 5-7 are the most frequently used formulas. The average value is rarely of practical importance. The same equations are used for current and for power in ac circuits.

PHASE

The current from an ac voltage source is ordinarily in step with the voltage. That is, when the voltage increases, so does the current, and when the voltage decreases, the current also decreases. We say that the voltage and the current are *in phase*. Their cycles start at the same instant.

In later chapters we'll learn how some components can throw the voltage and current *out-of-phase*. That is, one is delayed so that the two are no longer in step with each other.

MULTIPLE AC SOURCES

Remember that when we have two or more dc voltages in series (such as cells in a battery), we can find the total voltage simply by adding the component voltages from each individual source. Or, if we have two voltages of opposite polarity, we can simply subtract the smaller from the larger to find the total effective voltage in the circuit. With ac sources, however, the situation is much more complex. If the two voltage sources in a circuit like in Fig. 5-10 are in phase with each other, there is no problem. We can simply add the voltages, just as with dc. Or, if the voltage sources are 180° out of phase with each other (one full cycle equals 360°—see Fig. 5-11), we can just subtract the smaller from the larger voltage. Of course, if two equal ac voltages are 180° out of phase with each other, they will cancel each other out, leaving a net voltage of zero.

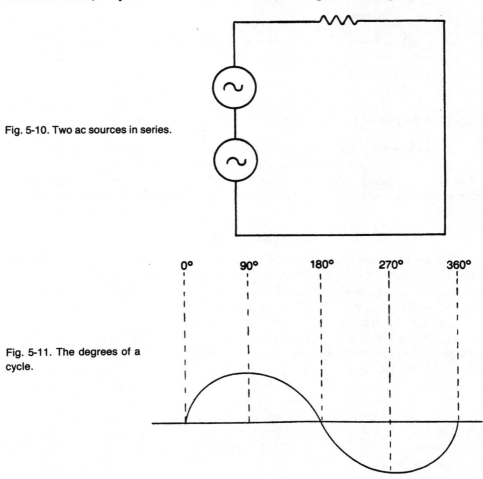

Fig. 5-10. Two ac sources in series.

Fig. 5-11. The degrees of a cycle.

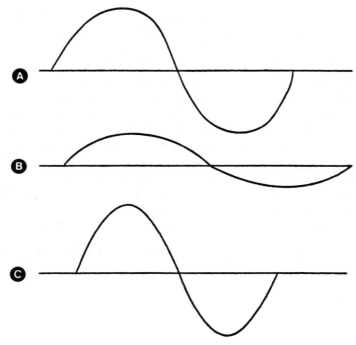

Fig. 5-12. Adding out-of-phase ac voltages.

To calculate any other phase relationship requires fairly complicated mathematics. For example, consider the graphs in Fig. 5-12. The voltage in graph B is 60° out of phase with the voltage in graph A. Graph C is the result of combining graphs A and B. Mathematically calculating the effective voltage shown in graph C would require numerous equations.

VECTOR DIAGRAMS

Fortunately, there is an easier way to solve combinations of out of phase ac signals. This is a method using *vector diagrams*. Vector diagrams may look and sound rather complicated, and the mathematical theory behind them is rather complex, but actually using these diagrams in practical situations is really quite simple. If you know how to use a ruler and a protractor, you should have no trouble.

Let's use a vector diagram to solve for the resultant voltage in Fig. 5-12. Let's say voltage A is 5 volts and voltage B is 3 volts. Voltage B is 60° out-of-phase with voltage A.

First draw a straight line to represent voltage A. You can use any scale you find convenient. For example, if you are using a scale of one inch equals one volt, line A should be five inches long. This step is illustrated in Fig. 5-13.

Starting at the point of origin of line A, draw a second line to represent voltage B. This line must be drawn to the same scale as line A. In our example, line B should be three inches long. You also have to use a protractor to draw line B at an angle equal to the phase difference. Naturally, line A is a 0°. In our example, line B should be at a 60° angle to line A. This is shown in Fig. 5-14.

Fig. 5-13. Step 1 of a vector diagram.

Fig. 5-14. Step 2 of a vector diagram.

Fig. 5-15. Step 3 of a vector diagram.

Fig. 5-16. Step 4 of a vector diagram.

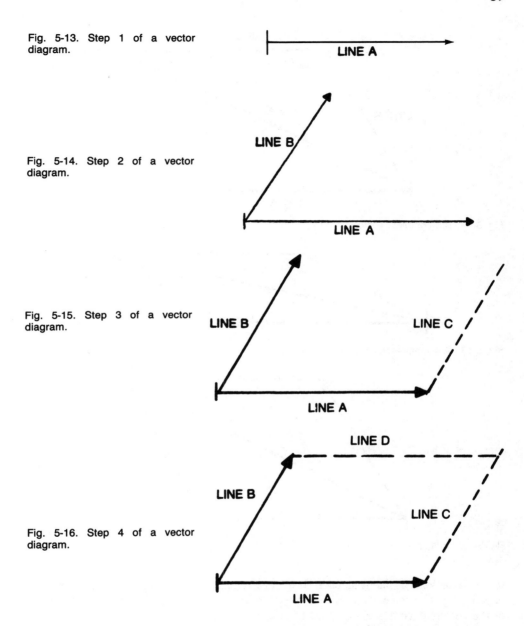

Draw a dotted line parallel to line B from the open end of line A. This new line (line C) should be the same length, or longer than line B. Line C is added to our vector diagram in Fig. 5-15. Draw a fourth line (dotted line D in Fig. 5-16) from the open end of line B to line C. This line will be parallel to line A.

Draw one last straight line from the original point of origin (the junction of line A and line B) to the junction of line C and line D. This is the line labeled line E in Fig. 5-17. If we now measure line E, its length will represent the resultant voltage, drawn

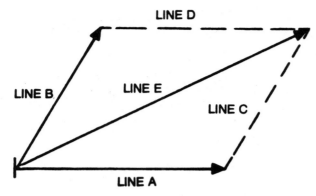

LINE D

LINE B

LINE E

LINE C

LINE A

Fig. 5-17. The complete vector diagram.

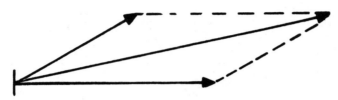

Fig. 5-18. Additional vector diagram.

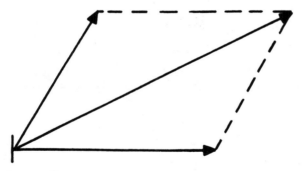

Fig. 5-19. Additional vector diagram.

to the same scale as line A and line B. In our example, it comes out to about 8.5 inches, or 8.5 volts. The angle of line E with respect to line A is the same as the phase difference of the resultant voltage (compared to voltage A).

Additional examples of vector diagrams are shown in Figs. 5-18 and 5-19. Also try drawing your own vector diagrams to solve for the resultant voltage for each of the combinations listed in Table 5-1.

COMBINING AC AND DC

What happens when you have a circuit with both a dc voltage source and an ac voltage source, like the one illustrated in Fig. 5-20? Let's assume the ac puts out a 5 volt

Table 5-1. Values for Practice Vector Diagrams.

VOLTAGE A	VOLTAGE B	PHASE
6	6	45°
8	3	90°
7	5	30°

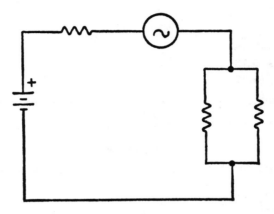

Fig. 5-20. A circuit with ac and dc voltage sources.

peak sine wave, and the dc source is 9 volts. At any given instant, the ac signal has a specific value. For instance, whenever the signal crosses the 0 line, it has an instantaneous value of 0 volts. At this instant, the ac source has no effect on the circuit—the total circuit voltage at this brief point in time is simply the 9 volts from the dc source.

However, at another instant in its cycle, the ac source is putting out its peak value of 5 volts. Meanwhile, the dc source is still producing 9 volts. These values are simply added, and the instantaneous value is 14 volts. When the ac voltage reaches its negative peak (−5 volts), the polarity opposes that of the 9 volts dc source. In this case we have −5 volts and +9 volts, so the instantaneous value is +4 volts.

If we continue figuring the instantaneous values at various points throughout the cycle, we will find we have a sine wave that varies between +4 volts and +14 volts. The graph of the resultant voltage is shown in Fig. 5-21. Notice that it is identical to an ordinary sine wave, except it varies above and below nine volts, rather than zero volts. We say we have 3.535 volts ac rms (5 volts peak) *superimposed* on 9 volts dc.

If the ac voltage peak is less than the dc voltage, circuit polarity will not be reversed at any time in the cycle. If, on the other hand, the ac voltage is larger than the dc voltage, then polarity will be reversed during at least some of the cycle. For example, 10 volts ac peak superimposed on 6 volts dc will produce a combined voltage that fluctuates between −4 volts and +16 volts.

In solving most circuit equations, the ac and dc elements are handled separately. In many cases, one or the other can be ignored without adversely affecting our understanding of circuit operation.

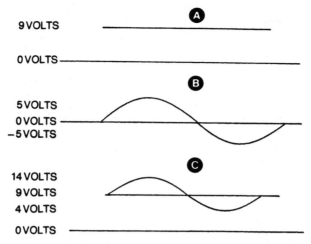

Fig. 5-21. Combining ac and dc voltages.

Self-Test

1. Which of the following is *not* a unit for measuring ac frequencies?

A *Cycles per second*
B *Waves per second*
C *Hertz*
D *Kilocycles*
E *Megahertz*

2. If an ac signal has a rms voltage of 55 volts, what is the peak voltage?

A *77.55 volts*
B *75 volts*
C *38.885 volts*
D *86.35 volts*
E *None of the above*

3. If an ac signal has a peak voltage of 90 volts, what is the average voltage?

A *63.63 volts*
B *99.9 volts*
C *126.9 volts*
D *57.24 volts*
E *None of the above*

4. When comparing rms voltages and average voltages, which of the following statements is true assuming sine waves?

A *The average voltage will always be greater than the rms voltage*
B *The rms voltage is always greater than the average voltage*

C *Either the rms or the average voltage may be larger*
D *There will always be a large difference between rms and average voltages*

5. A complete wave cycle consists of how many degrees?

A *360°*
B *180°*
C *90°*
D *720°*
E *None of the above*

6. If two ac signals of five volts each are combined, when voltage B is 180° out-of-phase with voltage A, what will the resultant voltage be?

A *5 volts*
B *0 volts*
C *10 volts*
D *2.5 volts*
E *None of the above*

7. In a circuit with both a dc voltage source and an ac voltage source, where the instantaneous voltage ranges from −2 volts to +10 volts, what is the peak ac voltage?

A *12 volts*
B *10 volts*
C *5 volts*
D *2 volts*
E *None of the above*

8. In the problem described in question 7, what is the rms value of the ac voltage?

A *4.242 volts*
B *6.66 volts*
C *13.32 volts*
D *5.4 volts*
E *None of the above*

9. In the problem described in questions 7 and 8, what is the dc voltage that the ac voltage is superimposed on?

A *+4 volts*
B *+6 volts*
C *0 volts*
D *−2 volts*
E *None of the above*

10. What voltage is applied to the third prong of a three prong ac plug?

A *110 volts*
B *220 volts*
C *77.77 volts*
D *0 volts*
E *None of the above*

6

Capacitance

It was mentioned in Chapter 3 that the most commonly used component in electronic circuits is the resistor. Probably the second most commonly used component is the *capacitor*.

WHAT IS CAPACITANCE?

If two metal plates are separated by an insulator (or *dielectric*), and a dc voltage is applied between the plates, current will not be able to cross the dielectric. But a surplus of electrons will be built up on the plate connected to the negative terminal of the voltage source, while there will be a shortage of electrons on the plate connected to the positive terminal. The voltage source will try to force electrons into one plate (negative terminal) and draw them out of the other (positive terminal).

At some definite point, these plates will be completely saturated. No further electrons can be forced into the negative plate, and no more electrons can be drawn from the positive plate. At this point, the plates have an electrical potential equal to that of the voltage source. In fact, the plates now act like a second voltage source in parallel with the first, and with the opposite polarity. Figure 6-1 shows the equivalent circuit. Naturally, since these opposing voltages are equal, they cancel each other out and no current can flow between the voltage source and the plates in either direction. The plates are said to be *charged*.

Now, if the voltage source is removed from the circuit, as in Fig. 6-2, the plates will stay charged, because there is no place for the electrons on the negative plate to

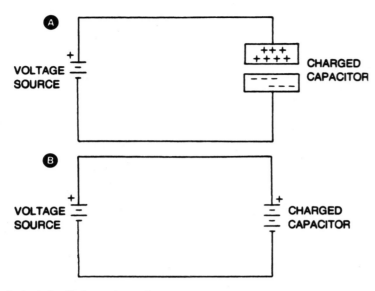

Fig. 6-1. Equivalent circuits for a charged capacitor.

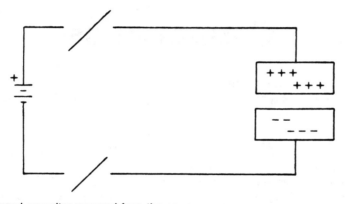

Fig. 6-2. Charged capacitor removed from the power source.

go. Similarly, there is no place for the positive plate to draw electrons from. The voltage is stored by the plates.

Replacing the missing voltage source with a resistor, as shown in Fig. 6-3 provides a current path for the excess electrons stored on the negative plate to flow to the positively charged plates. This will continue until both plates are returned to an electrically neutral state. This process is called *discharging* the plates.

Such a device (two conductive plates, separated by insulator) is called a *capacitor*. A capacitor is used to store electrical energy. At one time capacitors were known as *condensers*, but this term is somewhat misleading and has fallen largely into disuse. You may, however, run across it once in a while. Just remember, it is simply another name for a capacitor.

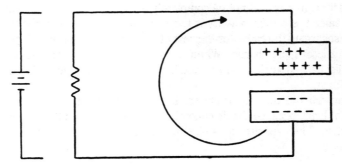

Fig. 6-3. Discharging a capacitor.

Fig. 6-4. Capacitor in an ac circuit.

A capacitor cannot hold a charge indefinitely. Even air can conduct some current, so the charge will slowly seep off into the air. This is a form of *leakage*. There will also be some leakage through the insulating dielectric. Of course, if all other factors are equal, the lower the internal leakage, the better the capacitor.

Now, let's consider what happens in a capacitor when an alternating current is applied to it. See Fig. 6-4. During the first part of the cycle, as the source voltage increases from zero, it will charge the plates of the capacitor in a manner similar to the dc circuit described above. The polarity of the charge capacitor opposes that of the source voltage.

The capacitor may or may not be completely charged by the time the applied voltage passes its peak and starts to decrease again (this will depend on the size of the plates, how much voltage is applied, and the frequency of the ac signal).

In either case, as the applied voltage decreases, a point will be reached when it is less than the charge stored in the capacitor. This will allow the capacitor to start discharging through the ac voltage source.

The capacitor may or may not be completely discharged when the ac voltage reverses polarity, but since the source polarity is the same as the capacitor polarity, the voltages aid, quickly discharging the capacitor the rest of the way, then charging it with the opposite polarity from the original charge. When the ac source voltage reverses direction, the capacitor is discharged again, and the entire process is repeated with the next cycle of the ac waveform.

If you constructed the circuit illustrated in Fig. 6-5 with a dc voltage source, the lamp would not light, because the dc current cannot flow through the circuit—it is blocked by the dielectric. The capacitor acts like an open circuit as far as direct current is concerned.

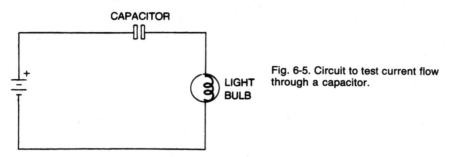

Fig. 6-5. Circuit to test current flow through a capacitor.

If, however, the same circuit is built with an ac voltage source, the lamp will light. See Fig. 6-6. This indicates that alternating current is flowing through the circuit. Of course, virtually no current (except the tiny leakage current) will flow across the dielectric itself. Remember, any given electron doesn't travel very far in an electric circuit. It merely moves far enough to disturb its neighbor. The process of charging, discharging, and recharging a capacitor from an ac voltage source, gives the same effect as if the current was actually flowing through the capacitor itself. Moreover, if we decrease the frequency of the ac source, the lamp will dim. Increasing the frequency will cause the lamp to burn brighter. A capacitor lets more current flow as the frequency of the source voltage is increased.

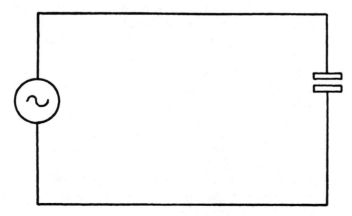

Fig. 6-6. Ac capacitor circuit.

İf we measured the dc resistance of a discharged capacitor, the meter's needle would show a sharp kick down to a moderately low resistance as the capacitor is charged. Then it will settle down to a very high resistance value. In an ideal capacitor we would have an infinite resistance, i.e., a completely open circuit.

However, we've already indicated that ac can flow through a circuit with a capacitor. The apparent resistance of a capacitor in an ac circuit is less than its dc resistance. This apparent ac resistance is called *capacitive reactance*. Its value decreases as the applied frequency increases. A capacitive reactance slows down voltage more than it does current, so the voltage lags the current by 90° (assuming a purely capacitive circuit).

Before we can present the formula for determining capacitive reactance, we need to know how capacitance is determined. The basic unit of capacitance is the *farad*. If one ampere of current flows when the applied voltage changes at a rate of one volt per second, we have one farad of capacitance (1 F).

In actual circuits, the farad is far too large a value. Instead, we generally use the *microfarad* which is one millionth of a farad. The microfarad can be abbreviated as μF. The abbreviation is generally preferred today. A still smaller unit is the *picofarad* (pF), which is one millionth of a microfarad. Sometimes a picofarad is referred to as a *micromicrofarad* ($\mu\mu$F).

Capacitance can be increased by making the metal plates larger, or by bringing them closer together or, in other words, making the dielectric thinner.

The specific dielectric used also has an effect on the capacitance of the unit. Air, for example, has a *dielectric constant* (K) of one. Mica, however, has a dielectric constant of six. Paper is between two and three, and titanium oxide can have a dielectric constant as high as one hundred and seventy.

Capacitance (in picofarads) can be calculated with the following formula:

$$C = \frac{0.0885 \times K \times A}{T} \qquad \textbf{Equation 6-1}$$

A represents the area of the side of one of the plates that is actually in physical contact with the dielectric. This area is measured in square centimeters for this equation. T represents the thickness of the dielectric (or the space between the plates), and is also measured in centimeters. K, of course, is the dielectric constant.

Let's assume we have a capacitor where A = 35 square centimeters, and T = 2 centimeters, and the dielectric is air (K = 1). Algebraically we have C = (0.0885 × K × A)/T = 0.0885 × 1 × 35)/2 = 3.0975/2 = 1.54875, or about 1.5 pF.

If, however, the dielectric is changed to mica (K = 6) and all other variables are the same, we find that C = (0.0885 × K × A)/T = 0.08854 × 6 × 35)/2 = 18.585/2 = 9.2925, or approximately 9 pF.

Now, let's try increasing T to 5. C = (0.0885 × 6 × 35)/5 = 18.585/5 = 3.717, or just under 4 pF. Increasing T decreases the capacitance, while increasing K increased the capacitance. The capacitance will also increase with an increase in A. For example, we can change A to 50 square centimeters. Then, C = (0.0885 × 6 × 50)/5 = 26.55/5 = 5.31, or just over 5 pF.

As one final example, let's see what happens when the dielectric constant is increased by a very large amount. We'll replace the mica (K = 6) with titanium oxide (K = 170).

Now the equation comes out to C = (0.0885 × K × A)/A/T = (0.0885 × 170 × 50)/5 = 752.25/5 = 150.45, or approximately 150 pF.

Now, assume you have a capacitor that uses paper as a dielectric (K = 2.5). A = 90 square centimeters, and T = 3 centimeters. What is the capacitance?

In actual practice, you will rarely need to use this formula, unless you are manufacturing capacitors, but it is helpful to understand how the variables interrelate in determining capacitance.

Returning to capacitive reactance, its value is determined by the capacitance and the frequency of the applied ac voltage. Reactance is stated in ohms, like ordinary dc resistance. The formula is as follows,

$$X_c = \frac{1}{2\pi FC} \qquad \textbf{Equation 6-2}$$

In this equation, F stands for the frequency (in hertz), and C represents the capacitance in farads (*not* in microfarads or picofarads). The symbol π is the Greek letter pi, and it is used in many formulas to represent a universal constant. The value of pi is always approximately 3.14, so $2 \times \pi$ = about 6.28. This means the formula can be written as $X_c = 1/(6.28 \times F \times C)$.

Notice that if the applied frequency is 0 (that is, dc) the value of the denominator becomes 0, regardless of the capacitance value, since zero multiplied by any number always equals zero. 1/0 is unsolvable—it equals infinity. Infinite resistance in a circuit, of course, acts like an open or incomplete circuit. This is exactly the way a capacitor behaves in a dc circuit.

Let's try a few examples with ac frequencies. We'll start with a 1 μF (1×10^{-6} farad) capacitor. The denominator equals $2\pi FC$, which we can rewrite as $6.28 \times (1 \times 10^{-6}) \times F$, or $6.28 \times 10^{-6} \times F$, so the reactance equation for a 1 μF capacitor is $X_c = 1/(6.28 \times 10^{-6} \times F)$.

If F is 10 Hz, we find that $X_c = 1/(6.28 \times 10^{-6} \times 10) = 1/(6.28 \times 10^{-5})$, or about 16,000 ohms (16 k). If F is increased to 50 Hz, $X_c = 1/(6.28 \times 10^{-6} \times 50) = 1/(3.14 \times 10^{-1})$, or approximately 3,200 ohms (3.2 k). If F is 600 Hz, then $X_c = 1/(6.28 \times 10^{-6} \times 600) = 1/(3.768 \times 10^{-3})$, or about 265 ohms.

You can clearly see that when the capacitance remains constant, and the frequency increases, the reactance decreases. Additional examples are given in Table 6-1.

Table 6-1. Capacitive Reactance Samples.

CAPACITANCE	FREQUENCY		
	10 Hz	100 Hz	1000 Hz
0.001 μF	16 megohms	1.6 megohms	160 kilohms
0.01 μF	1.6 megohms	160 kilohms	16 kilohms
0.1 μF	160 kilohms	16 kilohms	1600 ohms
1 μF	16 kilohms	1600 ohms	160 ohms
10 μF	1600 ohms	160 ohms	16 ohms
100 μF	160 ohms	16 ohms	1.6 ohms

OHM'S LAW AND REACTANCE

Ohm's law works with reactance in exactly the same way it does with regular dc resistance. The only difference that you must keep in mind is that the result of any specific equation is true only for a single, specific frequency.

For an example, we'll consider the circuit shown in Fig. 6-6, which consists of just a capacitor and an ac voltage source. For the sake of simplicity, we'll be considering only reactance. In this simple case, the effects of regular resistance can be considered minimal. In most practical circuits, however, both resistance and reactance would have to be considered. We'll see how this is done later.

We'll assume the capacitor has a value of 1 μF, so we can use the reactance values found in the examples in the last section. We'll also assume the ac voltage source generates 100 volts rms. According to Ohm's law, the current equals the voltage divided by the resistance, or, in this case, the voltage divided by the reactance. At 10 Hz, we know that a 1 μF capacitor has a reactance of about 16,000 ohms. I = E/X = 100/16,000 = 6.25×10^{-1} amperes, or 0.625 mA. This could also be written as 625 μA. When the frequency is increased to 50 Hz, the capacitive reactance drops to about 3,200 ohms. 100/3200 = 3.125×10^{-3} amperes, or about 3 mA. At 600 Hz, the reactance is a mere 265 ohms, so I = 100/265 = 0.038 amperes, or 38 mA. Notice that as the frequency goes up, the current also increases, while the reactance value goes down.

According to Ohm's law, current and resistance are always in an inverse relationship to each other. When one increases, the other decreases. Now, assume the voltage source generates 24 volts rms, and the capacitor is 0.2 μF (2×10^{-7}). How much current flows in the circuit at 10 Hz, 50 Hz, and 600 Hz?

TYPES OF CAPACITORS

In discussing dielectric constants, we mentioned that a number of different materials can be used as a dielectric in a capacitor, and the dielectric constant of each material helps to determine the capacitance of a device of a given size. Capacitors are generally classified according to their dielectric material. Some of the more common types of capacitors will be discussed in this section.

Regardless of the type of capacitor, the schematic symbol is generally the same. The two most common schematic symbols for capacitors are shown in Fig. 6-7. Most capacitor types are interchangeable.

Fig. 6-7. Schematic symbols for capacitors.

Ceramic Capacitors

Perhaps the most commonly used type of capacitor consists of a wafer of ceramic material between two silver plates. Leads are connected to the plates, and the entire assembly (except for the leads, of course) is encased in a protective plastic shield. See Fig. 6-8. This type is called a *ceramic capacitor*, or, because of its most common shape, a *ceramic disc*. Some ceramic capacitors, however, are enclosed in rectangular cases.

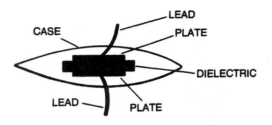

Fig. 6-8. Structure of a ceramic disc capacitor.

There is no electrical difference, and often the rectangular units are called ceramic discs anyway.

The values for ceramic capacitors usually range from about 10 pF to about 0.05 μF. Occasionally you'll also find ceramic capacitors with values up to 0.5 μF.

There are two important voltage ratings for capacitors. One is the *breakdown voltage*. This is the absolute maximum voltage the dielectric material can withstand without breaking down and rupturing. For most dielectrics the breakdown voltage is quite high (typically, more than 10 kilovolts), so you generally don't have to worry about it in practical circuits.

The second important voltage rating in a capacitor is the *working voltage*. This is the maximum dc voltage the manufacturer recommends to be placed across the plates safely. Exceeding this voltage can damage or destroy the capacitor, and perhaps other nearby components. The working voltage is usually much lower than the breakdown voltage.

One thing that is important to remember is that the working voltage is given in dc volts. Since ac ratings are usually given in rms units, you must find the peak voltage to determine if it can safely be applied to a given capacitor. For example, if a capacitor has a working voltage rating of 100 volts, you should not apply an ac voltage of more than 100 volts peak. In terms of rms voltage this is 100 × 0.707, or no more than 70.7 volts rms. If you put 100 volts rms across this device, the peak voltage would actually be 100 × 1.41, or 141 volts. This much overvoltage could easily damage the capacitor.

Working voltages for standard ceramic capacitors generally range from 50 to about 1600 volts. If a lower voltage unit can do the job, it will generally be less expensive, but a higher voltage rating can always be substituted. Capacitors with low working voltages also tend to be physically smaller.

Special high capacitance ceramic capacitors are also available with capacitances up to about 0.5 μF. Often this will be marked as 0.47 μF—capacitor values aren't precise enough for this small difference to matter. The working voltage for these large capacitance units is generally only about 10 to 100 volts.

The values are usually stamped right on the case of ceramic capacitors. This type of capacitor is very widely used because it is inexpensive, comparatively small, and has a low degree of power loss (leakage resistance). In radio frequency tuned circuits (very high frequencies) ceramic capacitors often prove to be unsatisfactory because of poor stability. But at frequencies up to about 100 kHz, they are generally an excellent choice.

Mica Capacitors

Another popular type of capacitor uses thin slices of mica sandwiched between a number of interconnected plates. See Fig. 6-9. Again, this type of capacitor is named

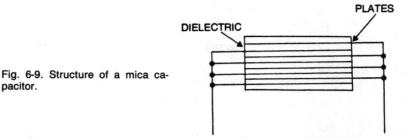

Fig. 6-9. Structure of a mica capacitor.

PLATES

DIELECTRIC

after its dielectric. It is called a *mica capacitor*. As a rule, mica capacitors are somewhat more expensive than ceramic discs, but they tend to be much more stable at radio frequencies. Their values typically range from 5 pF to 0.01 μF, and their working voltages are generally between 200 and 50,000 volts.

A special type of mica capacitor is the *silver-mica capacitor*. This is also sometimes called a *silvered mica capacitor*. In this type of capacitor a very thin coating of silver is applied directly onto the mica sheets, instead of the usual foil plates. These units are more expensive, naturally, but they can have capacitance tolerances as close as 1%. Most ordinary capacitors have a tolerance of only 10 to 20%. In most circuits a plus or minus error of 20% in the capacitance value really won't matter much, or can be compensated for by fine tuning a variable component. But in some special purpose circuits, precise values are needed, so silver-mica capacitors are used.

Sometimes a mica capacitor has its value stamped directly on its body, but other units are identified via a color code similar to the one used with resistors. Figure 6-10 shows the arrangement of color coding dots on a typical mica capacitor. Some indication, such as the arrow in the figure, will be given so that you will know which dot is which.

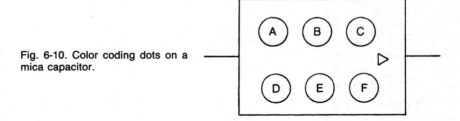

Fig. 6-10. Color coding dots on a mica capacitor.

The first dot (dot A) is generally either white or black. This simply indicates that the unit is a mica capacitor. Dot B is the first significant figure of the capacitance value, and dot C is the second significant figure. The value for each color is shown in Table 6-2. Notice that the values for dot B and dot C are identical to the resistor color code.

The multiplier is given by dot F, directly below dot C—not dot D at the beginning of the next row, as might be expected. Dot F corresponds to the third band on a resistor. The color values are basically the same as in the resistor color code, but capacitors don't have as wide a value range as resistors, so some colors aren't used as multipliers. Multiplying the significant figures by the multiplier gives the capacitance value in picofarads (pF). To convert to microfarads (μF), you must divide the result by 1,000,000.

Table 6-2. The Capacitor Color Code.

Color	Dot A	Dot B	Dot C
Black	Mica	0	0
Brown	--	1	1
Red	--	2	2
Orange	--	3	3
Yellow	--	4	4
Green	--	5	5
Blue	--	6	6
Violet	--	7	7
Gray	--	8	8
White	Mica	9	9
Silver	Paper	--	--

Color	Dot D	Dot E	Dot F
Black	±1000	±20%	X 1
Brown	±500	±1%	X 10
Red	±200	±2%	X 100
Orange	±100	±3%	X 1000
Yellow	−20 to +100	--	X 10000
Green	0 to +70	±5%	--
Gold	--	±5%	X 0.1
Silver	--	±10%	X 0.01

Dot D in the lower left hand corner gives the *temperature coefficient* for the capacitor. The capacitance will vary somewhat with changes in temperature. The temperature coefficient tells you the maximum extent of the fluctuation for a given change in temperature. It is given in parts per million per degree centigrade. For example, if dot D is red, the temperature coefficient is ± 200. This means if the temperature changes one degree centigrade, the value may increase or decrease by no more than 200 parts per million or 0.02%.

Notice that some capacitors can increase in value more than they decrease. In fact if dot D is green, the capacitance can only increase. The value marked is said to be the *minimum guaranteed value*.

Dot E simply gives the manufacturing tolerance of the capacitor. It corresponds to the fourth band on a resistor.

Let's suppose we have a capacitor with the following markings; Dot A is white, dot B is brown, dot C is black, dot D is orange, dot E is black, and dot F is brown. What is the value of this capacitor?

White at dot A simply indicates that this is a mica capacitor. Dot B is brown, which means the first significant figure is 1. Black at dot C means the second significant figure is 0. The value of the capacitor is 10 times the multiplier value. Dot D is orange so the capacitor has a temperature coefficient of plus or minus 100 parts per million (0.01%). The tolerance is given by dot E, which is black. This guarantees the actual value is within plus or minus 20% of the value marked. Dot F is the multiplier. It is brown on our sample resistor, so the multiplier is 10. The value of this capacitor is 10 × 10, or 100 pF plus or minus 20%, with a temperature coefficient of plus or minus 0.01%. Since the marked value has a tolerance of 20%, the actual value of the capacitor can range from 80 to 120 pF and still be correctly marked.

Let's assume the capacitance value is exactly as marked—100 pF, and the temperature changes 3 degrees centigrade. This will cause the value of the capacitor to change as much as 0.03 pF in either direction.

Now, let's suppose the temperature changes 50° centigrade. The value can change by plus or minus 0.5 pF. That is still not very much. You can see that the temperature coefficient can usually be ignored in practical circuits unless very wide fluctuations of temperature are expected. If dot A is black, it simply means the capacitor was manufactured to military specifications.

Paper Capacitors

Sometimes a higher capacitance is needed in a circuit than a ceramic disc or a mica capacitor can provide. Another type of capacitor can be formed by placing a strip of waxed paper (dielectric) between two long strips of tin foil (plates). These foil strips are often several feet long, but only about an inch wide. This assembly can be tightly rolled up to save space, and it is usually enclosed in a cardboard tube or plastic container. Such a device is called a paper capacitor or a tubular capacitor. See Fig. 6-11. Of course, leads are connected to the foil strips and brought out through the ends of the cardboard tube. In some units, the capacitor is sealed in a metal or plastic case to prevent moisture from getting into the capacitor.

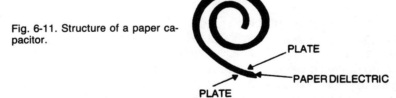

Fig. 6-11. Structure of a paper capacitor.

PLATE

PAPER DIELECTRIC

PLATE

The capacitance range for paper capacitors is from about 0.0001 μF to about 1 μF, with working voltages from 200 to 5000 volts.

Paper capacitors are relatively inexpensive, but they have a fairly low leakage resistance (a paper capacitor has a lower leakage resistance than, say, a mica capacitor), and they have some trouble giving optimum performance at the upper radio frequencies because of high dielectric losses. However, they are an excellent choice when a fairly large amount of capacitance is needed in a low or medium frequency circuit, especially when space is at a premium.

Paper capacitors generally have their values stamped directly onto their bodies, but sometimes they are color coded with the same system used for mica capacitors. For a paper capacitor, dot A is always silver.

Synthetic Film Capacitors

A similar type of capacitor can be made using a thin coating of synthetic film instead of waxed paper as the dielectric. Such a capacitor is usually identified by the specific type of film it uses. For example, there are *polystyrene capacitors*, *Mylar capacitors*, and *polyester capacitors*.

These synthetic films are usually thinner than the paper used in paper capacitors, so a film capacitor of a given value will tend to be somewhat smaller than a paper capacitor of the same capacitance.

Another advantage of these film capacitors is that they can operate over a wider temperature range than the paper units. In other words, they have a lower temperature coefficient. They also tend to be somewhat more precise, and closer to the marked value. That is, the tolerance is smaller. The capacitance of synthetic film capacitors ranges from 0.001 μF to about 2 μF. The working voltages are generally between 50 and 1000 volts.

Electrolytic Capacitors

The capacitors we have been talking about so far are *non-polarized*. That is, they can be placed in a circuit with the leads connected in either direction. Other capacitors are designed to work only in dc circuits and can only be hooked up one way. Such a component is said to be *polarized*.

The schematic symbol for a polarized capacitor is basically the same as for a regular capacitor, but a "+" is added at the positive terminal to indicate polarity. See Fig. 6-12. This terminal must be kept positive with respect to the other terminal.

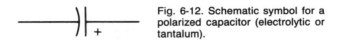

Fig. 6-12. Schematic symbol for a polarized capacitor (electrolytic or tantalum).

The most common type of polarized capacitor is the electrolytic capacitor. This type of capacitor consists of a pasty, semi-liquid electrolyte between aluminum foil electrodes, or plates. The positive plate is specially treated to form a thin oxide film on its surface. This film serves as the capacitor's dielectric. If the capacitor is subjected to a reverse polarity voltage, this thin layer of film could be punctured, ruining the capacitor.

Because the dielectric is such a thin film, very high capacitance values can be achieved in a reasonably small space. Typical values range from 0.47 μF to 10,000 μF (0.01 farad), and occasionally even higher. Working voltage ratings can be anywhere from 3 to 700 volts. The leads on an electrolytic capacitor can be placed in either an *axial* or a *radial* arrangement.

Internally an electrolytic capacitor resembles the construction of a paper capacitor, except, of course, the dielectric is considerably thinner.

One problem with electrolytic capacitors is that the semiliquid electrolyte can dry out, rendering the capacitor useless. This generally occurs when the capacitor is not in use—applying a voltage across it seems to prevent the electrolyte from drying out—so it is not a good idea to stock pile electrolytic capacitors in large quantities for possible future use.

Similarly, drying out can sometimes occur if the electrolytic capacitor is operated from too low a voltage. If the voltage is extremely low, then, as far as the capacitor is concerned, it is still just sitting on the shelf. Certainly a 15 volt capacitor could be used in a 5 volt circuit, but you wouldn't want to use a 500 volt unit. Besides, capacitors with higher working voltages are much larger, heavier, and more expensive than smaller units.

Preferably an electrolytic capacitor should be operated at a voltage between one third and two thirds of its rated maximum. This is enough to keep the electrolyte from drying out, but allows for unexpected over-voltage surges.

It might seem that the electrolytic capacitor would be of rather limited value because of its inability to function in an ac circuit, but actually it has a large number of important applications. These include power supply filtering, circuit coupling, audio frequency bypassing and achieving large time constants in RC circuits (described later in this chapter).

Electrolytic capacitors usually have a very wide tolerance. They might vary from their nominal value by as much as plus or minus 50 to 70%. Except in critical timing applications, the precise value is rarely important, so this loose tolerance is acceptable.

Tantalum Capacitors

Closely related to the electrolytic capacitor is the *tantalum capacitor* (see Fig. 6-13). Like the electrolytic capacitor, it can only be used in dc circuits. A dot on the body of the capacitor indicates the positive lead.

Fig. 6-13. A tantalum capacitor.

Tantalum capacitors offer a number of advantages over their electrolytic counterparts. They tend to be considerably smaller for a given capacitance value. They aren't as prone to drying out. They have less leakage. Their values tend to be more precise (lower tolerance) and they are much less susceptible to noise. Especially because of the last reason, they are generally preferred in computer applications. On the other hand, they tend to be more expensive than electrolytic capacitors, and they are limited to a much smaller range of capacitances and working voltages. Usually they range only from 0.5 μF to about 50 μF with working voltages that are rarely higher than 50 volts.

Variable Capacitors

Most capacitors have a single fixed value, but, just as there are variable resistors, there are *variable capacitors* too. One type uses a springy material for the plates. If they weren't held in place, the plates would fly apart. Between the plates is a piece of dielectric material, such as mica. The assembly is held together by a screw. By tightening or loosening this screw, the distance between the plates can be changed, thus altering the effective capacitance of the device. Figure 6-14 illustrates the structure of this type of capacitor. This kind of variable capacitor is called a *padder capacitor*, or a *trimmer capacitor*.

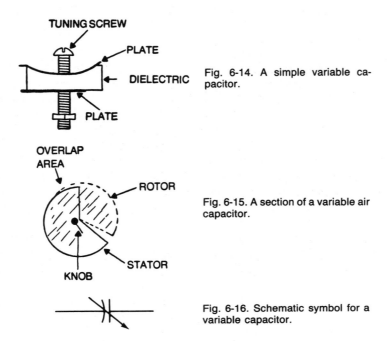

Fig. 6-14. A simple variable capacitor.

Fig. 6-15. A section of a variable air capacitor.

Fig. 6-16. Schematic symbol for a variable capacitor.

Another type of variable capacitor consists of a series of interweaved metal plates. One set (called the *stator*) is stationary. The other set of plates (called the *rotor*) can be moved by a knob. The amount of overlap between the plates determines their effective area. Moving the rotor electrically acts like the plates are being increased or decreased in area. Figure 6-15 illustrates the way this works. The dielectric in this type of capacitor is simply air. It is called a *variable air capacitor*. The schematic symbol for either type of variable capacitor is shown in Fig. 6-16.

RC TIME CONSTANTS

If a resistor and a capacitor are connected in series across a voltage, as in Fig. 6-17A, the capacitor will be charged through the resistor at a specific rate, determined by both

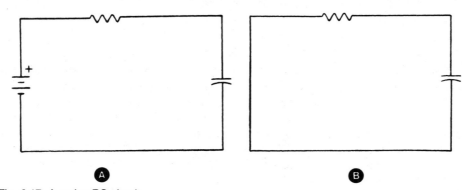

Fig. 6-17. A series RC circuit.

the capacitance and the resistance. The capacitance, of course, determines how many electrons the negative plate can hold when fully charged, and the resistance slows down the flow of electrons.

The time it takes for the capacitor to be charged to 63% of its full potential charge level is called the *time constant* of the combination. For obvious reasons, such a combination of a resistor and a capacitor is called an *RC circuit*.

Similarly, if the voltage source is removed, as in Fig. 6-17B, the capacitor will be discharged through the resistor. In this case, the time constant is defined as the time it takes the capacitor to drop down to 37% of its fully charged value. The charging time constant and the discharging time constant are always equal.

The time constant of a specific RC circuit can be found simply by multiplying the resistance and the capacitance;

$$T = RC \hspace{3cm} \text{Equation 6-3}$$

T is the time constant in seconds, R is the resistance in megohms (1 megohm = 1,000,000 ohms), and C is the capacitance in microfarads (μF).

For example, if the resistor has a value of 100,000 ohms (0.1 megohm) and the capacitor is 10 μF, the time constant will be equal to 0.1 megohm × 10 μF, 1 second. The same time constant could be achieved with other RC combinations. For example, 1 megohm and 1 μF, or 10,000 ohms and 100 μF would also result in a one second time constant.

As another example, let's take a 470,000 ohm (0.47 megohm) resistor and a 2.2 μF capacitor. The time constant in this circuit would equal 0.47 megohm × 2.2 μF, or 1.034 second.

If the resistor is 22,000 ohms (0.022 megohm), and the capacitor is 0.3 μF, the time constant is 0.022 × 0.3, or 0.0066 second (6.6 milliseconds).

Here are two more examples for you to work out on your own. What is the time constant for an RC circuit consisting of a 3.9 megohm resistor and a 0.68 μF capacitor? How about a 910,000 ohm resistor and a 15 μF capacitor?

FILTERS

A *filter* is a circuit that allows some frequencies to pass through it, while others are blocked. A capacitor is automatically a sort of filter by definition, since it allows higher frequencies to pass through it easily, while it blocks a dc signal (i.e., 0 Hz). A filter that passes high frequencies, but blocks low frequencies is called a high-pass filter.

The circuit for a practical, but simple high-pass filter is shown in Fig. 6-18. High frequencies are passed with little or no *attenuation*. That is, they are not reduced in

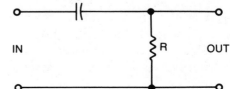

Fig. 6-18. Passive high-pass filter.   IN   R   OUT

level. But, at some specific frequency, the filter starts to impede the passage of the signal. This is the *cut-off frequency*.

It would be impossible to design a practical filter that would completely block all signals below the cut-off-frequency, and completely pass all frequencies above it. Instead, the level of the signal allowed to pass through the filter begins to drop off at a regular rate. This dropping off is called the *slope* of the filter. Obviously, the steeper the slope, the better the filter. For a simple filter, such as the one in Fig. 6-18, the slope is quite broad. A graph of the filter's frequency response is shown in Fig. 6-19.

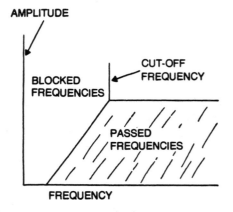

Fig. 6-19. A frequency response graph for a high-pass filter.

The formula for finding the cut-off frequency of this type of filter is,

$$F_c = \frac{158,000}{RC} \qquad \textbf{Equation 6-4}$$

Resistance (R) is in megohms, and capacitance (C) is in microfarads, so the resulting cut-off frequency (F_c) is in hertz.

Notice that the denominator in this equation is the formula for finding the time constant of an RC circuit. The formula could be rewritten as,

$$F_c = \frac{159,000}{T} \qquad \textbf{Equation 6-5}$$

T is the time constant in seconds.

Let's try a few examples with equation 6-4. Let's suppose the circuit in Fig. 6-18 consists of a 3.3 megohm resistor (3,300,000 ohms) and a 0.68 μF capacitor. The cut-off frequency would equal 159,000/(3.3 × 0.68) = 159,000/206.04, or approximately 772 hertz.

If we replace the resistor with a 820,000 ohm (0.82 megohm) unit, and keep the same capacitor, the cut-off frequency now becomes 159,000/(0.82 × 0.68) = 159,000/0.5576, or approximately 285,000 Hz.

Now let's keep the 0.82 megohm resistor and use a 0.002 μF capacitor. F_c = 159,000/(0.82 × 0.002) = 159,000/0.00164 = 96951220 Hz, or just under 97 MHz.

Notice that if either the capacitance or the resistance is decreased in value, the time constant decreases, and the cut-off frequency increases. Of course, it also works in the other direction. Increasing either the resistance or the capacitance will increase the time constant and decrease the cut-off frequency.

Try finding the cut-off frequency for the following combinations.

R	C
470,000 ohms	0.1 μF
3.9 megohms	22 μF
22 kilohms	150 μF

Now, what would happen if we reversed the positions of the resistor and the capacitor, as shown in Fig. 6-20?

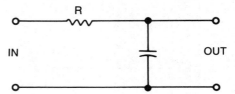

Fig. 6-20. Passive low-pass filter.

In this case the capacitor provides a *short-circuit*, or *shunt* across the signal source for high frequencies, but not for low frequencies and dc, which can't pass through the capacitor. Since current will tend to flow through the path with the least resistance (or reactance), very little high-frequency current will get through to the output.

Low frequencies, however, face a high reactance if they try to flow through the capacitor to ground, so they will appear at the output.

This filter, then, passes low frequencies, but blocks high frequencies. Of course, this circuit is called a *low-pass filter*. It operates as an exact mirror image of the high-pass version, as is indicated by its frequency response graph (Fig. 6-21). The formula for finding the cut-off frequency is identical for both circuits.

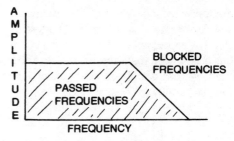

Fig. 6-21. A frequency response graph for a low-pass filter.

COMBINATIONS OF CAPACITORS

Now, let's examine what happens when we have more than one capacitor in a circuit.

80

Capacitors in Parallel

A circuit with two capacitors in parallel, as shown in Fig. 6-22, can be drawn more pictorially, as in Fig. 6-23. Since plates A and B are tied together, they are at the same electrical potential. We can think of them as a single plate. Similarly, plates C and D are electrically combined into an apparent single plate.

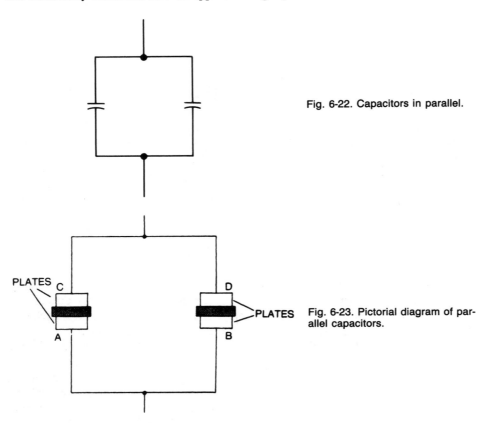

Fig. 6-22. Capacitors in parallel.

Fig. 6-23. Pictorial diagram of parallel capacitors.

Remember that the larger the surface area of the plates in a capacitor, the higher the capacitance will be. Obviously, combination plate A-B is going to be larger than either plate A or plate B separately. The same is true of combination plate C-D. So the total effective capacitance of multiple capacitors in parallel always increases. The total capacitance is larger than any of the separate, component capacitances.

In fact, we can simply add the capacitances of capacitors in parallel. That is, for n capacitors in parallel:

$$C_T = C_1 + C_2 + \ldots C_n \qquad \textbf{Equation 6-6}$$

Notice that this is the same as the formula for finding the total resistance of multiple resistors in series.

Fig. 6-24. Capacitors in series.

Capacitors In Series

Similarly, capacitors connected in series, as in Fig. 6-24, work against each other, reducing the total effective capacitance of the circuit. The formula for capacitors in series mirrors the formula for multiple resistors in parallel:

$$\frac{1}{C_T} = \frac{1}{C_1} + \frac{1}{C_2} + \cdots \frac{1}{C_n}$$ **Equation 6-7**

Therefore, two 0.1 μF capacitors in series would act like a single 0.05 μF capacitor. If the same capacitors were connected in parallel they would equal 0.2 μF.

Of course, both series and parallel combinations of capacitances can be included within a single circuit, just as with resistances.

For example, consider the string of capacitors in Fig. 6-25. We'll assume C1 is 0.1 μF, C2 is 0.033 μF, C3 is 0.0015 μF, and C4 is 0.22 μF. First solve for the series combination of C1 and C2. $1/C_T = 1/C1 + 1/C2 = 1/0.1 + 1/0.033 = 10 + 30.30303$ = about 40. Taking the reciprocal, we find the series combination of C1 and C2 is approximately 0.025 μF. This capacitance is in parallel with C3, so C_T = 0.025 + 0.0015 = 0.0265.

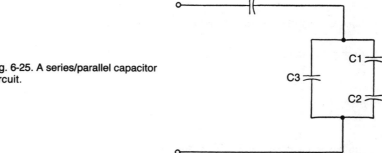

Fig. 6-25. A series/parallel capacitor circuit.

This effective capacitance is in series with C4, so the effective capacitance for the entire circuit equals $1/0.0265 + 1/0.22 = 37.7 + 4.5 = 42.2$—taking the reciprocal we conclude that the total effective capacitance is just under 0.024 μF.

Now, find the total effective capacitance for the circuit if C1 = 15 μF, C2 = 0.47 μF, C3 = 3.3 μF, and C4 = 2.2 μF.

STRAY CAPACITANCES

Since a capacitor is simply two conducting surfaces, separated by an insulator, small, unintentional capacitances can be formed by adjacent wires, or component leads. Generally, these *stray capacitances* are far too small to be of any real significance, but

82

in some very high frequency circuits (such as radio circuits) they can be very troublesome. These undesirable capacitances can allow signals to pass into portions of the circuit where they could hinder proper operation.

To prevent such stray capacitances in high frequency circuits, leads should be as short as possible reducing the effective plate area. Leads should also be *shielded* (i.e., enclosed in a conductor that is connected to ground—either earth ground or common ground) if more than just a few inches long.

Self-Test

1. A dielectric is which of the following?

A *A conductive plate in a capacitor*
B *A measurement of capacitance*
C *An insulator between two metal plates in a capacitor*
D *A charged particle*
E *None of the above*

2. Which of the following describes the action of a capacitor?

A *Opposes changes in current flow*
B *Converts ac into dc*
C *Creates a dc resistance*
D *Stores electrical energy*
E *None of the above*

3. A capacitor displays more resistance to what type of signals?

A *Low frequency signals*
B *ac signals*
C *High frequency signals*
D *Out-of-phase signals*
E *None of the above*

4. Assuming an ideal capacitor with no leakage, what will be the capacitive reactance of a 10 μF capacitor at dc (0 Hz)?

A *0 ohms*
B *1,000,000 ohms*
C *16,000 ohms*
D *63 ohms*
E *None of the above*

5. What is the capacitive reactance of a 22 μF capacitor at 500 Hz?

A *0 ohms*
B *144 ohms*

C *1,000,000 ohms*
D *14.5 ohms*
E *None of the above*

   6. What is the capacitive reactance of a 22 μF capacitor at 7,500 Hz?

A *0 ohms*
B *96 ohms*
C *1 ohm*
D *14.5 ohms*
E *None of the above*

   7. What is the capacitive reactance of a 0.033 μF capacitor at 4,300 Hz?

A *5 ohms*
B *1122 ohms*
C *10,000 ohms*
D *0 ohms*
E *None of the above*

   8. If two 0.25 μF capacitors are connected in series, what will the total effective capacitance be?

A *0.50 μF*
B *0.0625 μF*
C *0.125 μF*
D *2.5 μF*
E *None of the above*

   9. If two 0.25 μF capacitors are connected in parallel, what will the total effective capacitance be?

A *0.50 μF*
B *0.0625 μF*
C *0.125 μF*
D *2.5 μF*
E *None of the above*

   10. What is the cut-off frequency of a low-pass filter made up of a 0.56 μF capacitor and a 330 k resistor?

A *860 kHz*
B *5.4 Hz*
C *54 Hz*
D *850 Hz*
E *None of the above*

7

Magnetism and Electricity

Very closely related to the concept of electricity is the concept of magnetism. In this chapter we will study how these two phenomena interact.

WHAT IS A MAGNET?

Magnetism has been known to mankind for well over two-thousand years. The ancient Greeks discovered a peculiar lead-colored stone that had the mysterious ability to attract small particles of iron ore. Some time later, the Chinese found a practical use for this seemingly magical stone. They learned that if a piece of this stone was suspended on a string, or floated on a liquid, it would always try to point in one specific direction (north). Since they used this device to lead them through the desert, the stone came to be called *lodestone,* i.e., the leading stone.

We know now that the lodestone is a natural *magnet.* While, in some ways, magnetism is still rather mysterious, much is now known about its properties. Magic is not involved.

We can make magnets out of certain other materials, even though they aren't naturally magnetized. Lodestone is a fairly weak magnet, but stronger magnets can be made of iron, nickel, cobalt, or steel.

The two opposite ends of a magnet are called the *poles.* See Fig. 7-1. One pole will tend to point towards the earth's north pole if the magnet is floated or freely suspended. This north-seeking pole is called the *north pole* of the magnet. The other pole is referred to as the *south pole.*

Fig. 7-1. Poles of a magnet.

Remember that in an electrical circuit like charges repel and opposite charges attract. The same effect occurs with magnetic poles. If two magnets are brought together, north pole to north pole, they will try to repel each other. If, however, one of the magnets is turned around so that the north pole of one magnet is facing the south pole of the other, the magnets will exhibit a strong attraction towards each other.

If a bar shaped magnet is placed under a sheet of paper and some iron filings are spilled on top of the paper, then the paper is gently shook, the filings will tend to arrange themselves into a pattern like the one shown in Fig. 7-2.

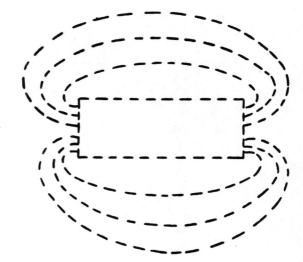

Fig. 7-2. Magnetic lines of force.

Notice that the iron filings arrange themselves in a set of parallel lines arcing from one pole to the other. These lines never cross or unite. They are an indication of the *magnetic lines of force,* or *flux.* The area they cover is the *magnetic field.*

The flux flows from the north pole to the south pole of the magnet, just as electrical current flows from the negative terminal to the positive terminal of a voltage source. The flux is produced by a force called *magnetomotive force.* Magnetomotive force is somewhat analogous to electrical voltage (which is also sometimes called *electromotive force*).

Just as certain substances conduct electrical current better than others, certain substances allow magnetic lines of flux to pass through them more readily than other substances. In other words, some materials present a greater resistance to the flux. The magnetic equivalent to resistance is called reluctance.

86

The similarities between magnetism and electricity are so strong, that Ohm's law applies to magnets too. In magnetic circuits flux equals magnetomotive force divided by reluctance, which directly reflects the electrical formula, current equals voltage divided by resistance (I = E/R).

PRODUCING MAGNETISM WITH ELECTRICITY

When an electric current passes through a conductor, such as a piece of copper wire, a weak magnetic field is produced. The magnetic lines of force encircle the wire at right angles to the current flow, and are evenly spaced along the length of the conductor. See Fig. 7-3. The strength of the magnetic field decreases at greater distances from the conductor. The size, and overall strength of the magnetic field is dependent on the amount of power flowing through the electrical circuit, but it is always fairly weak. The magnetic force surrounding the conductor can, however, be dramatically increased by winding the wire into a coil, so the lines of force can interact and reinforce each other.

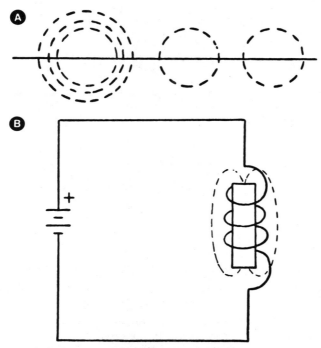

Fig. 7-3. The magnetic field surrounding an electric conductor.

An even greater magnetomotive force can be generated if the coil is wound around a piece of low reluctance material, such as soft iron.

Since the magnetomotive force vanishes as soon as the current stops flowing in the wire, we have a magnet that can be turned on and off. The strength of the magnet is also electrically controllable. Such a device is called an *electromagnet*.

PRODUCING ELECTRICITY WITH MAGNETISM

Since we can produce magnetism with an electrical current, it shouldn't be surprising that we can also produce electricity with a magnet.

Look at Fig. 7-4. It is basically the same set-up as in Fig. 7-3, but there is no electrical voltage source, and the material in the center of the coil (the *core*) is a permanent magnet.

Fig. 7-4. Producing electricity with a magnet.

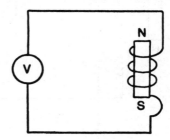

If we move the magnet up through the coil of wire, an electric current will start to flow through the wire. The strength of this induced current depends on a number of factors. These include the intensity of the magnetic field, how many lines of force are cut by the conductor, the number of conductors (each turn of the coil acts like a separate conductor in this case) cutting across the lines of force, the angle at which the lines are cut, and the speed of the relative motion between the magnet and the conductor.

This current will continue to flow until either the magnet is too far away for any of its lines of force to cut across the conductor, or the magnet stops moving.

If the magnet and the coil are stationary with respect to each other, no current is induced. Then, if we push the magnet back down through the coil (the direction of the movement is reversed) current will also flow, but it will have the opposite polarity. That is, it flows in the other direction.

The exact same effect can be achieved if the magnet is stationary and the conductor is moved. It is the relative motion between the components that is important.

All this might not seem terribly useful, since you have to keep moving the magnet or the coil back and forth to produce a continuing current. The current will keep reversing polarity each time the direction of movement is changed; but this is actually a very efficient method of producing electricity.

This is the way the power companies produce their high wattage ac power. Any of a number of mechanical means can be used to rotate a conductor between a magnetic north pole and south pole (see Fig. 7-5). It is usually more practical to rotate the conductor rather than the magnet. Since the conductor is rotating between the magnetic poles, the direction of its relative movement between the poles appears to alternate, so the induced current, as mentioned above, is an alternating current. Very large amounts of electrical power can be produced in this manner.

Self-Test

1. Which of the following materials can *not* be used to make a magnet?

A *Lodestone*
B *Cobalt*

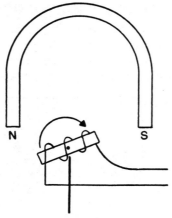

Fig. 7-5. Producing ac electricity with a magnet.

C *Carbon*
D *Iron*
E *Nickel*

2. What are the ends of a magnet called?

A *Poles*
B *Lodestones*
C *Ions*
D *Armatures*
E *None of the above*

3. What is another name for magnetic lines of force?

A *Magnetic field*
B *Flux*
C *Magnetic pole*
D *Lodestone*
E *None of the above*

4. If like poles of two magnets are brought near each other, what will happen?

A *They will attract each other*
B *The magnets will be damaged*
C *They will repel each other*
D *Nothing*

5. If a magnetic pole is freely suspended, which direction will the north pole of the magnet point to?

A *North*
B *South*

C *East*
D *West*
E *The direction will be random*

    6. What is the magnetic equivalent to electrical voltage?

A *Flux*
B *Magnetomotive force*
C *Reluctance*
D *Magnetic field*

    7. What is the magnetic equivalent to electrical current?

A *Flux*
B *Magnetomotive force*
C *Reluctance*
D *Magnetic field*

    8. What is the magnetic equivalent to electrical resistance?

A *Flux*
B *Magnetomotive force*
C *Reluctance*
D *Magnetic field*

    9. An electrical voltage can be induced with a coil and a magnet by which of the following?

A *Placing the coil at right angles to the magnetic field*
B *Placing the coil parallel to the magnetic field*
C *Holding both the magnet and the coil perfectly stationary*
D *Moving either the magnet or the coil*
E *None of the above—it can't be done*

    10. Rotating an armature in a magnetic field produces what type of electricity?

A *Static*
B *ac*
C *dc*
D *Pulsating dc*
E *None*

8

Inductance

In the last chapter it was pointed out that winding a wire carrying an electric current into a coil will increase the electromagnetic effect. Another result of passing current through a coil of wire is a phenomenon called *inductance*. Inductance is another important factor in electronic circuits.

WHAT IS INDUCTANCE?

Since an electric current through a coil of wire can create a magnetic field, and a magnetic field moving relative to a coil of wire can create an electric current, what happens when the current flowing through a coil changes?

As long as current flows through the coil at a steady, constant level, and in just one direction (dc), a non-moving magnetic field is generated. As long as the magnetic field and the coil are stationary in relation to each other, the magnetic field will have no particular effect on the current flow through the coil. But if the current through the coil starts to drop, the magnetomotive force generated by the coil will also be decreased, causing the magnetic lines of force to move in closer. Some of these moving lines of force will cut through some of the turns of the coil, inducing an electric current in the coil. This induced current will flow in the same direction (same polarity) as the original current.

Of course, this induced current passing through the coil will produce a magnetic field of its own. Plainly this means some finite period of time is required for this back and forth effect to die down. Current through a coil cannot be stopped or reversed in polarity instantly. Inductance tends to oppose any change in the current flow.

In some ways inductance is the opposite of capacitance. Capacitance offers very little resistance to high frequencies, but opposes low frequencies, or dc (constant current). Inductance, on the other hand, passes dc with practically no resistance, but opposes higher ac frequencies (changing current). This opposition to high frequencies is called *inductive reactance*.

Inductance is measured in *henries*. One henry is the inductance in a circuit in which the current changes its rate of flow by one ampere per second and induces one volt in the coil.

The henry is too large a unit for practical electronic circuits, so the millihenry (one thousandth of a henry) is more commonly used. This is abbreviated as mH.

INDUCTIVE REACTANCE

The formula for inductive reactance is:

$$X_L = 2\pi FL \hspace{4cm} \text{Equation 8-1}$$

where X_L is the inductive reactance in ohms, L is the inductance in henries (not millihenries), and F is the frequency in Hertz. 2π, of course, is a constant, equalling approximately 6.28.

Let's suppose we have a circuit with 100 millihenries of inductance (0.1 henry). If the frequency of the source voltage is 60 Hz, then the inductive reactance equals 6.28 × 60 × 0.1, or just under 38 ohms.

If the same circuit is used, but the applied frequency is increased to 500 Hz, the inductive reactance becomes 6.28 × 500 × 0.1, or 314 ohms.

Raising the ac frequency still further, to 2000 ohms, brings the inductive reactance up to 6.28 × 2000 × 0.1, or 1256 ohms.

If the frequency remains constant, but the inductance is increased, then the reactance will also be increased. For example, we've already found that 100 millihenries in a 60 Hz circuit results in an inductive reactance of about 38 ohms. If we increase the inductance to 500 mH (0.5 henry), and keep the frequency at 60 Hz, the inductive reactance comes out to 6.28 × 60 × 0.5, or approximately 188 ohms.

You'll notice the relationship between the frequency and the inductance to the inductive capacitance is just the opposite of that of the frequency and the capacitance to the capacitive reactance.

Determine the inductive reactance of a circuit consisting of 25 mH (0.025 henry) of inductance at 50 Hz, 300 Hz, and 4000 Hz. Then change the inductance to 300 mH (0.3 henry) and solve for the same three frequencies.

COILS

In electronics, the component called a *coil* is just that—a coil of insulated wire wound around some core. This core may be made of powdered iron, or some other magnetic material, or it may simply be air, or a small cardboard tube.

The inductance of a coil is determined by a number of factors; the width of the core, the diameter of the wire, the number of turns of the wire around the core, and the spacing between the turns of the coil, to name just a few of the more important factors.

The material the core is made of is also important. A core with low magnetic reluctance can increase the strength of the magnetic field, thereby increasing the strength of the induced voltage. Figure 8-1 shows the construction of a typical coil.

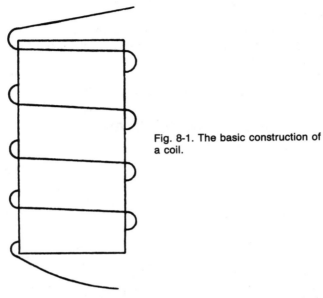

Fig. 8-1. The basic construction of a coil.

Some coils are adjustable. Usually the core is constructed so it can be moved slightly in and out of the center of the coil with a screw called a *slug*. This makes the core appear to be partially made of air, and partly of (usually) powdered iron, thereby altering the reluctance of the core, and thus, the inductance of the coil.

Many coils have the wires visibly exposed (but insulated, of course) but some are sealed in metal cans to avoid interaction with other components. Without this *shielding*, the magnetic field could induce a voltage in other nearby components. Obviously inducing a voltage where it's not intended can be quite detrimental to circuit operation.

The wires in a coil must usually be insulated, because they are generally wound quite closely together. If separate turns of uninsulated wire shift position and touch, allowing current to pass between them, a *short circuit* exists, making the coil appear to have fewer turns, as far as the current is concerned. See Fig. 8-2.

Figure 8-3 shows the most common schematic symbols for coils. A and B are fixed inductance coils, while the arrows through symbols C and D indicate that these coils are adjustable.

Sometimes special schematic symbols are used to indicate the core material of the coil. In this system, all of the symbols in Fig. 8-3 are for air-core inductors. If two dotted lines are drawn beside the symbol, as in Fig. 8-4A, it means the core is made of powdered iron. Two solid lines, as in Fig. 8-4B, indicate that the core is made of stacks of thin sheet iron.

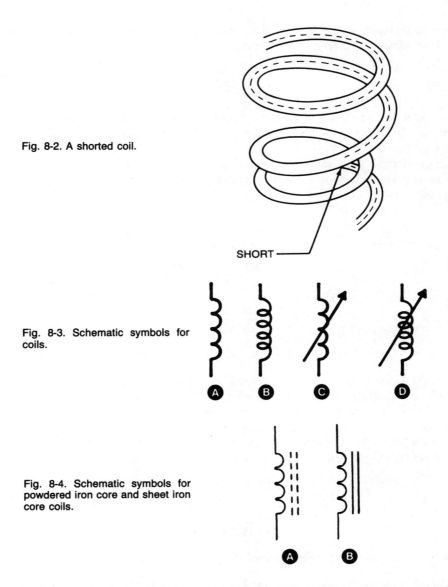

Fig. 8-2. A shorted coil.

SHORT

Fig. 8-3. Schematic symbols for coils.

Ⓐ Ⓑ Ⓒ Ⓓ

Fig. 8-4. Schematic symbols for powdered iron core and sheet iron core coils.

Ⓐ Ⓑ

Most schematic diagrams, however, simply use the standard symbols shown in Fig. 8-3, and the type of coil is specified in the parts list. Coils (and inductance values) are usually represented by the letter L, since I is commonly used for current.

WINDING COILS

A number of factors determine the inductance of a coil. These include the core material and diameter, the number of turns in the coil, and how closely they are spaced.

For a single layer coil (no over-lapping windings) on a non-magnetic core, the formula for determining inductance is as follows;

$$L = \frac{0.2d^2N^2}{3d + 9l}$$

where L is the inductance of the coil in millihenries (mH), d is the diameter of the coil winding in inches, l is the length of the coil winding in inches, and N is the number of turns in the coil.

For example, let's say we have a coil on a 0.75 inch diameter core, consisting of 150 closely wound turns of #32 enameled wire. The length of the coil is 1.2 inches. The inductance works out to:

$$L = \frac{0.2 \times 0.75^2 \times 150^2}{3 \times 0.75 + 9 \times 1.2}$$

$$= \frac{0.2 \times 0.5625 \times 22500}{2.25 + 10.8}$$

$$= \frac{2531.25}{13.05}$$

$$= 193.96552 \approx 194 \text{ mH}$$

If the number of turns (N) is increased, while the diameter (d) is held constant, the inductance (L) will be increased. The amount of increase will depend on whether the length is increased by the added windings, or if it is held constant to the original value of l, by squeezing the turns more tightly together.

Let's return to our earlier example and double the number of turns (2 × N). All of the other values, including length (l) will remain the same:

$$d = 0.75$$
$$l = 1.2$$
$$N = 150 \times 2 = 300$$

How does this affect the inductance of the coil? Let's use our formula to find out;

$$L = \frac{0.2 \times 0.75^2 \times 300^2}{3 \times 0.75 + 9 \times 1.2}$$

$$= \frac{0.2 \times 0.5625 \times 90000}{2.25 + 10.8}$$

$$= \frac{10125}{13.05}$$

$$= 775.86207 \approx 776 \text{ mH}$$

The change in inductance works out to:

$$\frac{776}{194} = 4$$

or, the square of the increase in the number of turns. That is, $2^2 = 4$.

Now, let's try the same problem, but assume doubling the original number of windings also doubles the length of the coil. That is, the spacing between the windings is not changed:

$$d = 0.75$$
$$l = 1.2 \times 2 \times = 2.4$$
$$N = 150 \times 2 \times = 300$$

$$L = \frac{0.2 \times 0.75^2 \times 300^2}{3 \times 0.75 + 9 \times 2.4}$$

$$= \frac{0.2 \times 0.5625 \times 90000}{2.25 + 21.6}$$

$$= \frac{10125}{23.85}$$

$$= 424.5283 \approx 424 \text{ mH}$$

When increasing the number of turns (N) increases the coil length (l), while the diameter (d) remains constant, the original inductance value (L) will be multiplied by a factor slightly greater than the multiple of l and N. In our example, l and N are multiplied by 2, and the inductance increases by a factor of:

$$\frac{424}{194} = 2.185567$$

In practical applications, we will probably know the desired inductance value (L), and will need to determine how to achieve that value. Start out by selecting a likely core form of a given diameter (d), and pick a reasonable length. Now the basic inductance formula can be algebraically rearranged to solve for the necessary number of turns (N):

$$N = \sqrt{\frac{L(3d + 9l)}{0.2d^2}}$$

As an example, let's say we need a 100 mH coil. We have a 1.2-inch diameter coil form. We'll set the coil length (l) at 1 inch. This means the number of turns in our coil should work out to:

$$N = \sqrt{\frac{100 (3 \times 1.2 + 9 \times 1)}{0.2 \times 1.2^2}}$$

$$= \sqrt{\frac{100\ (3.6\ +\ 9)}{0.2\ \times\ 1.44}}$$

$$= \sqrt{\frac{100\ \times\ 12.6}{0.288}}$$

$$= \sqrt{\frac{1260}{0.288}}$$

$$= \sqrt{4375}$$

$$= 66.14 = 66\ \text{turns}$$

Let's use the same coil form and length (d and l constant), and change the desired inductance to 250 mH. This time the required number of turns works out to:

$$N = \sqrt{\frac{250\ (3\ \times\ 1.2\ +\ 9\ \times\ 1)}{0.2\ \times\ 1.2^2}}$$

$$= \sqrt{\frac{250\ \times\ 12.6}{0.288}}$$

$$= \sqrt{\frac{3150}{0.288}}$$

$$= \sqrt{10937.5}$$

$$= 104.58 = 104\ \tfrac{1}{2}\ \text{turns}$$

Several magnetic materials are often used for coil cores. These include iron, powdered iron, ferrite, and nickel alloy.

When a magnetic core is used, a new factor enters into the equation for determining the inductance of the coil. This new factor is the permeability of the core material, and is represented by the Greek letter, μ.

The equation for determining the inductance of a coil with a magnetic core is as follows:

$$L = \frac{4.06N^2\ \mu\ A}{0.27_{(10^8)1}}$$

Once again, N stands for the number of turns, 1 is the coil length in inches. A is the cross-sectional area of the coil in square inches.

Let's try a typical example. We'll set N equal to 1000 turns, μ 5000, A = 0.4 inches, and 1 = 3 inches. This means the inductance of the coil will work out to:

$$L = \frac{4.06\ \times\ 1000^2\ \times\ 5000\ \times\ 0.4}{0.27\ \times\ (10^8)\ \times\ 3}$$

$$= \frac{4.06 \times 1000000 \times 5000 \times 0.4}{27000000 \times 3}$$

$$= \frac{8120000000}{81000000}$$

$$= 100.24691 \approx 100 \ \mu H$$

If we don't change the core, the values of μ, 1, and A will remain constant. This means we can change the inductance value (L) simply by changing the number of turns (N). Let's return to our earlier example, and increase the number of turns by a factor of 3:

$$N' = N \times 3 = 1000 \times 3 = 3000$$

Now let's see what this does to the inductance (L);

$$L = \frac{4.06 \times 3000^2 \times 5000 \times 0.4}{81000000}$$

$$= \frac{9000000 \times 8120}{81000000}$$

$$= \frac{73080000000}{81000000}$$

$$= 902.2222 \approx 900 \ \mu H$$

Increasing N by a factor of 3 increased the value of L by a factor of 9, or 3^2. The inductance varies as the square of the number of turns.

To design a coil for a specific inductance value (L), you can solve for the value of N (Number of turns) by rearranging the basic equation like this:

$$N = \sqrt{\frac{0.27 \ (10^8) \ L1}{4.06 \ \mu A}}$$

$$= \sqrt{\frac{6650246.3054L1}{\mu A}}$$

Let's try an example by using the same coil from the earlier problems:

$$A = 0.4$$
$$1 = 3$$
$$\mu = 5000$$

We'll say we need a 500 μH coil, and solve for the number of turns required (N):

$$N = \sqrt{\frac{0.27 \ (10^8) \times 500 \times 3}{4.06 \times 5000 \times 0.4}}$$

$$= \sqrt{\frac{40500000000}{8120}}$$

$$= \sqrt{4987684.7}$$

$$= 2233.3125 = 2233\ \tfrac{1}{4}\ \text{turns}$$

Another type of coil uses a toroidal core. This is a toroid of magnetic material, as shown in Fig. 8-5. A toroid is a ring, or doughnut shaped object.

Fig. 8-5. A toroidal core is a donut shaped piece of magnetic material.

This type of coil core offers a number of important advantages, which include small size and compactness, high Q, and, perhaps most important of all, self-shielding. Toroidal cores can be operated at extremely high frequencies.

The formula for determining the inductance (L) of a toroidal coil is as follows:

$$L = 0.011684 N^2\ \mu h \log_{10}\ (OD/ID)$$

where N is the number of turns, μ is the permeability of the core material, h is the height of the core in inches, OD is the outside diameter of the core in inches, and ID is the inside diameter of the core in inches.

As a typical example, let's assume the following values:

$$N = 60$$
$$\mu = 400$$
$$h = 0.15$$
$$OD = 0.75$$
$$ID = 0.25$$

In this case the inductance (L) will work out to:

$$
\begin{aligned}
L &= 0.011684 \times 60^2 \times 400 \times 0.15 \times \log_{10}\ (0.75/0.25) \\
&= 0.011684 \times 3600 \times 60 \times \log_{10}\ (3) \\
&= 2523.744 \times 0.4771213 = 1204.1319 \approx 1200\ \mu H
\end{aligned}
$$

Adding or removing turns in a coil on a given toroidal core (μ, h, OD, and ID remain constant) increases or decreases the inductance (L) by the square of the change in the number of turns. For instance, if the number of turns is increased by a factor of 2.5:

$$N' = N \times 2.5$$

the inductance value will become:

$$L' = L \times 2.5^2 = L \times 6.25$$

Similarly, if the number of turns is reduced by a factor of 1.75:

$$N' = \frac{N}{1.75}$$

the inductance value will be changed to:

$$L' = \frac{L}{1.75^2} = \frac{L}{3.0625}$$

The basic formula can be rearranged algebraically to determine the number of turns required for a specified inductance, using a specific toroidal core:

$$N = \sqrt{\frac{L}{0.011684 \; \mu h \; \log_{10} (OD/ID)}}$$

As an example let's solve for n, using the following values:

$$L = 250 \; \mu H$$
$$\mu = 400$$
$$h = 0.15 \text{ inch}$$
$$OD = 0.75 \text{ inch}$$
$$ID = 0.25 \text{ inch}$$

The value of N should be equal to:

$$N = \frac{250}{0.011684 \times 400 \times 0.15 \times \log_{10} (0.75/0.25)}$$

$$= \frac{250}{0.70104 \times \log_{10} (3)}$$

$$= \frac{250}{0.70104 \times 0.477}$$

$$= \frac{250}{0.33448}$$

$$= 747.42642 \approx 747 \; \frac{1}{2} \text{ turns}$$

If we want to double the inductance (L), the number of turns will have to be increased by a factor of $\sqrt{2}$, or approximately 1.41.

RL CIRCUITS AND TIME CONSTANTS

A circuit consisting of a coil and a resistor, like the one shown in Fig. 8-6 also has a definite associated time constant, just as resistance/capacitance (RC) circuits do (see Chapter 6).

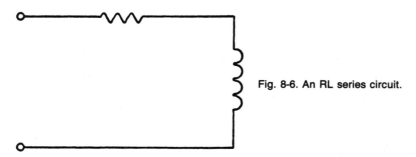

Fig. 8-6. An RL series circuit.

For *RL circuits,* the time constant (in seconds) is found by dividing the inductance (in henries) by the resistance in ohms:

$$T = L/R$$

Equation 8-2

In an RL circuit the time constant is the time required for the induced voltage to reach 63% of its full value.

For example, suppose the coil in Fig. 8-6 has an inductance of 100 millihenries (0.1 henry), and the resistor has a resistance of 1000 ohms. T = L/R = 0.1/1000 = 0.0001 second.

Increasing the inductance to 200 mH (0.2 henry), changes the time constant to 0.2/1000, or 0.0002 second. Increasing the inductance increases the time constant.

On the other hand, if the inductance is kept at 100 mH (0.1 henry), and the resistance is increased to 2000 ohms, the time constant becomes 0.1/2000, or 0.00005 second. Increasing the resistance decreases the time constant.

What would the time constant be for an RL circuit consisting of a 50 mH (0.05 henry) coil, and a 470 ohm resistor? What if the inductance is changed to 25 mH (0.025 henry)?

By the way, coils are also called *inductors.*

Self-Test

1. Which of the following characterizes inductance?

A *Tends to oppose changes in voltage*
B *Tends to oppose changes in current*
C *Tends to oppose dc*
D *Opposes all frequencies equally*
E *None of the above*

2. The reactance of a 25 mH coil at 5000 Hz is which of the following?

A 0.0013 ohms
B 785,000 ohms
C 785 ohms
D 13 ohms
E None of the above

3. What is the reactance of a 25 mH coil at 600 Hz?

A 785 ohms
B 94 ohms
C 0.011 ohms
D 94,000 ohms
E None of the above

4. What is the inductance of a single-layer coil on a 0.8 inch diameter nonmagnetic core with a length of 1.25-inch, and 320 turns of wire?

A 960 mH
B 3.8 mH
C 3.8 H
D 1200 mH
E None of the above

5. Assume we need a 150 mH coil on a 0.75-inch nonmagnetic diameter, 1-inch long. How many turns of wire will be required?

A 15,000
B 15
C 122 1/2
D 507 1/2
E None of the above

6. If we have a coil consisting of 500 turns on a magnetic core with a cross-sectional area of 0.35 inch, and a permeability rating of 750, and the coil is 1.5 inches long, what is the inductance?

A 6580 mH
B 6.6 mH
C 13 mH
D 100 mH
E None of the above

7. Assume we have a coil on a toroidal coil that is 0.22-inch high. The outside diameter is 0.8 inch, and the inside diameter is 0.3 inch. The permeability of the core

material is 500. What is the inductance, assuming there are 100 turns of wire around the core?

A *34,000 mH*
B *185 mH*
C *55 mH*
D *5500 mH*
E *None of the above*

8. How many turns would be needed to double the inductance of the coil described in question 7?

A *141*
B *400*
C *200*
D *50*
E *None of the above*

9. What is the time constant of a 250-mH coil and a 33,000-ohm resistor in series?

A *8.25 seconds*
B *0.0076 second*
C *0.0000076 second*
D *132 seconds*
E *None of the above*

10. In a RL circuit the time constant is the time required for the induced voltage to reach what percentage of its full value?

A *100%*
B *0%*
C *63%*
D *37%*
E *None of the above*

9

Transformers

If a number of shielded coils are placed in a circuit in series (see Fig. 9-1) their inductance values simply add, the same as with multiple resistances in series. That is:

$$L_T = L_1 + L_2 + \ldots L_n \qquad \text{Equation 9-1}$$

n, of course, represents the total number of inductors in the series circuit.

If the shielded coils are placed in a parallel circuit (see Fig. 9-2) the formula for finding the total effective inductance is similar to the resistance formula. The reciprocal of the total effective inductance equals the sum of the reciprocals of each of the parallel inductances. That is:

$$\frac{1}{L_T} = \frac{1}{L_1} \quad \frac{1}{L_2} + \ldots \frac{1}{L_n} \qquad \text{Equation 9-2}$$

You'll notice that I've specified shielded coils in these equations. This is because unshielded coils can interact. If the inductances are allowed to interact, the equations are complicated by a factor called the *coefficient of coupling*.

COEFFICIENT OF COUPLING

If two unshielded coils are brought within fairly close proximity of each other, their magnetic fields will interact. The magnetic lines of force from one coil will cut across

103

104

L3   L2   L1

Fig. 9-1. Coils in series.

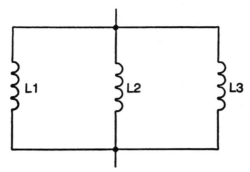

Fig. 9-2. Coils in parallel.

the turns of the other, so the coils will each induce a voltage in each other, as well as the voltage they induce in themselves. This is called *mutual inductance*. Previously, we have considered only *self-inductance*. The strength of the mutual inductance between coils is described as the coefficient of coupling.

If the coils are placed so that only a few of their magnetic lines of force interact, they are said to be *loosely coupled*. If, on the other hand, they are placed very close to each other so most of the lines of force from each coil cuts across the other, they are *closely coupled*.

This mutual inductance can either aid or oppose the self inductance of the coils, depending on the respective polarity of their magnetic fields. If both north poles or both south poles are facing each other, that is, each coil is wound in the opposite direction (see Fig. 9-3), the mutual inductance will be in opposition to the self inductance.

L1   L2

Fig. 9-3. Coils wound in opposite directions produce negative mutual inductance.

Fig. 9-4. Coils wound in the same direction produce positive mutual inductance.

If, however, a north pole is facing a south pole the coils are wound in the same direction (see Fig. 9-4) the mutual inductance will aid, or add to the self inductance.

Mutual inductance is represented algebraically by the letter M, and is measured in henries, or millihenries, the same as self inductance (L).

If two coils are in series and wound so that their mutual inductance aids their self inductance, the total effective inductance can be found with the following formula:

$$L_T = L_1 + L_2 + 2M$$ **Equation 9-3**

Similarly, if the mutual inductance is such that the magnetic fields of the two coils oppose each other, the formula is:

$$L_T = L_1 + L_2 - 2M$$

Equation 9-4

The same principle works with two coils in parallel. If their magnetic fields aid each other, the formula is:

$$\frac{1}{L_T} = \frac{1}{(L1 + M)} + \frac{1}{(L2 + M)}$$

Equation 9-5

If the magnetic fields of two coils in parallel oppose each other, the formula becomes:

$$\frac{1}{L_T} = \frac{1}{(L1 - M)} + \frac{1}{(L2 - M)}$$

Equation 9-6

The theoretical maximum amount of coupling between two coils is 100 %. Obviously, this is when all of the magnetic lines of force of one coil cut across all of the turns of the other, and vice versa. In practice 100% coupling can never be totally achieved. However, you can come quite close if both coils are wound upon a single, shared core. See Fig. 9-5.

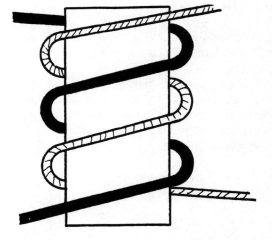

Fig. 9-5. Two coils wound on a single core.

TRANSFORMER ACTION

Interacting coils don't need to be in the same circuit. The magnetic field from the coil in one circuit can induce a voltage in the coil of a second, otherwise unconnected circuit. This process is called *transformer action*.

A *transformer* is essentially two (or sometimes more) coils wound upon a single core, and arranged so their mutual inductance is at a maximum.

The typical construction of a transformer is shown in Fig. 9-6. The schematic symbols for transformers are illustrated in Fig. 9-7. The symbol in Fig. 9-7A can be used as any

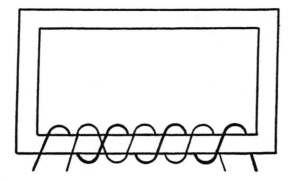

Fig. 9-6. Construction of a typical transformer.

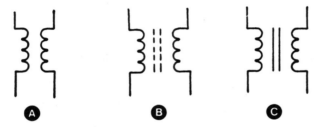

Fig. 9-7. Schematic symbols: A. air core transformer; B. powdered iron core transformer; C. sheet iron core transformer.

type of transformer, or as an air core transformer. Figure 9-7B represents a transformer with a powdered iron core, and Fig. 9-7C indicates the core consists of a stack of sheet iron wafers.

Let's examine what happens in an ideal, lossless transformer. We'll use the simple circuit shown in Fig. 9-8. Assuming that the ac source puts out 100 volts rms, the self induced voltage in the first (or *primary*) coil will also be 100 volts rms. Remember, we are assuming there are no losses in the system. If the coil consists of 100 turns of wire, 1 volt will be induced in each turn, so the total self induced voltage is 100 volts rms.

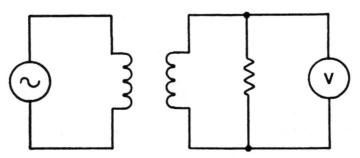

Fig. 9-8. Circuit demonstrating an ideal, lossless transformer.

Assuming the coils have 100% coupling, the same amount of magnetomotive force will cut across the turns of the other (secondary) coil. That is, one volt will be induced in each turn of the secondary.

If the secondary also consists of 100 turns, 100 volts rms will be provided to the load circuit. This kind of transformer is called an *isolation transformer* or a *1:1 transformer*, since the ratio between the two windings is one to one.

Such a device simply electrically isolates the load circuit from the power source. It is often used to reduce the risk of electrical shocks, or to prevent feedback (discussed in later chapters).

But what happens when the turns ratio is not exactly one to one? For example, what if the secondary has only 25 turns? The magnetic field of the primary will still be the same, of course, so one volt rms will still be induced in each turn of the secondary. But, since there are only 25turns in the coil, the total voltage available to the load circuit is only 25 volts rms. This kind of transformer is called a *step-down transformer*, since the output voltage is stepped down, or reduced from the input voltage.

However, the same amount of current is induced in the secondary coil, regardless of the number of turns. Remember, current is the same in all parts of a series circuit, so the current induced in one turn will flow through the entire coil. In other words, the power consumed by the secondary, is the same as the power consumed by the primary. When the voltage is stepped down, the current is stepped up.

Similarly, if the secondary has more turns than the primary, the voltage will be stepped up (increased) and the current will be stepped down (decreased). This device is called a *step-up transformer.*

For example, if the secondary of our imaginary transformer had 250 to the primary's 100 turns, the output voltage across the secondary coil would be 250 volts.

The 100 volt rms source and 100 turn primary were selected for these examples, for simplicity. Not all transformers induce one volt per turn. If the source voltage was increased to 230 volts, 2.3 volts would be induced in each turn. If the primary consisted of 200 turns, a 230 volt input would induce 1.15 volt per turn.

Changing the input voltage, changes the output voltage too. Let's say we have a transformer that steps down a 100 volt input to a 25 volt output. The turns ratio is 4:1. If 230 volts was applied to the input, the output would be 57.5 volts (all of these voltages are rms).

Suppose you have a transformer with 350 turns in the primary winding and 70 turns in the secondary. What is the output voltage with a 100 volt rms input? What is the output if the input is increased to 700 volts rms?

CENTER TAPS

Many transformers have an extra connection, or *tap* in the center of their secondary winding. The schematic symbol for a *center-tapped transformer* is shown in Fig. 9-9.

The secondary can be considered two coils connected at the center tap. Let's assume the secondary consists of 100 turns of wire. If the center tap is in the exact center of this coil, there will be 50 turns on either side of the tap. If the primary induces one volt per turn in the secondary, the full output from A to C is 100 volts rms (100 turns × 1 volt per turn).

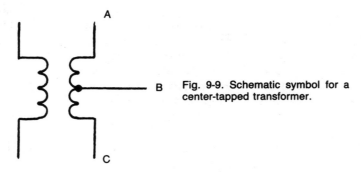

Fig. 9-9. Schematic symbol for a center-tapped transformer.

From A to B there is only 50 turns. 50 turns × 1 volt per turn = 50 volts rms from point A to point B. Similarly, there will be 50 volts across the 50 turn coil from B to C. Notice that AB + BC = AC

If the center tap is grounded, as in Fig. 9-10, the voltage across AB and BC will look like the graphs shown in Fig. 9-11. Notice that the two voltages are equal—50 volts rms, with respect to ground, but there are 180° out of phase. When voltage AB goes up, voltage BC goes down, and vice versa.

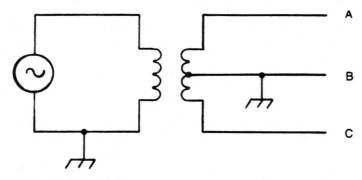

Fig. 9-10. A typical transformer circuit with a grounded center tap.

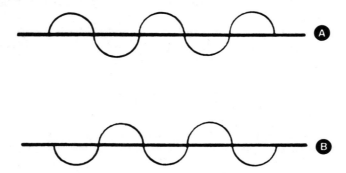

Fig. 9-11. Voltages in Fig. 9-10 are: A. voltage AB; B. voltage BC.

AUTOTRANSFORMER

An autotransformer is a unique form of transformer that uses a single coil winding as both the primary and the secondary, using the center tap principle. See Fig. 9-12.

Let's say there are 100 turns in the entire coil, and 100 volts rms is applied across AC. Because the entire voltage is distributed equally across the entire coil, the self induction will equal 1 volt rms per turn.

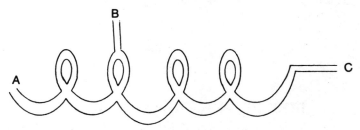

Fig. 9-12. Construction of an autotransformer.

Now, let's suppose the center tap (B) is placed so that there are 70 turns of wire between A and B, and only 30 turns from B to C. If we took the voltage across BC, we'd have a 30 turn coil with one volt rms induced across each turn. The output voltage would be 30 volts rms. The single coil acts like two very closely placed, but separate coils. An autotransformer can only be used in step-down applications, or step-up applications (the input voltage is applied across BC and the output is taken across (AC), but not for isolation. Since the primary and the secondary are the same coil, there is no electrical isolation between them. Usually even in step-down or step-up applications some degree of isolation is required. For this reason standard two coil transformers are more commonly used. The schematic symbol for an autotransformer is shown in Fig. 9-13. The center tap is often movable.

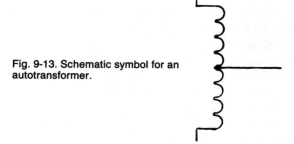

Fig. 9-13. Schematic symbol for an autotransformer.

Some transformers have more than one secondary. See the schematic symbol in Fig. 9-14. The primary coil, AB, induces a voltage in both CD and EF. The circuits connected to these two secondaries can be completely isolated from each other electrically.

In addition to voltage changing (step up and step down), and isolation, transformers are often used for impedance matching. This concept will be discussed in the next chapter.

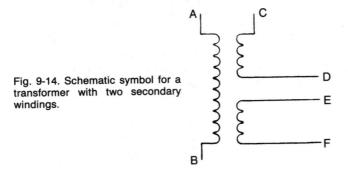

Fig. 9-14. Schematic symbol for a transformer with two secondary windings.

LOSSES IN A TRANSFORMER

So far we have been dealing with ideal, lossless transformers that waste no energy. In actual practice, of course, such perfection is never possible. You'll always have to put more power into a transformer (or any other electronic component, for that matter) because the transformer itself will use up, or lose some of the electrical energy.

Most of these losses can be made quite small, so they can usually be ignored if you're only interested in rough, ballpark figures. However, they can be quite significant in certain cases. For instance, a transformer with 100 turns in each winding and with 100 volts rms applied should have a nominal output of 100 volts rms. In an actual circuit, it may only put out about 98.5 volts rms. Another transformer with the same basic specifications, but greater internal losses might only put out 79 volts rms or so.

Generally, commercially available transformers are marked with their intended input and output voltages. These figures usually take most internal losses into account. You will rarely have to be concerned with the actual turns ratio of a transformer, just the voltage ratio. You should, however, understand some of the major causes of energy losses in practical transformers.

One limitation has already been mentioned—100% mutual inductance can never be achieved between two coils. Some of the magnetic lines of force will be lost into the surrounding air. However, it is possible to achieve a very high coefficient of coupling. Manufacturers can also largely compensate for the difference by adding a few extra turns to the secondary winding.

Since 100% coupling cannot be achieved, some of the magnetic field is leaked off, out of the circuit. These lost magnetic lines of force are called *leakage flux*, and the effect is called *leakage reactance*.

Leakage flux can be greatly reduced by winding the coils on an iron core or some other low reluctance substance.

The actual shape of the core can also have a significant effect on the leakage reactance. Figure 9-15 shows some of the most common core shapes. The *shell core* type (Fig. 9-15C) typically has the lowest leakage reactance. It is also the most expensive core to manufacture.

Another source of energy loss in a transformer is the dc resistance. Any practical conductor has some degree of resistance, which increases as the length of the conductor increases. A coil is essentially just a very long wire, wound into a small space. It acts

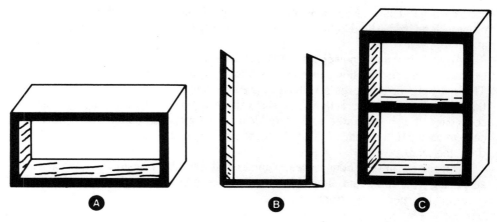

Fig. 9-15. Typical transformer core shapes.

like a very low value resistor. This dc resistance is usually very small—almost never as much as 100 ohms, but it does represent a small voltage loss.

The dc resistance between the separate windings of a transformer (from primary to secondary) should be virtually infinite, because there is no direct electrical connection between them. They are only connected magnetically. Of course, this would not be true of an autotransformer.

There is also the ac inductive reactance to be considered. Resistances and reactances cannot simply be added together. The next chapter will explain how these two similar, but different effects can be combined into a single specification called *impedance*.

Remember that inductive reactance increases as the applied frequency increases.

Since the coils are conductors, and they are separated by insulation, unwanted capacitances can be set up to add to the power loss.

In a transformer, dc resistance is sometimes referred to as copper loss, since the coils themselves are usually made of copper wire.

Another important type of energy loss in a transformer is *iron loss*. There are actually two forms of iron loss. The first of these is called *hysteresis loss*. Since the iron core is within a strong magnetic field (generated by the surrounding coil) it becomes magnetized itself. As the alternating current flows through the coil, the magnetic charge on the coil is forced to continually alternate its polarity. This process, naturally, uses up a certain amount of energy, because the iron core will oppose any change in its condition. This energy must be stolen from the electrical current, since it is the only energy source within the system. In other words, power is lost due to this effect.

Hysteresis loss can be reduced by constructing the core of some material which is very easy to magnetize and demagnetize. Some materials of this type are silicon steel, and certain other alloys.

The other form of iron loss is due to the electrical current that is induced in the conductive core by the fluctuating magnetic fields of the coils. This current (called an *eddy current*) is more wasted power that is not allowed to reach the desired load circuit.

To reduce eddy currents, transformer cores are often made of a pile of very thin metallic sheets (called laminations) rather than a solid mass of iron or whatever alloy is being used. These laminations are individually insulated from each other by a layer of varnish or oxide, so eddy currents cannot flow through the entire core and pick up strength.

Another inevitable power loss in transformers is due to an effect called *reflected impedance*. As current flows through the primary winding, a magnetic field is created. This magnetic field induces a voltage in the secondary winding. So far, this is simply the desired transformer action. But since there is now current flowing through the secondary, it also generates a magnetic field of its own, which induces an interferring current back into the primary. This energy is taken from the secondary winding's circuit (i.e., the load).

Despite all these various losses, commercially available transformers are generally surprisingly efficient. And, as we'll see in later chapters, these devices are used in a great many types of electronic circuits.

Self-Test

1. If three shielded 100 mH coils are connected in series, what is the total effective inductance?

A *33.333 mH*
B *300 mH*
C *100 mH*
D *67.777 mH*
E *None of the above*

2. If three shielded 100 mH coils are connected in parallel, what is the total effective inductance?

A *33.333 mH*
B *300 mH*
C *100 mH*
D *67.777 mH*
E *None of the above*

3. A transformer consists of which of the following?

A *A capacitor and an inductor*
B *An inductance and a resistance*
C *A parallel resonant circuit*
D *Two coils wound on a common core*
E *None of the above*

4. A transformer with 100 turns in the primary winding, and 25 turns in the secondary winding is which of the following?

A *An isolation transformer*
B *A step-down transformer*
C *A step-up transformer*

D *An autotransformer*
E *None of the above*

5. If 250 volts ac is fed to the primary coil of the transformer described in question 4, what will be the voltage induced across the secondary winding (assuming 100% coupling)?

A *125 volts*
B *500 volts*
C *2.5 volts*
D *250 volts*
E *None of the above*

6. In a step-up transformer, what is the effect on the current?

A *It is the same in both windings*
B *The current in the secondary is greater than the current in the primary*
C *The current is blocked at low frequencies*
D *The current in the secondary is less than the current in the primary*
E None of the above

7. If the center tap of a transformer is grounded, what will be the phase relationship between the two ends of the secondary winding?

A *In phase*
B *360° out of phase*
C *180° out of phase*
D *90° out of phase*
E *None of the above*

8. An autotransformer contains how many coils?

A *None*
B *One*
C *Two*
D *Three*
E *None of the above*

9. What is the name of the effect of some of the magnetic field leaking off due to less that 100% coupling?

A *Electromagnetic effect*
B *Eddy currents*
C *Leakage resistance*
D *Self-inductance*
E *None of the above*

10. Which of the following does *not* contribute to losses in a transformer?

A *Self-inductance*
B *Iron losses*
C *Stray capacitance*
D *Leakage resistance*
E *None of the above*

10

Impedance and
Resonant Circuits

In the previous chapter, we made a reference to *impedance*, and stated that it was a combination of reactance (inductive and/or capacitive) and dc resistance. In other words, impedance is the total effective ac resistance of a component or circuit.

In this chapter we will explore the concept of impedance, and discover how it can be determined mathematically. We'll also look at what happens when a circuit contains both inductive and capacitive components.

IMPEDANCE IN INDUCTIVE CIRCUITS

Any practical inductor has some dc resistance, as well as its inductive reactance. It also has some capacitance, but usually this value is small enough to be ignored in practical circuits, unless very high frequencies are involved. For now, we'll ignore this factor. So a circuit containing a resistor, a coil, and an ac voltage source can be redrawn as in Fig. 10-1. The resistor marked R_L does not exist as a separate component. It is the internal dc resistance of the coil itself. It is shown only to make the following discussion clearer.

For the following example, we'll assign these values to the components. L is 0.1 henry (100 mH), R_L is 22 ohms, and R_x is 1000 ohms. The ac voltage is 100 volts rms. We'll examine what happens as different frequencies are applied.

If the frequency of the ac source is 60 Hz, the inductive reactance is about 38 ohms. It might seem that we could find the total effective resistance of the circuit, simply by adding $X_L + R_L + R_x$, giving a total of 1060 ohms. Unfortunately, this simple approach

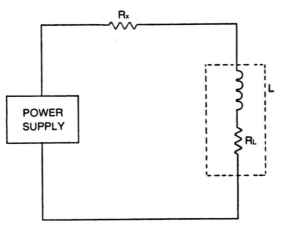

Fig. 10-1. Equivalent circuit for a series RL circuit.

will not give an accurate result. The reason is that dc resistance is 90° out of phase with inductive reactance.

An inductance offers more resistance to current than to voltage. You'll recall that ordinarily, the current and voltage from an ac source are in phase. After passing through a pure inductance, however, the voltage will lead the current by 90°. See Fig. 10-2.

When dc resistance is also in the circuit, the final phase difference between the voltage and the current will be proportionately less than 90°. In a purely resistive circuit, of course, the voltage and the current remain in phase.

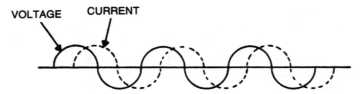

Fig. 10-2. Voltage leads current after passing through an inductance.

Another reason why resistance and reactance cannot simply be added together is that reactance is not a true resistance at all—it is only an apparent resistance. In a true dc resistance, power is consumed by the resistor (usually the energy is converted to heat), but no power is actually consumed by a reactance (either inductive, or capacitive).

We could find the combined value of reactance and resistance (that is, the impedance of the circuit) with a vector diagram, as in Fig. 10-3. But since the phase relationship between a resistance and inductive reactance is always 90°, we can solve the problem mathematically.

Notice that the vector diagram of the resistance and inductive reactance is always in the form of a right triangle. If we know the length of the sides connected at the right angle, we can solve for the third side with the algebraic formula, $c = \sqrt{a^2 + b^2}$. Or we can rewrite the formula, using the appropriate electrical terms;

$$Z = \sqrt{R^2 + X^2}$$ **Equation 10-1**

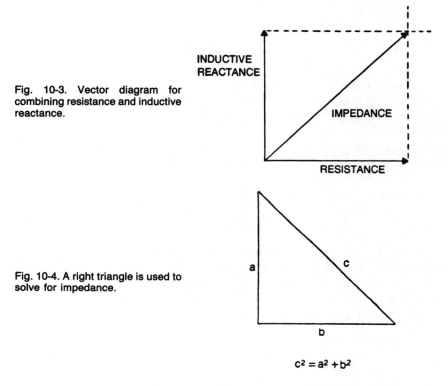

Fig. 10-3. Vector diagram for combining resistance and inductive reactance.

Fig. 10-4. A right triangle is used to solve for impedance.

$$c^2 = a^2 + b^2$$

where Z is impedance, R is resistance, and X is reactance. All three units are in ohms. See Fig. 10-4.

Returning to our example circuit, we find that the impedance equals $\sqrt{1022^2 + 38^2}$ = $\sqrt{1044484 + 1444}$ = $\sqrt{1045928}$ = about 1022.7 ohms.

For a second example, we'll keep the same components, but increase the ac frequency to 200 Hz. The dc resistance is still 1022 ohms, but the inductive reactance is now approximately 126 ohms. Therefore, $Z = \sqrt{1022^2 + 126^2}$ = $\sqrt{1044484 + 15876}$ = $\sqrt{1060360}$ = just under 1,030 ohms.

If the frequency of the ac source is now increased to 5000 Hz, the inductive reactance goes up to 3,142 ohms, so the impedance equals $\sqrt{1022^2 + 3142^2}$ = $\sqrt{1044484 + 9872164}$ = $\sqrt{10916648}$ = just slightly more than 3304 ohms.

You can see that as the frequency increases, both the inductive reactance and the total impedance of an RL circuit are also increased. The value of the dc resistance, of course, remains constant.

Now, change the value of R_L to 15 ohms, R_x to 2200 ohms, and L to 0.25 henry (250 millihenries) and find the circuit impedance at 60 Hz, 200 Hz, and 5000 Hz.

If you draw vector diagrams for each of these examples, you will find that the impedance phase angle in a circuit consisting of only resistance and inductance is always 45°.

When you have a parallel RL circuit, such as the simple one shown in Fig. 10-5, the situation is a bit more complicated.

The formula for a circuit of this type is:

$$Z = \frac{RX}{\sqrt{R^2 + X^2}}$$

Equation 10-2

Again, all three variables are in ohms.

Let's say we have 1000 ohms of resistance in parallel with an inductive reactance of 52 ohms. The impedance would be equal to $(1000 \times 52 / \sqrt{1000^2 + 52^2} = 52000 / \sqrt{1000000 + 2704} = 52000/\sqrt{1002704} = 52000/1001 =$ just under 52 ohms.

If the inductive reactance is increased to 890 ohms, the impedance becomes equal to $(1000 \times 890)/\sqrt{1000^2 + 890^2} = 890000/\sqrt{1000000 + 792100} = 890000/\sqrt{1792100} = 890000/1339 =$ about 665 ohms. The total impedance will always be less than either the dc resistance, or the inductive reactance.

You'll notice that we have been ignoring the dc resistance of the coil in these examples. To take this additional factor under consideration, the circuit becomes like the one in Fig. 10-6.

To solve the total impedance of this circuit, it is necessary to use vector diagrams. Usually the dc resistance of the coil can simply be ignored.

A circuit with both an inductance and a resistance in each leg of the parallel paths, can be solved mathematically, because the impedances are similar (i.e., in phase). Such a circuit is illustrated in Fig. 10-7.

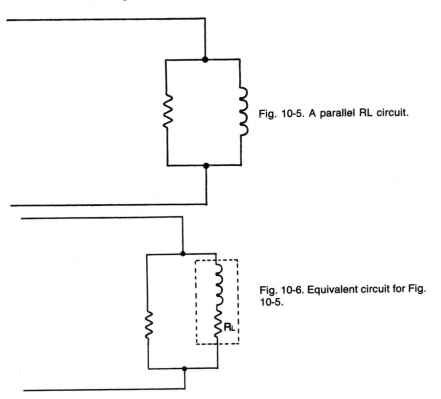

Fig. 10-5. A parallel RL circuit.

Fig. 10-6. Equivalent circuit for Fig. 10-5.

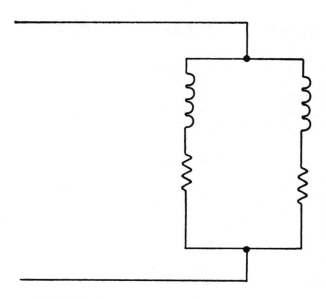

Fig. 10-7. Another parallel RL circuit.

To solve for the impedance of this kind of circuit, simply solve for each parallel path as a separate series RL circuit.

Let's suppose R1 equals 220 ohms, R2 equals 470 ohms, and L1 has an inductive reactance of 97 ohms, while L2 has an inductive reactance of 550 ohms.

The impedance in leg A equals $\sqrt{220^2 + 97^2} = \sqrt{48400 + 9409} = \sqrt{57809} =$ approximately 240 ohms. Z_B, on the other hand, equals $\sqrt{470^2 + 550^2} = \sqrt{220900 + 302500} = \sqrt{523400} =$ about 723 ohms.

We now have Z_A (240 ohms) in parallel with Z_B (723 ohms). Since impedances are essentially ac resistances, they can be combined in the same manner as dc resistances, as long as their phase angles are identical. In other words, impedances in series add:

$$Z_T = Z_1 + Z_2 + \ldots Z_n \qquad \textbf{Equation 10-3}$$

While impedances in parallel are solved by the formula:

$$\frac{1}{Z_T} = \frac{1}{Z_1} + \frac{1}{Z_2} + \ldots \qquad \textbf{Equation 10-4}$$

Dissimilar impedances (out of phase) cannot be combined in this manner.

In our example, we find that $1/Z_T = 1/Z_A + 1/Z_B = 1/240 + 1/723 = 0.004167 + 0.001383 = 0.00555$. Taking the reciprocal, we find that the total effective impedance of the entire circuit is about 180 ohms.

Change R1 to 1000 ohms, R2 to 100 ohms, X_{L1} to 550 ohms, and X_{L2} to 85 ohms, and find the total effective impedance of the circuit.

120

IMPEDANCE IN CAPACITIVE CIRCUITS

The situation with an RC circuit is basically similar. Capacitive reactance also consumes no real power. Ac flows in and out of a capacitor, but not through it.

In a capacitor the current leads the voltage by 90°, so the vector diagram of resistance and pure capacitive reactance would resemble Fig. 10-8. Notice that it is a mirror image of a resistance/inductive reactance combination. The resulting phase angle will always be 45° (or, 315° if the resistance is used as the 0° line).

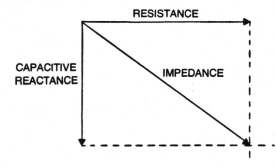

Fig. 10-8. Vector diagram for combining resistance and capacitive reactance.

Like an inductor, practical capacitors have some degree of dc resistance and a small (usually negligible) inductance.

The dc resistance through a capacitor is in parallel with its reactance, and is generally extremely high, so the effective impedance is essentially equal to the capacitive reactance. For this reason, the dc resistance is ordinarily ignored in all but the most critical calculations.

Combining dc resistance with capacitive reactance works in the same way as combining inductive reactance with dc resistance. For example, for a simple RC series circuit like the one in Fig. 10-9, the formula is:

$$Z = \sqrt{R^2 + X_c^2}$$ **Equation 10-5**

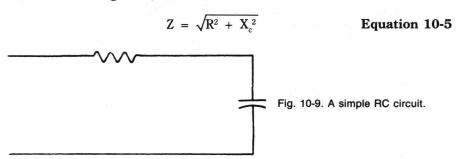

Fig. 10-9. A simple RC circuit.

This is the same formula used to solve for the impedance in RL series circuits. However, it is important to remember that capacitive reactance is 180° out of phase with inductive reactance. Also, since capacitive reactance decreases as the applied frequency increases, the circuit impedance of an RC series circuit also decreases with an increase in the applied ac frequency.

IMPEDANCE IN INDUCTIVE/CAPACITIVE CIRCUITS

Most practical electronic circuits have all three of the basic electrical elements: resistance, inductance, and capacitance. How would you go about finding the impedance of a circuit with a resistor, a capacitor, and a coil in series, like the one in Fig. 10-10?

Since the inductive and capacitive reactances are 180° out of phase we obviously can't simply add them together.

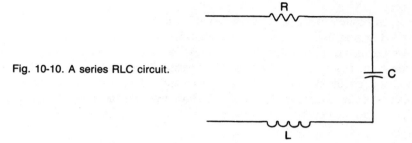

Fig. 10-10. A series RLC circuit.

Take a look at the graphs in Fig. 10-11. Signal B is smaller than signal A and 180° out of phase. Signal C is the result of combining these two waveforms. Notice that the output signal resembles (and is in phase with) the larger input signal (A), but has a lower amplitude. The difference between signal A and signal B is signal C.

In other words, when two similar signals are 180° out of phase, their combined value can be found by subtracting the smaller from the larger signal.

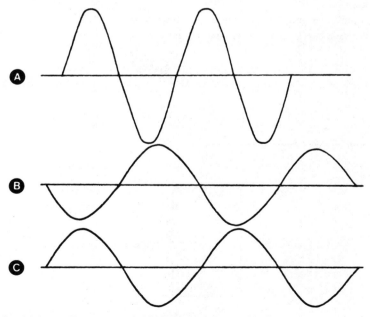

Fig. 10-11. Combining unlike reactances (inductive and capacitive).

We can obviously apply this same approach with combining inductive and capacitive reactances because they are, by definition, 180° out of phase with each other. The formula is:

$$X_T = X_L - X_C \qquad \textbf{Equation 10-6}$$

X_L is the inductive reactance, X_C is the capacitive reactance, and X_T is the total effective reactance.

If X_T is positive (as it will be whenever X_L is larger than X_C), then the resulting reactance behaves like an inductive reactance (increasing as the frequency increases). On the other hand, if X_T works out to a negative value (that is, X_C is larger than X_L), then the total reactance appears to be capacitive (decreasing as the frequency is increased). A circuit including both capacitance and inductance will act like an inductive reactance at some frequencies, and like a capacitive reactance at other frequencies.

X_T can then be combined with the dc resistance in the usual way to find the circuit impedance.

$$Z = \sqrt{R^2 + X_T^2} \qquad \textbf{Equation 10-7}$$

or

$$Z = \sqrt{R^2 + (X_L - X_C)^2} \qquad \textbf{Equation 10-8}$$

Of course, because the reactance is squared, the impedance will always have a positive value.

If an inductance and a capacitance are connected in parallel, as in Fig. 10-12, then the effective reactance of the combination is:

$$X_T = \frac{(X_L)(X_C)}{(X_L - X_C)} \qquad \textbf{Equation 10-9}$$

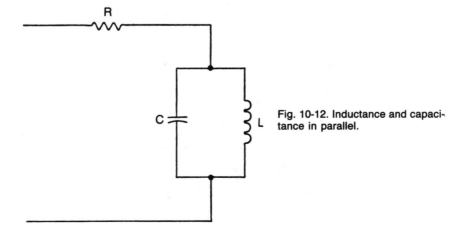

Fig. 10-12. Inductance and capacitance in parallel.

Again, if X_T is positive the effective reactance is inductance, but if X_T is negative, the reactance is effectively capacitive.

RESONANCE

Capacitive reactance decreases as the applied frequency increases, while inductive reactance increases along with the applied frequency. At some specific frequency for every combination of a capacitor and an inductor, the capacitive and the inductive reactances will be equal. This frequency is the *resonant frequency*, and it results in a unique circuit condition called *resonance*.

In a series connected LC circuit, the impedance equals $\sqrt{R^2 + (X_L - X_C)^2}$. If X_L equals X_C, these factors will cancel out leaving $Z = \sqrt{R^2 + O^2} = \sqrt{R^2}$, or simply $Z = R$.

At resonance, the impedance in a series LC circuit will be determined solely by the dc resistance. The impedance is at its minimum value at the resonant frequency.

Now, let's look at what happens at resonance in a parallel LC circuit? In this case

$$X_T = \frac{(X_L)(X_C)}{(X_L - X_C)}.$$

If $X_L = X_C$ then $X_L - X_C$ must be zero. Since this is the denominator of the equation, the value of X_T becomes infinite by definition, since any number divided by zero equals infinity. Clearly a parallel connected LC circuit has its maximum impedance at resonance.

Resonant circuits are extremely important in the study of electronics, and we'll learn how they are used in later chapters.

Any combination of an inductance and a capacitance will have a resonant frequency. Any given combination will be resonant only at a single frequency.

The formula for finding the resonant frequency for a specific inductance-capacitance combination (either series, or parallel connected) is:

$$F = \frac{1}{2\pi\sqrt{LC}} \qquad \text{Equation 10-10}$$

F is the frequency in hertz, L is the inductance in henries, C is the capacitance in farads, and 2π is approximately 6.28.

As an example, let's suppose we have a 0.1 henry (100 mH) coil, and a 10 μF (0.00001 farad) capacitor. The resonant frequency for this particular combination equals $1/(2\pi\sqrt{LC})$, or $1/(6.28 \times \sqrt{0.1 \times 0.00001})$. This works out to 1/0.00628, or a resonant frequency of approximately 159 hertz.

Try calculating the resonant frequency for a circuit with a 0.25 henry (250 mH) coil and a 100 μF (0.0001 farad) capacitor.

In a design situation you'll usually need to approach this equation from another angle. Probably, you'll know what frequency you want the circuit to resonate at, and you'll have to determine which components to use.

In this type of situation, you can arbitrarily select an inductance value and rewrite the resonance equation to solve for the capacitance as follows:

$$C = \frac{1}{4\pi^2 F^2 L} \qquad \text{Equation 10-11}$$

$4\pi^2$ is approximately 39.5.

Let's say we need a circuit that is resonant at 1000 Hz. We can arbitrarily decide to use a 0.1 henry (100 mH) coil. This will make the required capacitance equal to 1/(39.5 × 1000² × 0.1), or 1/(39.5 × 1000000 × 0.1), or 1/395000. Taking the reciprocal, we find that we need a capacitance of about 0.0000003 farad, or 0.3 μF.

Alternatively, we can choose a capacitance value, and solve for the inductance. The formula for this method is quite similar;

$$L = \frac{1}{4\,\pi^2 F^2 C} \qquad \textbf{Equation 10-12}$$

Again, we'll assume we need resonance at 1000 Hz, and we'll select a 1 μF (0.000001 farad) capacitor. L = 1 (39.5 × 1000² × 0.000001) = 1/(39.5 × 1,000,000 × 0.000001) = 1/39.5 = 0.025, or about 25 millihenries.

You can see that different combinations of components can produce the same resonant frequency.

Let's go back and check that last problem, by recalculating the resonant frequency from our component values, 25 mH (0.025 henry) and 1 μF (0.000001 farad). F = 1/(2 $\pi\sqrt{LC}$) =1/(6.28 × $\sqrt{0.025 \times 0.000001}$) = 1/(6.28 × $\sqrt{0.00000025}$) = 1/(6.28 × 0.0001581) = 1/0.000993 = about 1007 hertz. The seven hertz error came from rounding off the inductance value. This degree of accuracy is sufficient for most practical purposes.

Another thing to consider is what happens when one of the component values is increased. Will the resonant frequency increase, or decrease?

Using the last example, let's increase the inductance to 50 mH (0.05 henry). The resonant frequency now would equal 1/(6.28 × $\sqrt{0.05 \times 0.000001}$) = 1/(6.28 × $\sqrt{0.0000005}$) = 1/(6.28 × 0.0002236) = 1/0.0014043 = slightly over 712 hertz. When the inductance was increased, the resonant frequency was decreased.

Similarly, if we now increase the capacitance to 25 μF (0.000025 farad), the resonant frequency becomes 1/(6.28 × $\sqrt{0.05 \times 0.000025}$) = 1/(6.28 × $\sqrt{0.0000013}$) = 1/(6.28 × 0.001118) = 1/0.0070213 = approximately 142 Hertz. Increasing the capacitance also decreases the resonant frequency.

To practice using these formulas, design a circuit that resonates at 250 Hz, and one that resonates at 5000 Hz.

IMPEDANCE MATCHING AND POWER TRANSFER

Any dc power source, such as a battery, has some amount of internal resistance. Similarly, any ac power source has a certain degree of internal impedance. This impedance will be constant for any given frequency, while the impedance of the load circuit can be any impedance that the designer makes it.

In Fig. 10-13 the internal impedance of the power source and the load impedance of the circuit are shown as resistances for simplicity. Bear in mind that these are actually ac impedances.

If we assume the ac source has an internal impedance of 50 ohms, Table 10-1 shows the current and the power apparently consumed by the load at various values of the load impedance. Notice that the power available to the load circuit increases until it reaches

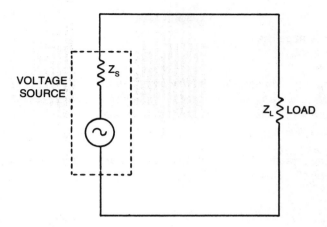

Fig. 10-13. Matching impedances.

Table 10-1. Matching Impedances.

E	Z_S	Z_L	Z_T	I	P_L
10	50	10	60	0.167	0.278
10	50	20	70	0.143	0.408
10	50	30	80	0.125	0.469
10	50	40	90	0.111	0.494
10	50	50	100	0.100	0.500
10	50	60	110	0.091	0.496
10	50	70	120	0.083	0.486
10	50	80	130	0.077	0.473
10	50	90	140	0.071	0.459
10	50	100	150	0.067	0.444

a maximum level, then the power transfer starts to drop back down again. This point of maximum power transference will always occur when the source impedance is equal to the load impedance. When these impedances are unequal, the source will not put out the maximum amount of energy it is capable of. This is called an *impedance mismatch*.

Often it isn't possible or practical to design the load circuit so that its impedance directly matches that of the signal source. In this case, some form of *impedance matching* is needed.

One of the simplest methods for this is with an *impedance matching transformer* (as mentioned in the last chapter). The inductance of the coils in such a transformer is carefully chosen to give the proper impedance ratio between the primary and secondary windings. Impedance can be either increased or decreased with such a transformer.

In practical circuits, the impedances will never match exactly, but a certain amount of leeway is usually acceptable. For instance, a 500 ohm load can generally be used with a 600 ohm single source with no problems. However, the circuit won't work very efficiently if it is driven by a 10,000 ohm source impedance. For best results, the source impedance and load impedance must be approximately equal.

Self-Test

1. What is impedance?

A *ac resistance*
B *An imaginary factor used to simplify circuit design*
C *A combination of capacitive reactance, inductive reactance, and dc resistance*
D *A combination of capacitive reactance and inductive reactance only*
E *None of the above*

2. Ignoring capacitive effects, what is the impedance of a 100 mH coil with an internal dc resistance of 45 ohms at 60 Hz?

A *59 ohms*
B *3500 ohms*
C *83 ohms*
D *105 ohms*
E *None of the above*

3. Ignoring capacitive effects, what is the total impedance of a 2700 ohm resistor and a 100 mH coil in parallel? The coil has an internal resistance of 30 ohms, and the operating frequency is 500 Hz.

A *314 ohms*
B *2750 ohms*
C *312 ohms*
D *97000 ohms*
E *None of the above*

4. What is the impedance of an RC series circuit made up of an 82,000 ohm resistor and a 0.022 μF capacitor at 300 Hz?

A *Infinite*
B *24114 ohms*
C *106,114 ohms*
D *85,472 ohms*
E *None of the above*

5. Ignoring dc resistance, what is the total reactance of a 350 mH coil and a 0.47 μF capacitor in series when the operating frequency is 1000 Hz?

A *2538 ohms*
B *1860 ohms*
C *50 ohms*
D *225 ohms*
E *None of the above*

6. In a series resonant LC circuit, what is the impedance at the resonant frequency?

A *Determined solely by the dc resistance*
B *The maximum impedance value*
C *Zero*
D *Infinity*
E *None of the above*

7. In a parallel resonant LC circuit, what is the impedance at the resonant frequency?

A *Determined solely by the dc resistance*
B *The maximum impedance value*
C *Zero*
D *Infinity*
E *None of the above*

8. What is the resonant frequency of an LC circuit made up of a 150 mH coil and a 0.22 μF capacitor?

A *36 Hz*
B *2196 Hz*
C *4,800,000 Hz*
D *876 Hz*
E *None of the above*

9. If a 0.1 μF capacitor and a 100 mH coil are connected in parallel, what is the total effective reactance at 2500 Hz, ignoring the effects of dc resistance?

A *1,000,000 ohms*
B *1070 ohms*
C *934 ohms*
D *637 ohms*
E *None of the above*

10. If we need an LC circuit to be resonant at 3000 Hz, and use a 250 mH coil, what should the capacitance value be?

A *0.01 μF*
B *0.1 μF*
C *0.22 μF*
D *4.7 μF*
E *None of the above*

11

Crystals

In the last chapter we learned how a resonant circuit can be built from a separate capacitor and coil. Resonant circuits can also be made with a single component called a *crystal* (often abbreviated as *XTAL*).

CONSTRUCTION OF A CRYSTAL

Figure 11-1 shows the basic structure of a crystal. A thin slice of quartz crystal is sandwiched between two metal plates. These plates are held in tight contact with the crystal element by small springs. This entire assembly is enclosed in a metal case which is *hermetically sealed,* to keep out moisture and dust. Leads connected to the metal plates are brought out from the case for connection to an external circuit. The schematic symbol for a crystal is shown in Fig. 11-2.

THE PIEZOELECTRIC EFFECT

A crystal works because of a phenomenon called the *piezoelectric effect.* Two sets of axes pass through a crystal. One set, called the *X axis* pass through the corners of the crystal. The other set, called the *Y axis* is perpendicular to the X axis, but in the same plane. Figure 11-3 shows two axes in a crystal, looking down through the top of the crystal.

If a mechanical stress is placed across a Y axis, an electrical voltage will be produced along the X axis.

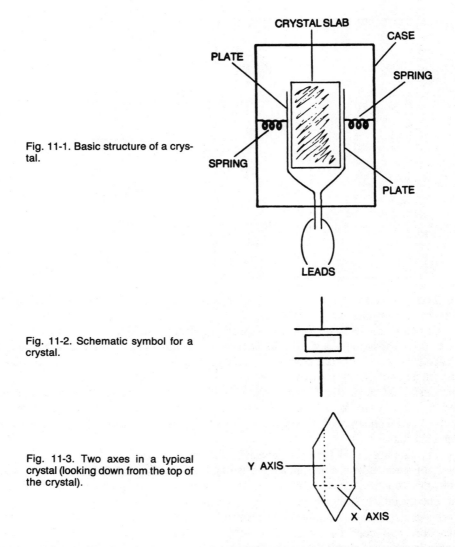

Fig. 11-1. Basic structure of a crystal.

Fig. 11-2. Schematic symbol for a crystal.

Fig. 11-3. Two axes in a typical crystal (looking down from the top of the crystal).

The reverse also holds true—if an electrical voltage is applied across an X axis, a mechanical stress will be created along the Y axis.

This piezoelectric effect can cause the crystal to ring (vibrate), or resonate at a specific frequency under certain conditions. In practice, only a thin slice of crystal is used. This slice can be cut across either an X or a Y axis.

USING CRYSTALS

Figure 11-4 shows the equivalent electrical circuit for a typical crystal. Depending on how the crystal is manufactured, it can replace either a series resonant (minimum impedance) or a parallel resonant (maximum impedance) capacitor-coil circuit. Generally, a crystal designed for series resonant use, cannot be used in a parallel resonant circuit.

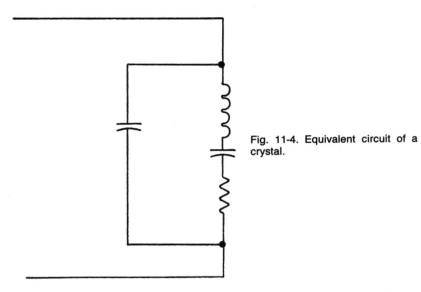

Fig. 11-4. Equivalent circuit of a crystal.

The resonant frequency of a crystal is determined primarily by the thickness and size of the crystal slice.

Crystals can also be made to resonate at integer multiples of its main resonant frequency. These multiples are called *overtones*, or *harmonics*. The primary frequency is called the *fundamental*. For example, a crystal designed to resonate at 15,000 Hz (the fundamental) will also resonate (but to a lesser degree) at 30,000 Hz (second harmonic), 45,000 Hz (third harmonic), 60,000 Hz (fourth harmonic), and so forth. Notice that no first harmonic is mentioned, since one times the fundamental equals the fundamental frequency. The resonance effect will become steadily less pronounced at higher harmonics.

Crystals are usually more expensive than separate capacitors and coils, but their resonant frequency is much more precise and stable. Capacitor-coil resonant circuits often drift off frequency (that is, the the components change their values slightly), particularly under changing temperature conditions.

Crystals are also somewhat temperature sensitive (although not as much so as capacitors and coils) so when very high accuracy is required (as in broadcasting applications), the circuit is enclosed in a *crystal oven*, which maintains a constant temperature.

Reliability is another advantage of crystals. Their failure rate tends to be somewhat lower than capacitors and coils. However, they can be damaged by high over-voltages, or extremely high temperatures. Also, a severe mechanical shock (such as being dropped onto a hard surface) could crack the delicate crystal element. If this happens, the entire crystal package must be replaced. There is no way to replace just the crystal slice.

Occasionally the hermetic seal develops a small leak, allowing contaminants (like dust, or water) to get inside the metal case. This can interfere with the electrical contact between the metal plates and the crystal element. All of these defects are relatively rare, but they do occur frequently enough to warrant mentioning them here.

The most important disadvantage of crystals is that the resonant frequency generally cannot be changed. In a capacitor-coil circuit, one or both components can be made variable, but there is no such thing as a variable crystal. Sometimes special external circuitry can be added for a limited degree of fine tuning, but generally a crystal resonant circuit is fixed. The only way the frequency can be changed is by physically replacing the crystal. For this reason crystals are usually inserted into sockets, rather than actually being permanently soldered into the circuit. The leads of the crystal plug into holes in the socket which make electrical contact with the external circuit. Crystals can easily and quickly be removed and replaced. Using sockets also sidesteps the potential problem of thermal damage resulting from soldering the crystal's leads directly.

Crystals are available, cut for many different frequencies. The most common type of crystal is tuned for a frequency of approximately 3.58 MHz. This frequency is used for the standard color burst signal in a color television transmission, so there is a very large market for such crystals. Many other circuits, which often have nothing at all to do with television, are designed to use the same frequency, simply to take advantage of the wide availability and low cost of 3.58 MHz crystals.

Self-Test

1. What material are crystals made out of?

A *Nickel alloy*
B *Quartz*
C *Iron*
D *Glass*
E *None of the above*

2. What is the purpose of the hermetically sealed housing used with crystals?

A *To stabilize the temperature*
B *To protect the fragile crystal from being cracked*
C *To shield the crystal from external magnetic fields*
D *To keep out moisture and dust*
E *None of the above*

3. What is the principle behind the operation of a crystal?

A *The piezoelectric effect*
B *The photoelectric effect*
C *The electromagnetic effect*
D *The ionization effect*
E *None of the above*

4. What is the result of placing a mechanical stress along the Y axis of a crystal?

A *An electrical voltage will appear across the X axis*
B *The crystal may be damaged*

C *An electrical voltage will appear across the Y axis*
D *The mechanical motion will be amplified across the X axis*
E *None of the above*

5. A crystal can be used in place of what type of circuit?

A *An RC circuit*
B *A series resonant circuit*
C *A parallel resonant circuit*
D *Either a series or a parallel resonant circuit*
E *None of the above*

6. What determines the resonant frequency of a crystal?

A *External components*
B *The hermetic seal*
C *The size and thickness of the crystal material*
D *The temperature of the crystal*
E *None of the above*

7. If a crystal has a resonant frequency of 47,000 Hz, what is the third harmonic?

A *50,000 Hz*
B *141,000 Hz*
C *188,000 Hz*
D *300,000 Hz*
E *None of the above*

8. When should a crystal oven be used?

A *In very cold environments*
B *Always*
C *When extremely high accuracy is required*
D *When the crystal is being operated at a high harmonic of its resonant frequency*
E *None of the above*

9. What is the biggest disadvantage of using crystals in resonant circuits?

A *Size*
B *Cost*
C *The resonant frequency cannot be easily changed*
D *Poor accuracy*
E *None of the above*

10. What is the biggest advantage of using crystal in resonant circuits?

A *Size*
B *Greater accuracy and stability*
C *Less fragile*
D *Cost*
E *None of the above*

12

Meters

In practical electronics work it is very often necessary to measure the basic parameters in a circuit, i.e., voltage, current, and resistance.

We can calculate the nominal values, but component tolerances can throw the actual values off quite a bit. For example, a circuit containing three resistors, each with a tolerance of $\pm 20\%$ can be off value by as much as $\pm 60\%$ (actually, it's very rare for the values to be that far off, since some of the component errors will probably cancel each other out; but it is possible.)

Also, the calculations for complex circuits can be extremely tedious, time-consuming, and difficult to keep track of.

Various defects can cause components to change their values drastically, so for troubleshooting a method of actually measuring the circuit values is absolutely essential. It is the only way to pinpoint the defective component.

Fortunately, there is a relatively simple way to measure voltage, current, and resistance. These measurements are made with devices called *meters*.

THE D'ARSONVAL METER

By far, the most common type of meter is the *D'Arsonval meter*. The construction of this kind of meter is shown in Fig. 12-1.

A permanent magnet in the shape of a horseshoe is positioned around a coil of wire wound on a piece of soft iron. This coil—which is called the *armature*, or the *movement* of the meter—is on a pivot that allows it to move freely.

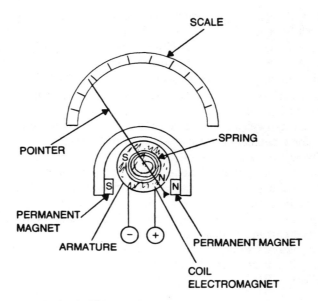

Fig. 12-1. Construction of a basic D'Arsonval meter.

When an electrical current flows through the coil with the polarity shown, the movement develops a magnetic field. Its north pole will be facing the north pole of the permanent magnet, and its south pole will be facing the south pole of the permanent magnet. Since like poles repel, the movement will turn on its pivot so its poles will no longer face the poles of the permanent magnet. Just how far the movement will turn will depend on the relative strength of the magnetic fields involved.

The magnetic field of the permanent magnet is constant, of course, but the magnetic field of the armature will depend on the strength of the electrical current flowing through it. Therefore, the motion of the armature will be directly proportional to the amplitude of the applied electrical current.

A *pointer* can be attached to the center of the armature, so that the amount of applied current is indicated on a *calibrated scale*. This scale is marked in equal units spaced so that the current value can be read directly.

A small spring opposes the rotation of the armature. This small amount of physical opposition is easily overcome by the applied current and its resulting magnetic field, but when current stops flowing through the coil, the spring forces the movement (and thus, the pointer) to return to the zero position.

The current applied to this type of meter must have the correct polarity. If the polarity of the current is reversed, the south pole of the electromagnet (armature) will be facing the north pole of the permanent magnet, and vice versa. Since unlike poles attract, the armature will not move.

Some meters can be adjusted so that the zero position is in the center of the scale. This will allow the pointer to move in either direction, thus indicating current of either polarity. Most meters, however, do not have this capability.

A meter's movement is quite delicate and fragile. Meters are usually enclosed in protective cases (made of transparent plastic), but they can still be permanently damaged if dropped, or otherwise mishandled. Also, the coil in the armature is made of very fine wire so it will be light and move easily. This means the wire cannot carry much current. If too much current is applied directly, the armature can be quickly ruined. Fortunately, there are methods of decreasing the current actually applied to the meter, while still allowing the meter to give an accurate reading. This will be discussed later.

With a reasonable amount of care, a D'Arsonval meter movement is sturdy enough for practical use, and can last for years.

This type of meter (which is sometimes also called a *moving-coil movement meter*) is quite popular for a number of reasons. It is relatively inexpensive, highly accurate and *sensitive* (that is, very small currents can be readily measured). The scale is uniform and easy to read, because the movement of the pointer is directly proportional to the applied current, resulting in evenly spaced calibration markings. A D'Arsonval drains very little current from the circuit being tested, so it is quite efficient, and doesn't significantly affect the values being measured. Finally, the D'Arsonval meter can easily be adapted to read current, voltage, or resistance.

This type of meter is so commonly used, that whenever a meter is mentioned in electronics work, it can generally be assumed to be a D'Arsonval type unit unless otherwise specified.

AMMETERS

When a meter is used to measure electric current, it is referred to as an *ammeter* (from AMpere METER).

The standard schematic symbol for a meter is shown in Fig. 12-2A. Letters are generally placed within the circle to indicate exactly what kind of meter is being used. For example, the letter "A" in the symbol in Fig. 12-2B means the unit is an ammeter.

Since the ampere is a rather large unit for an electronic circuit, *milliammeters* (Fig. 12-2C), or *microammeters* (Fig. 12-2D) are frequently used instead. All three types of ammeters work in exactly the same way—the only difference is the range of values they are capable of measuring.

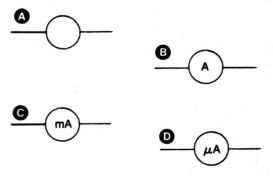

Fig. 12-2. Schematic symbols: A. any meter; B. ammeter; C. milliammeter; D. microammeter.

To measure current, the meter must actually be inserted into the circuit itself. In other words, the meter is placed in series with the circuit. This is shown in Fig. 12-3. Remember that in a series circuit the current is equal at all points, so if an ammeter

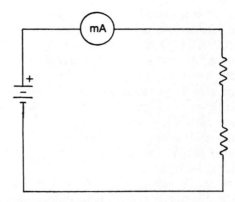

Fig. 12-3. An ammeter in series with the circuit to be measured.

is placed in series with the rest of the circuit, the same current that flows through the circuit will flow through the meter.

Because the meter is in series with the circuit, its internal resistance must be as low as possible, to avoid upsetting the normal current flow. For example, let's suppose we have a circuit with a total resistance of 10,000 ohms, and is powered by a 10 volt source. The nominal current will be equal to E/R = 10/10,000 = 0.001 ampere (1 mA).

When the meter is inserted in the circuit, its internal resistance will add to the resistance of the circuit. For instance, if the meter's internal resistance is 5000 ohms, the total resistance in the circuit becomes 15,000 ohms (10,000 + 5,000), so the current changes to 10/15,000 or 0.00067 ampere, (0.67 mA). The current value is dropped 3%. Obviously, that is a significant difference.

On the other hand, if the meter's internal resistance is only 50 ohms, the total circuit resistance will be 10,050 ohms, and the current will be equal to 10/10,050, or 0.000995 ampere (0.995 mA). This is within 0.5% of the correct nominal value. Clearly, for an accurate current reading, the ammeter's internal resistance must be as low as possible.

Because an ammeter must be used in series with the circuit it is testing, one of the connections in the circuit has to be physically disconnected, so the ammeter can be inserted. Often this requires desoldering.

Quite often it is necessary to measure a current that is larger than the available meter will handle. This problem can be taken care of with a *shunt resistance*, that is, a resistor in parallel with the meter movement. See Fig. 12-4.

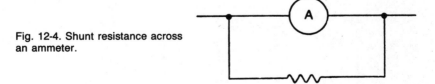

Fig. 12-4. Shunt resistance across an ammeter.

By carefully selecting the proper ratio between the shunt resistance and the internal resistance of the meter itself, we can measure virtually any amount of current flow. Most ammeters are actually milliammeters or microammeters with an appropriate shunt resistance, because the tiny coil in a meter movement usually can't carry a full ampere without damage.

The shunt resistance is generally quite small. As an example, let's suppose we have a meter with an internal resistance of 50 ohms, and which has a *full scale reading* of 1 mA (or 0.001 ampere). This means that a current of 1 mA will cause maximum deflection of the armature and pointer. A greater current might damage the meter.

Now, suppose we need to use this meter to read currents up to 100 mA (0.1 ampere). The meter itself can handle only 1% of the desired full scale reading, so, obviously, the shunt resistance will have to carry the other 99%.

Using Ohm's law we can find the full scale voltage dropped through the meter. E = IR, or, in this example, 0.001 ampere × 50 ohms, or 0.05 volt. For a full scale reading of 0.1 ampere, the shunt will have to carry a current of 0.099 ampere (0.001 + 0.099 = 0.1). Since the shunt resistance is in parallel with the meter, the voltage drop will be the same. Therefore, R = E/I = 0.05/0.099, or approximately 0.5 ohm. Notice that this drops the apparent resistance of the meter to 1/50 + 1/0.5, or 0.495 ohm.

It should be obvious from the above example that the full scale reading of an ammeter cannot be increased by too large a factor, or the shunt resistance will become impractically small. In fact, this particular example would probably be rather impractical, because resistance values below 10 ohms are fairly rare.

Of course, when the range of a meter is changed, the calibration markings on the scale face will no longer be accurate. However, when the increased full scale reading is an exact multiple of 10 (as in our example) the same calibration markings can be used, and the appropriate number of zeroes can be added mentally. For instance, if the meter in the example gave a reading of 0.5 milliamps on its old scale, it would mean that the actual current was 0.5 × 100, or 50 mA (0.05 ampere).

Of course, all shunt resistors should have the tightest tolerance possible. 1% tolerance resistors are generally essential, but 5% resistors can be adequate in some uncritical applications. The tolerance should never be more than 5%.

VOLTMETERS

To measure the current flowing in a circuit, the meter is placed in series with the circuit. If, on the other hand, a meter is connected in parallel across a resistance within a circuit (such as in Fig. 12-5), the deflection of the pointer will be proportional to the voltage drop across the resistance.

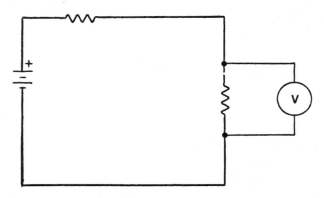

Fig. 12-5. A voltmeter in parallel with the circuit to be measured.

Actually, the meter is still measuring the current flowing through it, but since current is determined by voltage and resistance (I = E/R), making the resistance constant, will cause the current through the meter to vary in step with the voltage. The meter's scale can easily be calibrated so that the reading is given directly in volts. Such a device is called a *voltmeter*.

In an ammeter, the resistance should always be kept as low as possible, because it is in series with the circuit being tested, because a large resistance would produce a large voltage drop that would not be present if the meter was not in the circuit.

In a voltmeter, on the other hand, the resistance should be as large as possible. Generally, a rather large fixed resistor called a *multiplier* is connected in series with the meter, as shown in Fig. 12-6. This multiplier resistance serves two important functions—it protects the meter movement from excessive current, and it helps prevent *circuit loading*.

Fig. 12-6. Adding a multiplier resistance to a voltmeter.

Suppose we want to take measurements in the simple resistor/voltage source circuit shown in Fig. 12-7A. Let's assume each of the two resistors has a value of 1000 ohms, and the voltage source is 10 volts. The total resistance in the circuit is 1000 + 1000, or 2000 ohms. The current in the circuit equals 10/2000, or 0.005 ampere (5 mA). This means the voltage drop across each resistor equals 1000 × 0.005, or 5 volts.

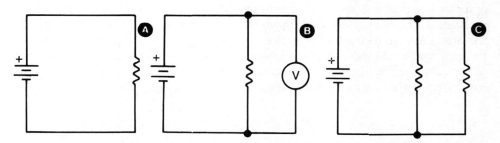

Fig. 12-7. A simple circuit: A. the basic circuit; B. the circuit with a voltmeter added; C. equivalent circuit of the circuit and the meter.

Now, when we connect the meter, as in Fig. 12-7B, it places a resistance in parallel with the resistor. The equivalent circuit is shown in Fig. 12-7C.

If the meter has an internal resistance of 50 ohms, the parallel combination of the meter and R1 is found by the formula, $1/R_T = 1/R_1 + 1/R_m = 1/1000 + 1/50 = 21/1000 = $ approximately 48 ohms.

Now, the total effective resistance in the circuit is just 48 + 1000, or 1048 ohms. The current is changed to 10/1048, or about 0.01 ampere (10 mA). The voltage drop across the R1-Rm combination is 0.01 × 48, or a mere 0.48 volts. This will be the reading shown on the meter. Quite obviously, this is extremely inaccurate.

Now, suppose we include an extra series resistor (i.e., a multiplier) inside the meter, so that the total resistance of the meter is 20,000 ohms. In this case, the parallel combination of R1 and Rm can be calculated as $1/R_T = 1/1000 + 1/20,000 = 21/20,000$, or about 952 ohms. This is much closer to the original value of 1000 ohms.

If we increase the meter resistance further to 1,000,000 ohms (1 megohm), the parallel combination value comes even closer to the nominal value of R1 alone—just slightly over 999 ohms.

For the most accurate readings a meter should be essentially non-existent as far as the circuit's operation is concerned. For a voltmeter, increasing the resistance of the meter will increase the accuracy of the measurement.

However, if you're comparing voltages between similar circuits (as in servicing, when you compare a defective unit to a working standard), the meters used in each circuit should have the same internal resistance, or identical voltages will not produce the same readings. If the difference is large, you may not be able to compare the readings in any meaningful way.

OHMMETERS

The third type of common meter circuit measures dc resistance and is called an *ohmmeter*. Of course, the meter itself is actually responding to the current flowing through it, but by passing a known voltage through an unknown resistance, we can measure the current and use Ohm's law to calculate the resistance. You don't have to actually do any calculations when you use an ohmmeter—the scale is calibrated to read the resistance directly in ohms.

The basic ohmmeter circuit is shown in Fig. 12-8. The resistor labeled Rm is included to calibrate the meter. For example, if the battery is three volts, and the total resistance of the meter circuit is 3000 ohms, and we short the test leads together (0 ohms external resistance), the current flowing through the meter will equal 3/(3000 + 0), or 0.001 ampere (1 mA). If the meter has a full scale reading of 1 mA, the pointer will move all the way over to the high end of the scale, which is marked 0 for an ohmmeter. Rm is used to aim the pointer directly at the zero mark when the leads are shorted. As the battery ages, the required resistance to do this will change.

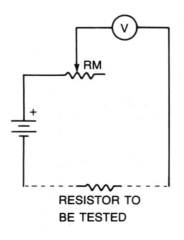

Fig. 12-8. A basic ohmmeter circuit.

On the other hand, if the test leads are disconnected (infinite external resistance) no current at all will flow through the meter, because there isn't a complete circuit path available. Of course, the pointer will remain at its rest position (low end of the scale). On an ohmmeter this position is labeled ∞, or infinity.

Now, if we connect a 3000 ohm resistor between the test leads, the total circuit resistance will equal 3000 (internal resistance) + 3000 (external resistance), or 6000 ohms. Therefore, the current flowing through the meter must equal 3/6000, or 0.0005 ampere (0.5 mA). The pointer will move the center of the scale.

You'll notice that the lower the resistance between the test leads, the farther the pointer moves up the scale. This, of course, is the exact opposite of what happens with voltmeters and ammeters, where the pointer moves further for higher values.

There is another major difference in the way the scale of an ohmmeter must be calibrated. This is indicated in Table 12-1 which compares the current flow with the test resistance. Unfortunately, the current through the meter does not change in a direct linear fashion with the external resistance. This is because the internal resistance is, of necessity, a constant value.

The internal resistance can be altered to change the total range of the meter (if the internal resistance is 10,000 ohms, a mid-scale reading would indicate an external resistance of 10,000 ohms), but within a given range, the resistance is fixed as far as the actual testing is concerned.

Table 12-1. Why an Ohmmeter's Scale Is Not Linear.

E	R_M	TEST RESISTANCE	R_T	I (mA)
3	3000	0	3000	1.00
3	3000	500	3500	0.86
3	3000	1000	4000	0.75
3	3000	1500	4500	0.67
3	3000	2000	5000	0.60
3	3000	2500	5500	0.55
3	3000	3000	6000	0.50
3	3000	3500	6500	0.46
3	3000	4000	7000	0.43
3	3000	4500	7500	0.40
3	3000	5000	8000	0.38
3	3000	5500	8500	0.35
3	3000	6000	9000	0.33
3	3000	6500	9500	0.32
3	3000	7000	10000	0.30
3	3000	7500	10500	0.29
3	3000	8000	11000	0.27
3	3000	8500	11500	0.26
3	3000	9000	12000	0.25
3	3000	9500	12500	0.24
3	3000	10000	13000	0.23
3	3000	10500	13500	0.22
3	3000	11000	14000	0.21
3	3000	∞	∞	0.00

AC METERS

An ohmmeter can only be used to test dc resistance—not reactance or impedance. It might seem that an ac ohmmeter could be built with an ac power source, using a circuit something like the one shown in Fig. 12-9. However, any measurement made with such a device would be meaningful only at the specific frequency of the source voltage used in the test. The voltage source could be made so that the frequency is variable, but such a circuit would add greatly to the instrument's cost and complexity. Testing would be a long, drawn-out process. Besides, such a device would give no indication of phase relationships.

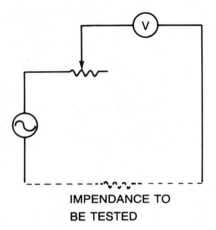

Fig. 12-9. A theoretical ac ohm-meter.

IMPENDANCE TO
BE TESTED

Unfortunately, there is no easy way to directly measure impedance. Usually it must be calculated from voltage and current measurements.

Fortunately, meters can be constructed to read ac voltages and currents, with the help of a device called a *diode*, which only lets current pass through it if the polarity is correct. If the polarity is reversed, the diode blocks the current. Fig. 12-10 shows what happens to an ac signal as it passes through a diode. Only half of the actual ac waveform is applied to the meter itself. Since the two halves of an ac waveform are mirror images of each other, the total value can easily be derived. Diodes will be discussed in detail in a later chapter.

Ac meters generally measure rms values of sine waves. If the waveform is something other than a sine wave, the reading will not be equal to the true rms value.

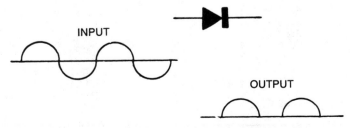

INPUT

OUTPUT

Fig. 12-10. What happens to an ac signal when it passes through a diode.

VOMS

Probably the most commonly used piece of equipment in electronics work is the *VOM*. VOM stands for *volt-ohm-milliammeter*. Various resistors are switched in and out of series or parallel to set up the meter for each type of measurement. An internal battery is included for resistance measurements. Some VOMs do not have the capability to measure current directly, and many are designed for dc use only.

Usually any of a number of different value resistances can be switched into the circuit, so measurements can be made within different ranges. Whichever range is easiest to read for a specific value can be easily selected, 10 millivolts would probably be very difficult to detect on a meter scale that went up to 100 volts.

The scale face of a VOM has a number of sets of calibration markings, so each of the metering functions can be read directly. Usually the different ranges are multiples of 10 of the basic range, so converting the reading to the appropriate range is simply a matter of mentally adding the correct number of zeroes. For example, a reading of 1.5 on a × 100 range would indicate a value of 150.

Closely related to the VOM is the VTVM, or *vacuum-tube voltmeter*. This device is operated by an ac power supply and has an extremely high input impedance, and therefore it has a very high degree of accuracy. The disadvantages of the VTVM are its greater cost and complexity, and the fact that it must be plugged into an ac wall socket, which limits its portability.

A VTVM's sensitivity and high impedance can be simulated by a special type of VOM that is built around a device called a *field-effect transistor*, or *FET*. This component will be dealt with in a later chapter.

In the last decade or so, meter-type VOMs have been largely (but not completely) replaced by *digital multimeters*, or *DMM*s. These devices are not built around a mechanical meter. Instead, the value of the measured signal is displayed directly in numerical form using LEDs (Light Emitting Diodes), or LCDs (Liquid Crystal Displays). These display devices will be covered later in this book.

For most applications, DMMs and VOMs or VTVMs are pretty much interchangeable. DMMs usually have very high input impedances, so they are quite accurate. In addition, the numerical display is easier to read than a meter pointer. There is no need to worry about any error from looking at the meter face at an angle. So why haven't DMMs replaced mechanical meters altogether? For some applications they really don't work very well. This is especially true when measuring values that change over time (for example, the charge on a capacitor). While the movement of a meter pointer is easy to follow either up or down (or even back and forth), a DMM will just display an unreadable and meaningless blur of rapidly changing numbers. A well-stocked modern electronics workbench will have both a VOM (or VTVM) and a DMM.

Self-Test

1. Which of the following can *not* be easily measured with a simple meter circuit?

A *Resistance*
B *Current*
C *Impedance*

D *Voltage*
E *None of the above*

2. What is the most common type of meter movement?

A *D'Arsonval*
B *Fixed oil*
C *Digital*
D *Farad*
E *None of the above*

3. What type of meter is used to measure current?

A *Ohmmeter*
B *Wattmeter*
C *Moving coil meter*
D *Ammeter*
E *None of the above*

4. How much internal resistance should an ammeter have?

A *As much as possible*
B *A variable amount determined by Ohm's law*
C *As little as possible*
D *It doesn't matter*

5. To increase the capacity of an ammeter, what should be added to the circuit?

A *A series resistance*
B *A shunt resistance in parallel with the meter*
C *A shunt capacitance in parallel with the meter*
D *A series inductance*
E *None of the above*

6. How is a voltmeter used?

A *It is placed in parallel across the component being measured*
B *It is placed in series with the component being measured*
C *It is placed within the magnetic field of the component being measured*
D *None of the above*

7. Which type of meter requires its own power source?

A *A voltmeter*
B *An ohmmeter*
C *A wattmeter*

D *An ammeter*
E *None of the above*

8. What is a VOM?

A *A combination voltmeter/ohmmeter*
B *A voltage only meter*
C *A combination ohmmeter, milliammeter, and voltmeter*
D *A measurement of the movement of a meter's pointer*
E *None of the above*

9. For the greatest accuracy, what should the input impedance of a VOM be?

A *50,000 ohms per volt*
B *1,000 ohms per volt*
C *As large as possible*
D *As small as possible*
E *1,000,000 ohms per volt*

10. What do ac voltmeters measure?

A *The peak voltage of a sine wave*
B *The rms voltage of any waveform*
C *The average voltage of a sine wave*
D *The rms voltage of a sine wave*
E *None of the above*

13

Other Simple Components

We have been concentrating solely on passive components. Later on, we will be looking at active components, such as tubes and transistors. An active component is one that is capable of amplification. A passive component does not amplify and does not need a power source other than the signal passing through it.

The most important types of passive components have already been covered. These are resistors (Chapter 3), capacitors (Chapter 6), and inductors (Chapters 8 and 9). This chapter will discuss a few additional passive devices that are used in many electronic circuits. We will mostly be working with switching devices. The last portion of this chapter will turn to motors and lamps.

WHAT IS A SWITCH?

In the circuits described so far, the only way to stop the current from flowing is to physically disconnect one of the components. Obviously this would be impractical in most types of equipment. Yet, a continuous current flow is generally undesirable—power is wasted, continuous operation may be annoying, or even dangerous, and it might interfere with the operation of other circuits. Also, it is often necessary to selectively allow current to pass through various sub-circuits or components at different times. The solution to these problems is a simple device called a *switch*.

TYPES OF SWITCHES

A switch is simply a device that makes and/or breaks an electrical connection, thereby completing or opening a current path.

The simplest type of switch is the *knife switch*, which is illustrated in Fig.13-1. When the handle of the switch is in the position shown in Fig. 13-1A, no current can flow through the attached circuit, because there is not a complete path for the current to follow. If the handle is moved into the position shown in Fig. 13-1B, the metal handle completes the circuit and current can flow.

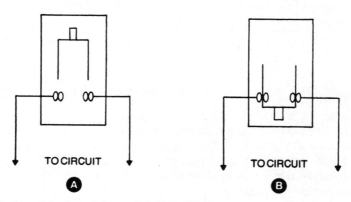

Fig. 13-1. A knife switch: A. switch opened; B. switch closed.

A switch like this is quite simple to make and use, but it is rarely used in modern circuits because it is quite bulky, and the electrical connections are exposed, which could mean a serious shock hazard.

A more common type of switch is shown in Fig. 13-2. This kind is called a *slide switch*. When the *slider* (moveable portion) is in the position shown in Fig. 13-2A, the circuit is open and no current can flow. But, when the slider is moved into the position shown in Fig. 13-2B, the metal strip on the bottom of the slider touches the two connections to the external circuit, thus completing the current path.

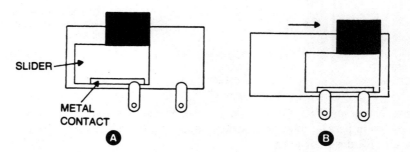

Fig. 13-2. Construction of an SPST slide switch: A. switch opened; B. switch closed.

Another popular type of switch is the *toggle switch*, illustrated in Fig. 13-3. It operates in a similar manner to the slide switch, but the slider is in the shape of a ball that rolls in and out of position.

Still another commonly used switch type is the *pushbutton*. This device will be dealt with later in this chapter.

148

Fig. 13-3. A toggle switch.

SPST Switches

The switches we have been talking about so far have only two connections. Because the slider, or *pole*, makes contact in only one position, this form of switch is said to be a *single-pole/single-throw switch*. This is generally abbreviated as *SPST*.

The schematic symbol for a SPST switch is shown in Fig. 13-4. This switch is usually shown in its open position for clarity. The same symbol is used for most kinds of switches (i.e., slide, toggle, or whatever).

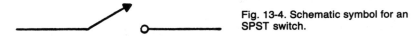

Fig. 13-4. Schematic symbol for an SPST switch.

When the switch is open, there is a break in the circuit. No current flows. When the switch is closed it acts like a simple piece of wire, and current can flow through it. See Fig. 13-5.

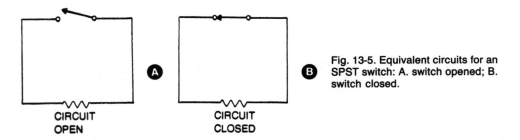

Fig. 13-5. Equivalent circuits for an SPST switch: A. switch opened; B. switch closed.

Another way an SPST switch might be used is shown in Fig. 13-6. This circuit has two resistors in parallel, but if the switch is open, current can flow through R1 only. Electrically, R2 does not exist.

However, when the switch is closed, current can flow through both resistors. R1 and R2 act like any parallel resistance combination. Referring again to Fig. 13-6, if the battery generates 6 volts, and each resistor has a value of 100 ohms, the milliammeter will indicate 60 mA when the switch is open. However, when the switch is closed, connection R2 in parallel with R1, the total effective resistance in the circuit drops to 50 ohms, so the current increases to 120 mA.

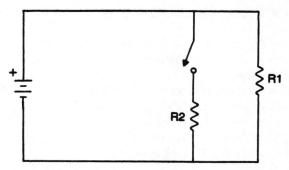

Fig. 13-6. SPST switch in a parallel circuit.

SPDT Switches

A single switch can be used to control two separate circuits if a third connection is added, as in Fig. 13—7. When this switch is in the position shown in Fig. 13-7A, the current can flow between connector 1 and connector 2. Connector 3 will be open (disconnected). Moving the slider into the position shown in Fig. 13-7B will allow current to flow between connector 2 and connector 3, but connector 1 will now be left open.

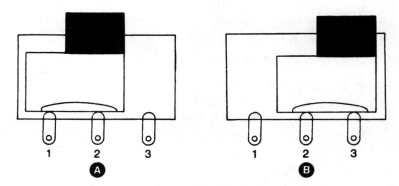

Fig. 13-7. An SPDT slide switch: A. switch opened; B. switch closed.

Since this type of switch has a single pole and two possible positions, it is called a *single-pole/double-throw* or *SPDT* switch.

The schematic symbol for this kind of switch is shown in Fig. 13-8. The slider may be shown in either position. Notice that connector 2 (Fig. 13-7) is common to either

Fig. 13-8. Schematic symbol for an SPDT switch.

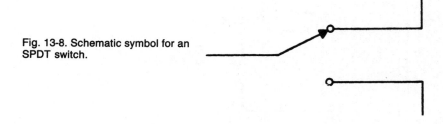

circuit, regardless of the slider position. The connector numbers are not shown in actual schematic diagrams.

A simple circuit using an SPDT switch is shown in Fig. 13-9. R2 and R3 are in parallel with R1, but current will flow through only one resistor at any given time, depending on the position of the slider.

In Fig. 13-9A, current can flow through R1 and R2, but not R3. In Fig. 13-9B, the situation is reversed. Current can flow through R1 and R3, but not R2.

If we assume the battery puts out 6 volts, R1 is 100 ohms, R2 is 220 ohms, and R3 is 4,700 ohms, the milliammeter will read 87 mA (0.087 ampere) in Fig. 13-9A, while in Fig. 13-9B it will read 98 mA (0.98 ampere).

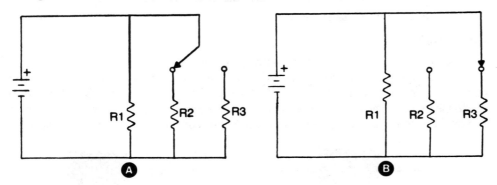

Fig. 13-9. An SPDT switch in a parallel circuit.

Some SPDT switches have a *center off* position. In this case, the slider would be positioned as illustrated in Fig. 13-10. Neither circuit 1-2, nor circuit 2-3 is complete, so no current flows through the switch at all.

An SPDT switch could be used as an SPST unit by just leaving one of the end connectors unused. See Fig. 13-11.

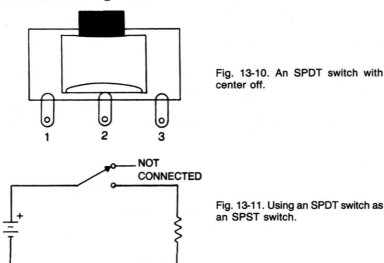

Fig. 13-10. An SPDT switch with center off.

Fig. 13-11. Using an SPDT switch as an SPST switch.

DPST

A similar type of switch is the *double-pole/single-throw* switch *(DPST)*. The schematic symbol for this device is shown in Fig. 13-12. Notice that it is actually two entirely separate (electrically) SPST switches with a common slider. The slider has two separate metal strips to complete two separate circuits with no common connections. Of course, these two SPST switches always work in unison. They are either both off, or both on. A DPST switch can be used as an SPST switch by just using one set of contacts.

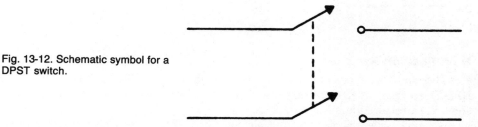

Fig. 13-12. Schematic symbol for a DPST switch.

Actually, DPST switches are fairly uncommon. When a DPST function is required, a DPDT switch (see below) is usually used.

DPDT Switches

Double-pole/double-throw (DPDT) switches have six connection terminals (see Fig. 13-13) and are essentially two SPDT switches with a common slider, just as a DPST switch is essentially two SPST switches with a common slider.

BOTTOM VIEW

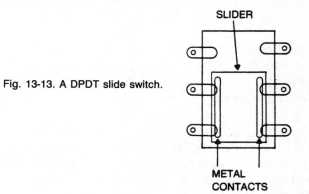

Fig. 13-13. A DPDT slide switch.

Obviously a DPDT switch can be substituted for any of the other switch types (SPST, SPDT, DPST, or a combination of an SPST and an SPDT) simply by using only the appropriate terminals. By using all six contacts, we can achieve two electrically separate SPDT actions with the throw of a single switch. The schematic symbol for a DPDT switch is shown in Fig. 13-14. Like the SPDT switch, DPDT switches are often equipped with a center-off position which leaves all six terminals open.

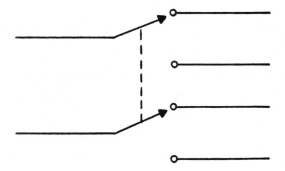

Fig. 13-14. Schematic symbol for a DPDT switch.

Multi-Position Switches

The four basic types of switches discussed above (i.e., SPST, SPDT, DPST, and DPDT) are the most commonly used configurations, but other configurations are sometimes required in specific circuits.

Generally slide switches are available only up to DP4T (two poles with four positions each). If more poles, and/or positions are needed, a *rotary switch* is used. This type of switch gets its name from the fact that a knob is rotated to control the position of the slider(s).

The schematic symbol for a rotary switch varies somewhat, of course, depending on the number of poles and positions in the specific switch. Figure 13-15A shows the schematic symbol for an SP12T rotary switch, while Fig. 13-15B is the symbol for a 3P6T unit. Many other combinations are also possible.

There are two basic varieties of rotary switches. The *nonshorting type* disconnects the circuit at one position completely before the connection of the next position is made.

The other kind of rotary switch is the *shorting,* or *make-before-break* type. With this type of switch, in switching from position A to position B, the switch makes contact with both position A and position B for a brief instant, before the connection at position A is broken.

In most circuits it doesn't really matter which type is used, but some specialized circuits require one type or the other.

Momentary Contact Switches

Sometimes it is necessary to open or close a circuit connection only briefly, then return it to its original condition. In many such cases, manually moving the switch back and forth is inconvenient and/or impractical. For this sort of situation a *momentary action* or *momentary contact* switch is used. This kind of switch is loaded with a spring that always returns the slider to a specific rest position, unless it is manually held in the other position. The rest position can be either *normally open(N O)* or *normally closed(N C),* depending upon the requirements of the specific circuit.

While momentary slide switches and toggle switches are available, this type of switch action is usually in the form of a *pushbutton switch*.

Pushbutton switches can be SPST, SPDT, or DPDT (or, occasionally DPST), but the most common configuration for momentary action switches is SPST.

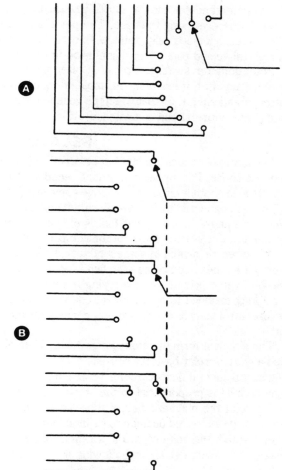

Fig. 13-15. Schematic symbols: A. SP12T rotary switch; B. 3P6T rotary switch.

If it is a momentary action type, this will be indicated in the parts list, or in a note on the schematic. Notice that a double-throw, momentary-contact switch has one set of contacts that are normally open, and another set of contacts that are normally closed.

Most pushbutton switches are of the momentary action type, but some work as regular switches that can be left in either position. A pushbutton switch that does not have momentary contacts is called a *push on/push off switch*.

Potentiometer Switches

Another type of commonly used switch fits onto the back of a potentiometer. This kind of switch is usually of the SPST type.

The switch is controlled by the potentiometer's knob. When the potentiometer is in its maximum resistance position, the switch is open (off). But as soon as the control knob is advanced from this extreme position, the switch is clicked shut, turning the controlled circuit on. From then on the potentiometer operates normally.

The potentiometer and the switch will usually be used in the same part of the circuit, but this isn't necessarily the case, The potentiometer and the switch are mechanically tied together, but electrically distinct. They could be used in two entirely separate circuits, although this could cause some confusion in operation.

Potentiometer switches are most often used to turn the main power supply of a circuit on and off. For example, an amplifier might have a switch connected to its volume control potentiometer, so that when the volume is turned all the way down to its minimum setting, the entire circuit is switched off.

RELAYS

In many applications it might not be practical to use any of the basic manual switches discussed so far. For instance, we might need a circuit that must be switched on when the voltage in another circuit rises above some specified level. Of course, we could watch a voltmeter connected to the second circuit and manually flick a switch in the first circuit at the appropriate moment, but that is obviously a highly impractical and inconvenient approach to the problem. Some sort of automatic switching would clearly be more efficient.

An automatic switch of some kind would also be necessary in a remotely controlled circuit. Often the circuit being switched is not readily accessible. It would be inconvenient at best to run a pair of wires carrying the full power supply voltage or electrical data over a long distance to a convenient control point. It would be far better to just send a small control voltage over light-duty connecting wires to a remote-controlled electrical switch.

The simplest form of automatic switching in an electrical circuit is through a device called a *relay*. A relay basically consists of two parts—a coil and a magnetic switch. When an electrical current flows through a coil, a magnetic field will be created around it. This magnetic field is proportional to the amount of current flow through the coil. At some specific point the magnetic field will be strong enough to pull the switch's slider from its rest, or *de-energized* position to its momentary, or *energized* position. If the electrical power through the coil drops, the strength of the magnetic field will drop off to zero, releasing the switch slider and allowing it to spring back to the original de-energized position.

The switch section of a relay can be any of the basic switching types discussed earlier in this chapter (SPST, SPDT, DPST, or DPDT). For single-throw units the switch contacts may be either normally open or normally closed. Of course, with an SPDT or a DPDT switching arrangement, you have both a normally open and a normally closed contact simultaneously.

The schematic symbols for an SPST and an SPDT relay are shown in Fig. 13-16. Relays are usually identified in schematic diagrams and parts lists by the letter *K*.

Fig. 13-16. Schematic symbols: A. SPST relay; B. SPDT relay.

A relay's coil and switching contacts are virtually always used in electrically isolated circuits. That is, the current through one circuit controls the switching of another circuit. Relays vary greatly in size, depending primarily on the amount of power that they can safely carry. Separate ratings are usually given for the coil and the switch contacts, because they are generally used in separate circuits. Relays range from tiny units intended for 0.5 watt transistorized equipment to huge megawatt (millions of watts) devices used in industrial power generating plants. At either extreme, the principle of operation is precisely the same.

Of course, the most important rating for a relay is the voltage required to make the switch contacts move to their energized position. This is called *tripping* the relay. Typical trip voltages for relays used in electronic circuits are 6, 12, 24, 48, 117, and 240 volts. Both ac and dc types are available.

The controlling voltage through a relay coil should be kept within about ±25% of the rated value. Too large a voltage could burn out the delicate coil windings. On the other hand, too small a control voltage could result in erratic operation of the relay.

Sometimes it may be necessary to drive a relatively high power circuit with a rather low power control signal. This can be done with the type circuitry illustrated in Fig. 13-17. High voltage supply B is operated only when the relay is energized, minimizing power consumption.

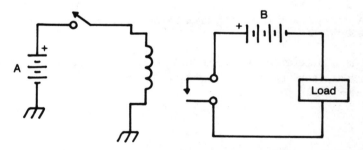

Fig. 13-17. A relay is used so that a low voltage can control a higher voltage load.

Occasionally the available control signal will not be sufficient to drive a large enough relay for the circuit to be controlled. In a case like this, the solution may be to add an additional medium-power relay to act as an intermediate stage, as shown in Fig. 13-18.

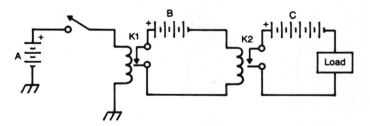

Fig. 13-18. A small relay can be used to drive a heavier relay.

The coil winding could self-destruct if the current through it changes suddenly, perhaps due to opening a series switch as shown in Fig. 13-19A. The voltage drops from V + to 0 in a fraction of a second. This causes the magnetic field around the coil to collapse rapidly. This abrupt change in the magnetic field will induce a brief high-voltage spike in the relay. This could be high enough to damage or eventually destroy the switch contacts.

A diode is often placed in parallel with the relay coil to suppress such high-voltage transients. This is illustrated in Fig. 13-19B. With this arrangement, the diode limits the voltage through the relay coil to the power supply voltage (unless, of course, the spike is large enough to damage the diode itself).

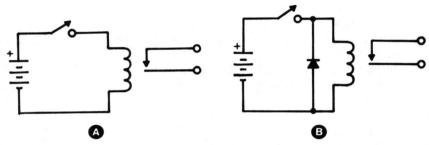

Fig. 13-19. A parallel diode is often used to protect a relay's coil from high-voltage spikes which can occur during switching.

A transistor amplifier is often used to drive a high-current relay from the low-current source, such as a small battery. This limits undue power drain. A typical circuit of this type is shown in Fig. 13-20.

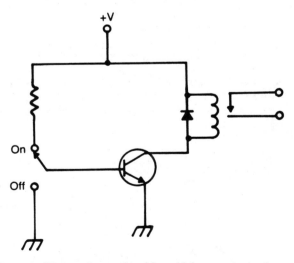

Fig. 13-20. A transistor amplifier can be used to drive a high-current relay from a low-current source.

Some relays are of a latching type. One control pulse closes the switch contacts, which will remain closed, even if the control pulse is removed. A separate control pulse opens the switch contacts. Each time the relay is triggered, it latches into the appropriate position until the next trigger signal is received. Not surprisingly, this type of device is called a *latching relay.*

FUSES AND CIRCUIT BREAKERS

Another kind of automatic switching is used specifically for circuit protection. The voltage through a circuit can be controlled by the design of the power supply, except for relatively rare transients. But, the current drawn through a circuit depends on the resistance and impedance factors within the load circuit. If the resistance drops because of a short circuit, or some other defect, the current could rapidly rise to a level that can damage or destroy some of the components. What is needed is a way to disconnect the power supply from the load circuit before the current reaches a dangerous level. This is most often done with a special device called a *fuse.* A fuse is basically just a thin wire that is carefully manufactured so that it will melt when the current passing through it exceeds a specific value.

The schematic symbol for a fuse is shown in Fig. 13-21, and Fig. 13-22 illustrates a simplified circuit using a fuse. If the current drawn by the load circuit exceeds the current rating of the fuse for any reason (such as changing the value of R1), the fuse will blow, opening the circuit. No further current will flow through the circuit.

Fuses are usually enclosed in glass (or sometimes metal) tubes for protection. The fuse wire is very thin, and could easily be damaged. Figure 13-23 shows a typical fuse.

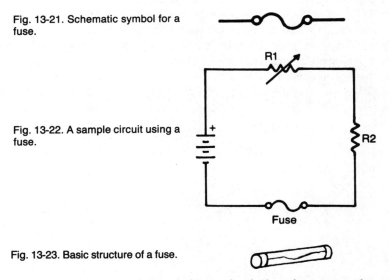

Fig. 13-21. Schematic symbol for a fuse.

Fig. 13-22. A sample circuit using a fuse.

Fig. 13-23. Basic structure of a fuse.

Sometimes fuses are soldered directly into a circuit, but since once a fuse element has been melted, it must be replaced before the circuit can be reused, this is rather impractical in most applications. For more convenient fuse replacement, some kind of socket is generally used for fuses. Most commonly the fuse is held between a set of spring clips. Another frequently used method is to fit the fuse into a special receptacle with a screw cap.

Frequently replacing fuses can be a nuisance, so sometimes a component called a *circuit breaker* is used. This is a special switch that will automatically open if the current through it exceeds some specific amount. To close the switch again, you just manually push a *reset* button.

Occasionally *transients* (brief irregular signals) can cause a fuse to blow, or a circuit breaker to open even if there is no defect in the load circuit at all. But if a new fuse immediately blows, or if the circuit breaker repeatedly opens when it's reset, it indicates something is wrong and repairs are needed. *Never* replace a fuse with a higher rated unit. You could end up blowing some expensive electrical components to protect the fuse, and that certainly doesn't make much sense.

MOTORS

Motors aren't switching devices, but they warrant a brief discussion here. Rather than include a separate short chapter, I decided to make this chapter something of a potpourri.

Basically, a motor is a *transducer*, a device which converts one form of energy into another form of energy. Additional transducers will be discussed in Chapters 19, 22, and 30. In the case of a motor, electrical energy is converted into mechanical energy. That is, an electrical signal can cause something to physically move.

A motor is a practical application of the electromagnetic fields discussed back in Chapter 7. There are many different types of motors; some are extremely tiny, others are huge. Some can only move very small weights, while larger motors can move tons; some run on dc and others run on ac. Regardless of these differences, all motors are basically the same, at least in their operating principles.

An electric current is fed through a set of coils, setting up a strong magnetic field. The attraction of opposite magnetic poles and the repulsion of like magnetic poles results in the mechanical motion of the motor.

A simplified cut-away diagram of a typical motor is shown in Fig. 13-24. Notice that there are two sets of coils. One is stationary, and is known as the field coil. The other

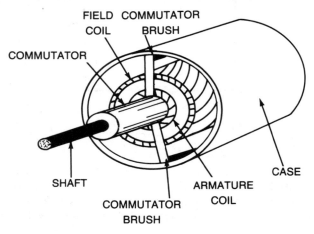

Fig. 13-24. Simplified cut-away diagram of a typical motor.

coil, which is known as the armature coil, can freely rotate within the magnetic field of the field coil. The motor shaft is connected directly to the movable armature coil. As the armature coil moves, the motor shaft rotates.

The commutator reverses the polarity of the current with each half-rotation of the armature and shaft. This keeps the armature coil constantly in motion. In Fig. 13-25A, the armature coil is positioned so that its magnetic poles are lined up with the like poles of the field coil. The like magnetic poles repel each other, forcing the armature coil to rotate, as shown in Fig. 13-25B. At some point the attraction of unlike poles will take over, pulling the armature into the position shown in Fig. 13-25C.

The commutator reverses the polarity of the current, so once again the like poles of the armature coil and the field coil are lined up. The whole process repeats for the second half-rotation (Fig. 13-25D), bringing us back to the position illustrated in Fig. 13-25A, and a new cycle begins.

Increasing the current through the coils (assuming everything else remains equal), will increase the torque of the motor. That is, it can turn a larger load if a larger current is supplied.

Using a given motor to move a load, the heavier the load is, the more current the motor will be forced to draw from the power supply.

Some motors are designed to operate at a constant speed; others will change their rotation speed with changes in the current and/or voltage applied to them. The size of the load can also affect the motor's rotation speed. Obviously, heavier weights will slow the motor down, because it has to work harder to turn the load. This is especially true if a constant current source is driving the motor. Some motors are specifically designed for the load to control the actual rotation speed.

Dc motors and ac motors are generally not interchangeable. Using the wrong type of power source could damage or destroy the motor. Excessive loads can also damage small motors.

LAMPS

A lamp is another common type of transducer. In this case, electrical energy is converted into high energy. It isn't too hard for most people to understand how a simple lamp or light bulb works. Most people are surprised to find out that there are also transducer devices that convert light energy into electrical energy. Such devices will be covered in Chapter 22.

Incandescent Lamps

The most common and familiar type of lamp is the incandescent lamp. This is the type of light bulb we're all acquainted with and use in our homes every day. Similar lamps are used in many electronics circuits. The only real difference is size.

An incandescent lamp is a vacuum enclosed glass bulb. The bulb is air-tight to maintain the vacuum. Within the bulb is a short length of a special resistive wire. The two ends of this wire are brought out separately to the bulb's metallic socket.

When a sufficient electrical current at the correct voltage is passed through this thin wire, or "filament", it will heat up and start to glow, giving off a great deal of light. The filament must be contained in a vacuum to prevent it from burning out too quickly.

160

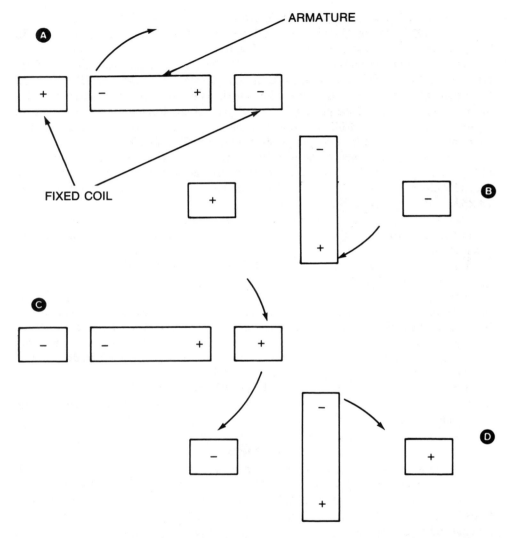

ARMATURE

FIXED COIL

Fig. 13-25. The commutator reverses the polarity of the current with each half-rotation to keep the armature in constant motion.

When the filament wire burns through or breaks from some other cause, the bulb must be replaced.

Contrary to popular belief, Edison did not really invent the incandescent lamp. He just came up with the first practical device of this type. He was the first to devise a suitable filament that would last.

Incandescent lamps vary widely in size and shape, as well as power requirements. You can find incandescent lamp bulbs designed to operate at almost any voltage ranging from under one volt up to several hundred volts.

In electronics work, we generally use only small, low voltage bulbs. These lamps are often called flashlight bulbs, because they are commonly used in that application. Small incandescent lamps are used in many electronics circuits (especially older designs) as indicator devices. Often, the bulb will be painted with translucent paint, or a translucent plastic cap will be placed over the bulb to give the light a specific color, such as red or green.

At one time, small incandescent lamps were just about the only practical choice for low-power indicator devices. Today they are largely being replaced by LEDs (see Chapter 19) because the lamps are bulky and fragile. They are also quite inefficient in terms of power. They generate more heat than light. Since many modern semiconductor components are heat sensitive, this is not just wasteful; it could actually cause harm.

Some other types of lamps are filled with a specific gas, rather than being evacuated. A fluorescent tube is a common example.

Neon Lamps

A number of electronic circuits (again, mostly older designs) use neon lamps as indicating devices. The construction of a neon lamp is illustrated in Fig. 13-26. We have an air-tight glass bulb, just like in the lamp discussed above, but this bulb does not contain

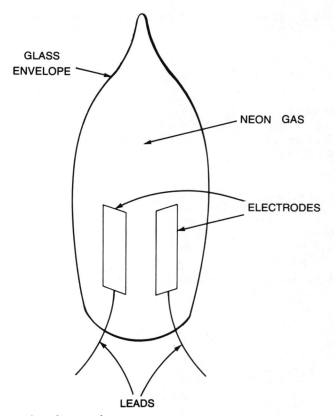

Fig. 13-26. Construction of a neon lamp.

a vacuum; it contains neon gas. Instead of a filament, there are two separated electrodes. If a voltage is connected across these electrodes, the neon gas between them will become ionized and start to glow. Neon glows with a characteristic orange color.

A high voltage is normally required to light a neon lamp; typically, close to 100 volts is required.

Because of the way the neon lamp suddenly switches on or "fires" when its threshold level (minimum operating voltage) is exceeded, this device is often used for triggering and voltage regulation applications, as well as an output indicator.

Self-Test

1. What is the simplest type of switch?

A *SPDT switch*
B *Toggle switch*
C *Knife switch*
D *Relay*
E None of the above

2. How many circuits can a DPDT switch simultaneously control?

A *One*
B *Two*
C *Three*
D *Four*
E *None of the above*

3. Which of the following is *not* a standard switch type?

A *SPST*
B *SPDT*
C *DPST*
D *DPDT*
E *None of the above*

4. If R1 in the circuit shown in Fig. 11-17 if 2200 ohms, and R2 is 3900 ohms, what is the circuit resistance when the switch is closed?

A *2200 ohms*
B *3900 ohms*
C *6100 ohms*
D *1407 ohms*
E *None of the above*

5. What is the term for a switch that disconnects the circuit completely at one position before the connection of the next position is made?

A *Rotary*
B *Shorting*
C *Nonshorting*
D *Make-before-break*
E *None of the above*

6. What happens when a momentary action switch is released?

A *Nothing*
B *The switch contacts latch into the new position*
C *The switch contacts return to their normal rest position*

7. What is contained in a relay?

A *An RC circuit*
B *A coil and a fuse*
C *A coil and a diode*
D *A coil and a set of switch contacts*
E *None of the above*

8. Why is a diode placed across the coil portion of a relay?

A *To increase the current flow*
B *To speed up the switching*
C *To protect the relay from high-voltage transients when the magnetic field collapses*
D *To protect the relay against incorrect polarity*
E *None of the above*

9. What is the name of a device that protects a load circuit from excessive current flow?

A *Relay*
B *Suppression diode*
C *SPDT switch*
D *Fuse*
E *None of the above*

10. When should a fuse be replaced with a higher rated unit?

A *If it blows*
B *Never*
C *When the original value is not available*
D *When fuses of the original value blow as soon as they are replaced*
E *None of the above*

14

Reading Circuit Diagrams

As the old cliché says, a picture is worth a thousand words. This is as true in electronics as in anything else, maybe even more so. It's impossible to imagine working in the electronics field without using diagrams and drawings.

There are several basic types of diagrams commonly used in the electronics field. For the most part, these various diagram types can be grouped into three broad categories;

Pictorial Diagrams, Block Diagrams, Schematic Diagrams

Each of these will be discussed in this chapter.

PICTORIAL DIAGRAMS

The most obvious type of diagram frequently encountered in electronics work is the pictorial diagram. While obvious, this is probably the least useful, although it can be helpful in certain cases.

A pictorial diagram is simply a drawing of the way a circuit or piece of equipment should look. This can be useful for hobbyists building a project, or in repairing a piece of equipment that has been modified from its original design.

A typical pictorial diagram is shown in Fig. 14-1. It tells us nothing more than a photograph of the circuit would tell us.

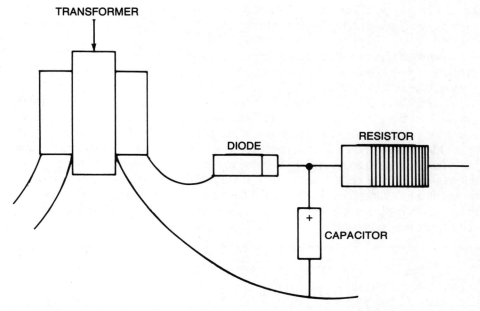

Fig. 14-1. Pictorial diagram.

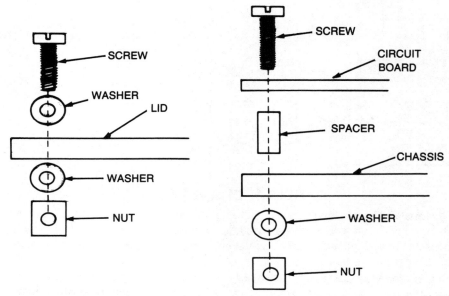

Fig. 14-2. Exploded diagram.

A slightly more sophisticated form of the pictorial diagram is the exploded diagram. An example is shown in Fig. 14-2. In an exploded diagram, the various parts are shown in their relative positions to one another, but are moved apart so they can be seen better. Lines show how the parts are interconnected. Generally, exploded diagrams are not used for circuits, per se. They are used to show how circuit boards and bulky components (such as heavy power transformers) are mounted within a case and how the case is assembled.

Component placement diagrams are often used with projects that are built on printed circuit boards. If you are not familiar with printed circuits, don't worry about it for now. They will be covered in Chapter 15.

A component placement diagram shows the correct position for each component mounted on the printed circuit board. This is important, because if a component is misplaced (with one or more leads in the wrong hole), the circuit will not function properly. Off-board components, or connections to other boards are also indicated. Usually just a labelled lead line coming off the board at the appropriate point is shown. To keep things simple and to avoid wasted effort, the actual off-board device(s) is generally not drawn, although it might be shown in some diagrams, if the technician drawing the diagram thinks this will offer more information to someone using the diagram later. A typical component placement diagram is shown in Fig. 14-3.

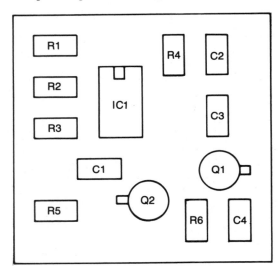

Fig. 14-3. Component placement diagram.

BLOCK DIAGRAMS

Pictorial diagrams show how a circuit is physically constructed, but tell us nothing about how the circuit works, or even what its intended purpose is. *Block diagrams* are more useful for understanding circuit operation. Some technicians refer to block diagrams as *functional diagrams*.

Except for very simple circuits, most electronic systems are made up of multiple sub-circuits. Each sub-circuit performs a specific function or set of related functions. The various sub-circuit functions work together to achieve more powerful and versatile results.

In a block diagram each sub-circuit is shown simply as a block. The actual components used to make up the sub-circuit are ignored. We are only concerned with what the circuit does, not how it does it. Each sub-circuit is considered a "black box". We know what the ''black box'' is supposed to do, but we use it as if we have no way of looking inside. If the input is A, then the output is B. The function of each sub-circuit is written in the appropriate block.

Occasionally a few components are shown separately in a block diagram. This usually is done when the component is not part of any specific sub-circuit, and, in essence, functions as a sub-circuit by itself. *Feedback components* that connect several blocks, or stages, are usually drawn separately, as are inter-stage switches, plugs, and jacks. A typical block diagram is shown in Fig. 14-4.

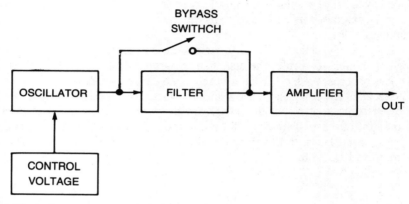

Fig. 14-4. Block diagram.

In some block diagrams special shapes are used to indicate certain functions. For example, circles are normally used to indicate oscillators or signal sources, while triangles are used to represent amplifiers. The diagram of Fig. 14-4 is redrawn using this system in Fig. 14-5. There is no functional difference in the two types of block diagrams. Some

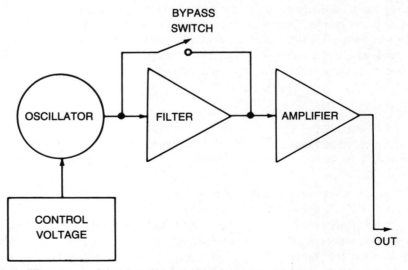

Fig. 14-5. A different way of drawing the block diagram of Fig. 14-4.

people find the varied shapes of the second version easier to read at a glance. Generally, which system you use will be strictly a matter of personal preference.

SCHEMATIC DIAGRAMS

The most important and most frequently used type of diagram in electronics work is the schematic diagram. A schematic diagram (sometimes known as a wiring diagram) shows all of the components in the circuit and how they are electrically interconnected. The position of any given component in a schematic diagram doesn't necessarily correspond to its actual position in the physical circuit. The arrangement of the components in a schematic diagram is influenced more by the clarity of the diagram than by any specific construction details.

Straight (usually) lines are used to represent interconnecting wires and leads between components. As a rule, the schematic diagram should be drawn so that as few lines as possible cross each other where there is no electrical connection. Unfortunately, in most circuits of any complexity, a few line crossings are inevitable. Certain conventions are followed to prevent confusion. There are three commonly used standards.

In the first of these systems, a dot is used to indicate an electrical connection between two crossing leads. If there is no dot, there is no electrical connection. This system is illustrated in Fig. 14-6.

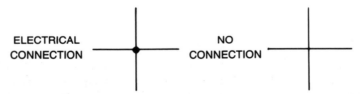

Fig. 14-6. First system used to indicate crossed wires.

Some technicians prefer to use the second system, illustrated in Fig. 14-7. Here, a crossing of any two lines assumes that there is an electrical connection between them. If there is to be no electrical connection, and the lines just cross in the diagram, a small loop is made in one of the crossing lines. This indicates that the one wire "jumps" over the other, without touching (making electrical contact).

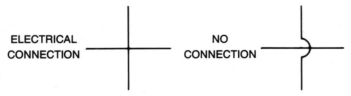

Fig. 14-7. Second system used to indicate crossed wires.

The third system is a combination of the first two. I prefer this third system because it leaves the least room for error. As shown in Fig. 14-8, a dot is used to indicate an electrical connection between a pair of crossing lines. When there is no electrical connection, the "jumping" loop is used. No two lines ever simply just cross each other,

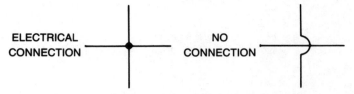

Fig. 14-8. Third (and least ambiguous) system used to indicate crossed wires.

because this is potentially ambiguous. This convention will be employed in all of the schematic diagrams in this book.

Standardized schematic symbols are used to indicate various component types. These symbols are introduced in the appropriate chapter for each type of component. Some of the most important (and most commonly used) component symbols for the components we have learned about so far are shown again in Fig . 14-9.

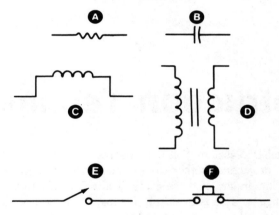

Fig. 14-9. Schematic symbols: A. resistor; B. capacitor; C. coil; D. transformer; E. SPST switch; F. push-button switch.

Many schematic diagrams also indicate certain electrical parameters that can be measured at specific test points within the circuit. The indicated parameters are most commonly voltages, currents, or waveforms. If the circuit is functioning properly, you should be able to measure the same values as those indicated in the schematic diagram. It is usually important to use test equipment with the same specifications as the equipment used to determine the original values shown in the schematic, or you might not get the same results.

15

Construction Techniques

I think it's reasonable to assume that most people reading this book will want to build some of their own projects, whether from plans in a magazine, book, from their own design, or from a kit. In any of these cases, you will need to know something about construction techniques. This is even true for many commercial kits that have rather terse instructions, assuming the kit-builder knows what he's doing. (A notable exception is the Heathkit Company. Their manuals are generally excellent, leading the first-time project builder step by step through the construction process, without really talking down to the old-hand electronics hobbyist.)

Obviously, constructing a circuit involves more than just gathering the components together and tossing them into a box. Some sort of electrical connection must be made between the component leads, according to the pattern of the schematic diagram.

There are several popular types of construction in wide use today. They will be briefly discussed in this chapter.

We will start with the construction technique known as breadboarding, it is the one best suited to the experiments in the next chapter (and in Chapters 23 and 31). Again, it is strongly recommended that you perform each of the experiments yourself. You will learn more this way than if you just read about them.

BREADBOARDING

A breadboarding system is a method of temporarily hooking up electronic circuits for purposes of testing and experimenting.

Of course, you could hard-wire and solder each test circuit, as if it was intended for permanent use. But this can rapidly become extremely expensive.

You could cut costs somewhat by desoldering each circuit when you're through with it and reuse the components. But desoldering tends to be very tedious, time consuming, and quite inconvenient. Besides, the repeated heating and reheating of their leads can damage some components.

Fortunately, there is a much easier way to set up temporary circuits—the breadboard. In its simplest form, this is merely a solderless socket that the various component leads and wires can quickly be plugged into, or pulled out of. A typical solderless socket is shown in Fig. 15-1.

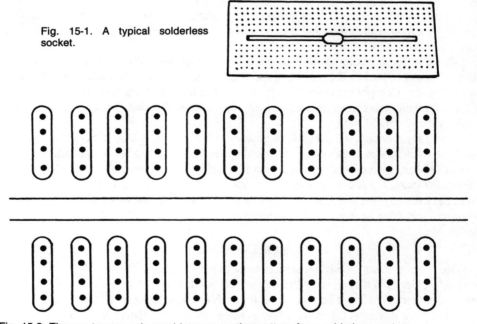

Fig. 15-1. A typical solderless socket.

Fig. 15-2. The most commonly used interconnection pattern for a solderless socket.

The various holes in this type of socket are electrically interconnected. The most frequently used pattern is shown in Fig. 15-2.

These sockets can make experimentation and circuit design much, much easier, but they are even more useful as part of a complete breadboarding system.

These systems consist of a solderless socket and various commonly used sub-circuits, such as *power supplies* (which produce a desired dc voltage from the standard ac house current, thus saving battery costs) and *oscillators* (which produce an ac signal, usually with a variable frequency). These sub-circuits can be separate, self-standing units used along with a simple solderless socket, but it's generally more convenient to have them grouped together within a single, compact unit. At any rate, these sub-circuits will be needed far too often to make breadboarding them each time they're needed reasonable. The way these (and other) basic circuits work will be described in detail in later chapters.

If you don't have a power supply available, you can perform the experiments in the next chapter with dry cell batteries. There will be no difference in circuit operation.

Unfortunately, there is really no substitute for the variable frequency oscillator. If you do not have such a device, you won't be able to perform Experiments 7 through 9 at this time. Just read them over carefully for now. Later in this book you'll learn how to build an oscillator. At that point you can return to this chapter and perform these experiments.

You should have no problem performing the rest of the experiments in this chapter.

Finally, most breadboarding systems have one or more potentiometers and switches handily available for use in experimental circuits.

Once you have designed your circuit, breadboarded the project and gotten all the bugs out, you will probably want to rebuild some circuits in a more permanent way. Solderless breadboarding sockets are great for testing and experimenting with prototype circuits, but they really aren't much good when it comes to putting the circuit to practical use.

Breadboarded circuits, by definition, have nonpermanent connections. In actual use, some component leads will easily bend and touch each other, creating potentially harmful short circuits. Components can even fall out of the socket altogether when the device is moved about. Interface signals can easily be generated and/or picked up by the exposed wiring.

Generally, packaging a circuit built on a solderless socket will be tricky at best. They tend not to fit well in standard circuit housings and boxes. Besides a solderless socket is relatively expensive. It is certainly worth the price if it is repeatedly re-used for many different circuits. But if you tie it up with a single permanent circuit, you are only cheating yourself. Less expensive construction methods are available that are more reliable, more compact, and offer better overall performance.

SOLDERING

Most permanent circuit construction methods involve soldering. A special metal (called "solder") is melted over the connection point of two or more leads, binding the leads together, creating a strong, reliable mechanical and electrical connection.

There are different types of solder available for various purposes. Some of these are summarized in Table 15-1. The most common type used in modern electronics work is 60-40 solder. This type of solder is composed of 60% tin, and 40% lead. It has a rosin core. For electronics work, you must use rosin core solder only. Never, ever use acid core solder on any electronic circuit. The acid is highly corrosive and will eat through many of the components. Acid core solder is used only for metal bonding applications.

Some solder is sold with no core at all. It is used with a separately applied rosin paste. The rosin helps make a good electrical connection between the leads being soldered together.

Before soldering, the leads must be clean. A few quick rubs with some fine sandpaper will be sufficient in most cases. The leads must be arranged in a mechanically solid connection before soldering. If the mechanical connection is weak, you will most certainly end up with a poor solder joint.

Table 15-1. Typical Types of Solder.

Metal Used	Core	Melting Point F	C	Applications
*(TL) 50-50	Rosin	430	220	electronic
*(TL) 60-40	Rosin	370	190	electronic low-heat
*(TL) 63-37	Rosin	360	180	electronic low-heat
*(TL) 50-50	Acid	430	220	metal bonding NOT FOR ELECTRONIC USE
silver	- -	600	320	high heat high current
				*(TL) = tin / lead

For most modern electronics work, you should use a low-power soldering iron. Generally a 20 to 30 watt unit will be best. Many electronic components are very sensitive to heat. This is especially true of semiconductors (discussed in later chapters). Even with a low-powered soldering iron, do not apply heat near sensitive components for too long a time or you will damage them.

Do not use a high powered soldering iron or soldering gun. Most low-power soldering irons are shaped like the one shown in Fig. 15-3. This is sometimes called a soldering pencil. You can also find some low-powered soldering guns, with an easy to hold, pistol-like handle.

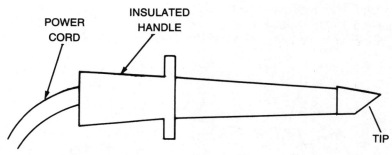

POWER CORD

INSULATED HANDLE

TIP

Fig. 15-3. Most low wattage soldering irons use the "pencil" shape.

Better soldering irons are grounded. This reduces problems with static electricity, which can damage some electronic components (especially CMOS ICs—see Chapter 28). A grounded soldering iron usually isn't absolutely necessary, but it is desirable.

The soldering iron should be fully warmed up before you begin soldering. For most soldering irons, this means you need to plug it in for about five to ten minutes before you actually start soldering. For maximum safety, the hot soldering iron should be placed in a soldering iron stand when not actually being used. A typical soldering iron stand is illustrated in Fig. 15-4. Such stands can be bought for just a few dollars, and they can reduce considerably the potential hazards of accidental burns, or fire. They are very cheap insurance. Don't scrimp on safety.

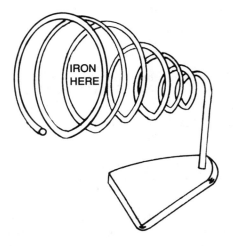

Fig. 15-4. A soldering iron stand can reduce the risk of burns or fire.

The soldering iron's tip should be cleaned periodically on a damp sponge. If too much rosin and miscellaneous build-up accumulates on the tip of the soldering iron, the heat transfer will be significantly reduced. The odds of producing cold solder joints (explained shortly) will increase enormously.

Before actually soldering, the tip of the soldering iron should be *tinned*; melting a little bit of solder (not too much) over the tip. Use just enough to coat the tip with solder.

Good soldering is normally a two-handed process. (Some specialized soldering aids will let you solder one-handed, but this is by far the exception, rather than the rule.) Hold the handle of the soldering iron in one hand and the solder in the other. Apply both to the joint being soldered, as shown in Fig. 15-5. Do *not* apply the iron directly to the

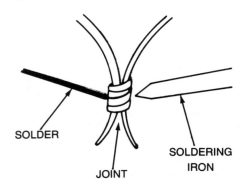

SOLDER

SOLDERING
IRON

JOINT

Fig. 15-5. Apply heat and solder separately to the joint.

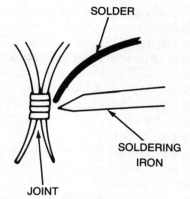

Fig. 15-6. Do not apply the soldering
iron directly to the solder.

solder, as illustrated in Fig. 15-6. You do not want to melt the solder and let it drip over the joint. The idea is to heat up the joint so that the solder will flow evenly over it.

Make the solder joint as quickly as possible, remove the unused solder and the soldering iron from the joint without moving or jarring the joint before the melted solder has a chance to cool and set. This takes a couple of seconds. If the joint is moved before the solder hardens, you will likely wind up with a cold solder joint.

Cold solder joints are one of the most common causes of problems in newly built electronic circuits, especially those built by beginners. Examine all solder joints very carefully. A small, high intensity lamp and a magnifying glass can be very helpful. A good solder joint will look smooth and shiny. If any solder joints look rough or grainy, they are probably cold solder joints. The cure is simple enough—just reheat the connection to re-melt the solder and let if flow more smoothly over the joint.

In a cold solder joint, there is not a good electrical connection between the soldered leads. In some cases, the mechanical connection is pretty weak too.

The term *cold solder joint* really refers to a number of possible problems. One of the easiest to visualize is a bubble within the solder joint. This kind of problem is illustrated in Fig. 15-7. From the outside, the leads appear to be soldered together, but they aren't.

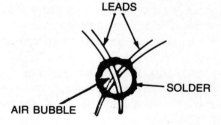

Fig. 15-7. A cold solder joint could
be the result of an air bubble.

Cold solder joints are not always apparent to visual examination. Even an experienced expert will not be able to identify all cold solder joints. If the completed project does not work properly, a cold solder joint might be the problem. Sometimes, if the project is not too complex, it is worthwhile just to reheat all of the solder connections rather than try to track down the specific trouble spot.

When soldering, be very careful not to create any *solder bridges*. A solder bridge is an undesired connection made by excess solder between two connections or adjacent printed circuit traces. Solder bridges are most likely to be a problem when a printed circuit board (discussed in the chapter) is used. Integrated circuits (or ICs), with their closely spaced pins are also frequent candidates for solder bridges. ICs will be discussed in later chapters. A careful visual examination of all solder joints can help save you a lot of grief. Check for solder bridges after each connection is made. Then when all soldering is complete, go back and carefully re-examine all the solder joints again before applying power to the circuit. Remember, a solder bridge is a form of short circuit. If a power line or a signal gets into the wrong part of the circuit, the circuit will not work properly, and, in some cases, some components (usually the more expensive ones) can be damaged or destroyed.

Do not use too much solder. Use enough to cover the joint thoroughly, but don't fill your project with great, ugly lumps of solder. Besides looking unattractive and unprofessional, too much solder can be the source of many problems. The more solder you use, the greater the chances of solder bridges, and the greater the likelihood that it all won't melt thoroughly, and cold solder joints will result. If you use too much solder, you will have to apply more heat longer to each joint, increasing the chances of damaging delicate semiconductors. Use what you need, but don't resort to overkill.

DESOLDERING

If you work in electronics (whether professionally, or as a hobby), sooner or later you're going to have to do some desoldering. I won't try to kid you—this is not the fun part of electronics. It's not awful, but it tends to be a tedious and obnoxious task at best. I have yet to meet an electronics technician or hobbyist who says they enjoy desoldering. It is sometimes necessary, and can save you a lot of money.

Desoldering is just the opposite of soldering, as the name suggests. The joint is heated and the solder is removed. This permits you to remove and replace a component in the circuit.

Suppose you make a mistake and solder the wrong two leads together, or perhaps you happened to get a bad component. Perhaps, if a project or other circuit has been in use for awhile, and you'd now like to modify the circuit in some way. In any of these instances, desoldering is the only alternative to junking the whole thing and starting over from scratch, not only an extremely inelegant solution, it tends to be ridiculously expensive.

The tricky part of desoldering is removing the old solder. Don't be deceived by advertising claims, no device is going to make desoldering truly easy or convenient. Some devices make the job a little less of a chore, but it's still something you'll want to avoid as much as possible.

Aside from the sheer nuisance value, desoldering can overheat temperature sensitive components (especially semiconductors). Desoldering almost always takes longer than soldering.

Basically, there are two main approaches to removing old solder in a desoldering operation;

> capillary action
>
> suction

We will discuss each of these.

Capillary Action

The *capillary action* approach to desoldering uses a braided cable. This cable is pressed against the joint as the solder is reheated. Thanks to some principles of physics (which we don't need to go into here), the melted solder will tend to be drawn up into the braid. This effect is known as capillary action.

Desoldering braid is available from many electronics dealers, including the ever-present Radio Shack chain. Unfortunately, some people seem to have a lot of difficulty with this method. They can never seem to get enough solder sucked up by the braid. Others find the capillary action method such a snap, they use ordinary stranded wire in place of the special desoldering braid.

If you can get capillary action to work for you, it tends to be the most efficient method. But, if after numerous tries you can't get it to work, don't get too upset. You're not alone. It seems to be a specialized skill that some people have a natural knack for, and others don't.

Suction

Melted old solder can also be removed with suction. Some sort of vacuum device pulls the molten solder up into some type of container. Almost anybody can use these suction devices, although, as a general rule, they are not quite as efficient as desoldering braid. A good suction device can be fairly expensive. They often tend to spit out tiny globules of solder that can cause short circuits if you are not careful. They are also prone to clogging when portions of sucked-up solder harden in the intake nozzle.

The simplest and least expensive suction based desoldering devices are simple rubber bulbs with a non-stick nozzle. Such a device is illustrated in Fig. 15-8.

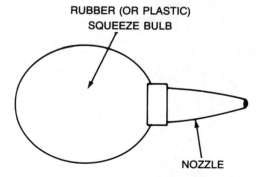

RUBBER (OR PLASTIC)
SQUEEZE BULB

Fig. 15-8. A simple rubber bulb with a non-stick nozzle can be used for desoldering.

NOZZLE

The bulb is squeezed and held, forcing most of the air out of it. The nozzle is then brought into position over the melted solder, and the bulb is released. There is now a minor vacuum within the bulb, because of the expelled air. Nature abhors a vacuum, so a strong suction force will appear at the nozzle until the pressure outside and inside the bulb is equalized. (This just takes a fraction of a second.) If the nozzle is properly positioned over molten solder, the liquid solder will be pulled up into the bulb. It usually takes several repetitions of this process to remove a sufficient amount of solder from the joint to allow removal of the component.

The next step up in suction based desoldering devices uses a spring loaded piston. In the set-up position, the spring is compressed, and a plastic plug blocks the nozzle. Some sort of triggering mechanism is included on the device. When the trigger is activated, the spring is released, and the plug/piston is pulled quickly back out of the way, creating a momentary vacuum and suction force. Again, the liquified solder is pulled up into the nozzle. There are many variations on this type of device. Some work a lot better than others, of course.

If you frequently have to do a lot of desoldering, you might want to purchase an electric desoldering pump. These rather expensive devices work in a manner very similar to the common vacuum cleaner.

POINT-TO-POINT WIRING

Let's discuss some actual permanent construction techniques for electronic circuits. For fairly simple circuits using just a handful of components, you could use point-to-point wiring. No real base is used. The leads of the components are simply soldered together. Often solder terminals are used. A solder terminal is a strip of bakelite, or other plastic with a screw-down foot for mounting onto the case of the project. One or more metal loops are provided as connection points for the leads to be soldered. Some typical solder terminals are shown in Fig. 15-9. The leads to be soldered are mechanically connected to the metal loop, or terminal, as shown in Fig. 15-10. The connection is then soldered.

This construction method is only suitable for very simple circuits. For a circuit of any complexity, you can very easily run into problems with "rats nest" wiring. Rats nest is a fairly self-explanatory name for jumbled wiring that goes every which-way, full

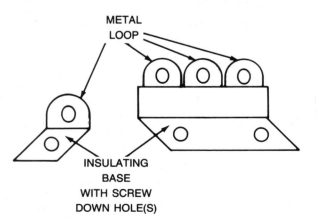

METAL
LOOP

INSULATING
BASE
WITH SCREW
DOWN HOLE(S)

Fig. 15-9. Solder terminals are often used for point-to-point wiring.

Fig. 15-10. The component leads are mechanically connected to the loop of the solder terminal.

of tangles. Such jumbled wiring is next to impossible to trace if any error is made, or if the circuit needs to be serviced or modified at a later date.

Loose, hanging wires can create their own problems, such as stray capacitances and inductances between them, allowing signals to get into the wrong portions of the circuit, resulting in erratic, or incorrect operation. Rats nest wiring is just begging for internal breaks within the connecting wires, and short circuits between them. Quite a bit of mechanical stress can be placed on some of the connecting wires. Momentary, intermittent shorts might not cause permanent damage in all cases, but can result in some strange circuit performance that can be maddeningly frustrating to diagnose and service.

PERF BOARDS

Simple to moderately complex circuits can be mounted on *perf board*. A "perf board" is a non-conductive board with a regular pattern of holes or perforations drilled through it. Component leads are mounted directly through these holes, or special clips can be mounted in the holes, and the component leads connected to the clips. The clips are commonly known as *flea clips*.

Perf board construction is essentially point-to-point wiring on a fixed base. The potential problems of rats nest wiring can also show up in this construction method.

In any minimally complex circuit, one or more jumper wires can be required. A jumper wire is used to make a connection from one part of the board to another. If a jumper wire crosses any other wire or component, it must be insulated.

Take your time and experiment with component placement before you start soldering. Try to find an arrangement that will minimize the number of jumper wires and crossings used. Use straight line paths for jumpers whenever possible.

Another reason to experiment with the component placement before you start soldering is to make sure all of the components will actually fit on the board. If you start soldering without checking this out first, you could find yourself facing some unpleasant surprises, like ending up with no place to mount that big filter capacitor.

PRINTED CIRCUITS

For moderate to complex circuits, or for circuits from which a number of duplicates will be built, a printed circuit board gives very good results.

A non-conductive board is used as a base. Copper traces on one side (or, in very complex circuits, on both sides) of the board act as connecting wires between the components. Very steady, stable, and sturdy connections can be made, since the component leads are soldered directly to the supporting board itself.

Great care must be taken in laying out a PC board to eliminate wire crossings as much as possible. Obviously, two copper traces cannot cross over each other (unless on opposite sides of the board). If a crossing is absolutely essential, an external wire jumper must be used.

Normally, components are mounted on the opposite side of the board from the copper traces. The component leads are fitted through holes drilled in the board, and soldered directly to the copper pad on the opposite side of the board. The excess lead is snipped off to reduce the chances of a short circuit, and to create a more professional appearance. A typical PC board solder joint is illustrated in Fig. 15-11.

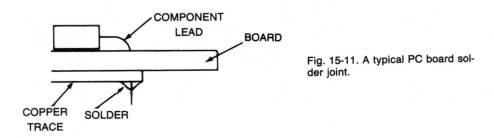

Fig. 15-11. A typical PC board solder joint.

A printed circuit board type of construction results in strong mechanical connections, and very short leads. Lengthy leads actually get in the way. Short leads can help minimize interference and stray capacitance problems.

Stray capacitances between adjacent traces can adversely affect circuit performance. In some critical circuits, a guard band between traces can help reduce this potential problem. Especially in circuits using ICs (integrated circuits—see Chapters 24 through 29), the copper traces are often placed very close to one another. You must be extremely careful to avoid solder bridges. Only small amounts of solder should be used. If too much solder is used, it will flow and bridge across adjacent traces. Also watch out for short circuits from other causes. A small speck of loose solder, or a piece of a component's excess lead could easily bridge across two (or more) adjacent traces, creating a short.

When soldering to a printed circuit (or PC) board, be very careful not to use too much heat, or to apply heat for too long a time. Excessive heat can cause the copper foil trace to lift off the board. The unsupported trace is very fragile and will break.

Tiny, near invisible hairline cracks in the copper traces can also be problematic, if you're not careful. Generally, fairly wide traces that are widely spaced are the easiest to work with and the most reliable. However, this isn't always practical with all circuits— especially where ICs are used.

Blank boards for use as printed circuits are widely available. These are non-conductive boards with one side (sometimes both sides) completely covered with copper foil. The desired pattern is put onto the board using a special resist ink. This can be done either by photographic methods, or it can be drawn on directly with a special resist pen. The board is then soaked in a special acid solution, which eats away the exposed copper, but the resist protects the portions of the copper foil it covers. The board is removed from the acid solution, and washed. The resist ink is removed, and the desired copper traces are left on the board.

UNIVERSAL PC BOARDS

There is a fairly recent variant form of printed circuit construction. Designing and etching a customized printed circuit board is a time-consuming, and somewhat tricky job. Now you can buy various universal PC boards. These boards have a generalized pattern of copper traces, and can be used for many different circuits.

WIRE-WRAPPING

There is one type of permanent circuit construction that does not require soldering. This is the wire-wrap method. It is used primarily in circuits using large numbers of integrated circuits.

In a wire-wrapped circuit, a thin wire (typically 30 gauge) is wrapped tightly around a square post. The edges of the post bite into the wire, making a good electrical and mechanical connection without soldering. Components are fitted into special sockets that connect their leads to the square wrapping posts.

This form of construction is best suited to circuits made up primarily of a number of ICs. If just a few discrete components (resistors, capacitors, etc.) are used, they can be fitted into special sockets, or soldered directly, while the connections to the ICs are wire-wrapped. If both soldering and wire-wrapping are used, it is known as hybrid construction. In circuits involving many discrete components, the wire-wrapping method of construction tends to be rather impractical.

Wire-wrapped connections can be made (or unmade) quickly and easily, without risking potential heat damage to delicate semiconductor components. Moreover, it usually is not difficult to make changes or modifications in the circuit.

Manual and electric wire-wrapping tools are available. The tool is needed to wrap the wire tightly enough around the square post. Most wire-wrapping tools can also be used for unwrapping. Many of these tools can also cut the wire and strip off the insulation.

There are some disadvantages to this type of construction. The use of discrete components is problematic and awkward at best. The thin wire-wrapping wire is very fragile and easily broken. It can carry only very low-power signals. In complex circuits, the wiring can be difficult to trace.

When many integrated circuits are involved (some advanced circuits require several dozen), wire-wrapping can be a very convenient construction method. Integrated circuits of various types will be discussed in Chapters 24 through 29.

16

Experiments 1

This chapter is a collection of simple experiments intended to illustrate and give you some practical experience with some of the concepts discussed in the previous chapters. Performing these experiments is, of course, optional, but it is highly recommended, because actually participating and seeing the principles in action is usually a more effective learning experience than simply reading about them.

Table 16-1 lists all of the equipment and parts you will need to perform the experiments in this chapter. All of the items in this list have been described in the previous chapters.

EXPERIMENT #1 OHM'S LAW

For this experiment build the circuit shown in Fig. 16-1 . As you can see, this is about as simple as a circuit can get—it consists only of a resistor and a dc voltage source with a 3 volt output.

With some breadboarding systems only fixed dc voltages are available. You may have to use a 5 volt supply. This will alter your results, of course, but the principles are the same. And, of course, if you don't have a dc power supply, you can make a 3 volt battery from two 1.5 volt dry cells in series.

In any case, carefully measure the dc voltage with a voltmeter before starting the experiment. Be sure the voltmeter range can handle more than the maximum voltage you expect to measure. If you're using a 3 volt supply, the voltmeter should be able to handle at least 4 volts full scale. For a 5 volt supply, a 6 volt full scale meter is about the minimum.

**Table 16-1. Equipment and Components
Needed for the Experiments in This Chapter.**

1	100 ohm resistor
2	1000 ohm resistors
1	10,000 ohm resistor
1	0.01μF disc capacitor
1	0.1μF disc capacitor
1	10μF electrolytic capacitor
1	100μF electrolytic capacitor
1	0.1 mH coil
1	power transformer—primary = 120 VAC: secondary = 6.3 VAC with center tap
1	6 volt SPDT relay
1	10,000 ohm potentiometer(preferably with a linear taper)
VOM	(including DC voltmeter, DC milliammeter, ohmmeter, and AC voltmeter)
Breadboarding system	(including solderless socket, power supply and oscillator)

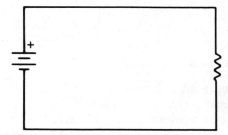

Fig. 16-1. Circuit for Experiment #1—Ohm's Law.

Table 16-2. Worksheet for Experiment #1—Ohm's Law.

R	E	I (measured)	I (calculated)
100 ohms 1000 ohms 10,000 ohms	——— ——— ———	——— ——— ———	——— ——— ———

Premeasuring the voltage is especially important if you are using batteries because the voltage can vary a great deal, depending on the age and condition of the cells. Two 1.5 volt cells in series will generally have a combined voltage somewhere between 2.5 and 3.25 volts. Enter the measured voltage of the source in the position labeled "SOURCE VOLTAGE" in Table 16-2.

Now use a 100 ohm resistor (marked brown-black-brown) to complete the circuit shown in Fig. 16-1. This can be either a ½ or a ¼ watt unit. The tolerance (fourth band) does not matter.

Measure the voltage drop across the resistor, as shown by the dotted lines in the schematic. Enter this value in the column labeled "E", beside the heading "100 ohms."

Remove the 100 ohm resistor and replace it with a 1000 ohm unit (brown-black-red). Again measure the voltage drop across the resistor and enter the result in the appropriate space in Table 16-2.

Next, repeat the procedure with a 10,000 ohm resistor (brown-black-orange) and enter the measured value.

The voltage dropped by each of the three resistors should be identical to the source voltage. This will always be true of any circuit, regardless of the resistance value. The voltage dropped by the total resistance of a circuit will always be equal to the source voltage.

Now, adapt the circuit by inserting a milliammeter between the voltage source and the resistor, as shown in Fig. 16-2. This meter should be able to measure up to 5 milliamps (0.005 ampere, or 5000 microamps). If the full scale value is too large, however, you won't be able to get accurate readings of very small currents. A 5 mA or a 10 mA meter would be ideal.

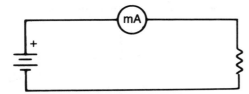

Fig. 16-2. Circuit from Fig. 16-1 with an ammeter added.

Return the 100 ohm resistor (brown-black-brown) to the circuit, and record the current read on the meter in the column labeled "I (measured)" in Table 16-2.

Now repeat the procedure with the 100 ohm resistor (brown-black-red), then with the 10,000 ohm resistor (brown-black-orange). Be sure to enter each value as accurately as possible. Disconnect the circuit and calculate the current for each resistor using Ohm's law (I = E/R). Enter the results in the column marked "I (calculated)". Try to ignore the measured values while you do this.

When you have finished, compare the "I (measured)" values with the "I (calculated)" values. They should be quite close, although there will probably be some minor variations due to measurement errors (no meter is ever 100% accurate) and the probability that one or more of the resistors is not precisely its nominal value.

If you like, you can calculate the exact resistance values with the formula R = E/I (be sure to use the measured values for I). Or, you could measure each of the resistors with an ohmmeter to find their true value.

Table 16-3 shows the results I got from performing this experiment.

EXPERIMENT #2   RESISTORS IN SERIES

The next circuit we'll be working with is shown in Fig. 16-3. This is quite similar to the circuit used in Experiment #1, except this time there are two resistors instead of just one. The resistors are connected in series.

Table 16-3. Author's Results for Experiment #1—Ohm's Law.

Source Voltage (E) 3.0 volts			
R	E	I (measured)	I (calculated)
100 ohms 1000 ohms 10,000 ohms	3.0 volts 3.0 volts 3.0 volts	2.7 mA 0.31 mA 0.03 mA	3 mA 0.3 mA 0.03 mA

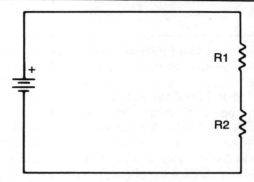

Fig. 16-3. Circuit for Experiment #2—resistors in series.

Use the same voltage source you used in the last experiment. R1 will be a 1000 ohm resistor (brown-black-red) for all the steps of this experiment.

In this first step use a 100 ohm resistor (brown-black-brown) for R2. Measure the voltage dropped across R1, then measure the voltage dropped across R2, and finally, the voltage across both resistors together. See Fig. 16-4. Enter all three measurements in the appropriate spaces in Table 16-4.

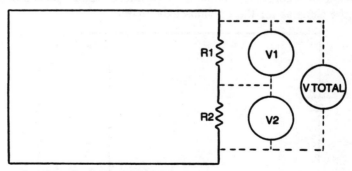

Fig. 16-4. Measuring the voltages from Fig. 16-3.

Now insert your milliammeter at point A and record the current. Then do the same for points B and C.

Finally, disconnect the voltage source, and use an ohmmeter to measure the actual value of the two resistors together. Enter this figure in the column labeled "R(total)".

Table 16-4. Worksheet for Experiment #2—Series Resistances.

Source Voltage (E) _____			
R1	1000 ohms	1000 ohms	1000 ohms
E (R1)	_____	_____	_____
R2	100 ohms	1000 ohms	10,000 ohms
E (R2)	_____	_____	_____
E (total)	_____	_____	_____
I (A)	_____	_____	_____
I (B)	_____	_____	_____
I (C)	_____	_____	_____
R (total)	_____	_____	_____

Repeat the previous steps with a 1000 ohm resistor (brown-black-red) for R2. Then substitute a 10,000 ohm resistor for R2 and repeat all of the above measurements.

There are several things you should notice in the chart of your results. First, in each circuit "E(total)" should equal the source voltage. Also, in any given circuit I(A), I(B), and I(C) should all be exactly equal. The same amount of current flows through all portions of a series circuit. Finally, R(total) should be approximately equal to R1 + R2. There will probably be some variation because of individual component tolerances.

If you use the nominal value of R1 + R2 as R, you can use Ohm's law to calculate the nominal current flowing through the circuit (I = E/R). You should come up with a figure that is close to the measured current value.

If you used the measured value of R(total) in the equation, you should come very close to the measured current value. Any minor error is due to imprecision in the measurement.

EXPERIMENT #3   RESISTORS IN PARALLEL

Now we'll experiment with the parallel resistance circuit shown in Fig. 16-5.

R1 again will be a constant 1000 ohms (brown-black-red). The source voltage will also remain the same as for the previous experiments.

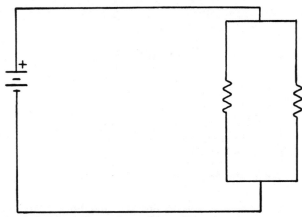

Fig. 16-5. Circuit for Experiment #3—resistors in parallel.

Table 16-5. Worksheet for Experiment #3—Parallel Resistances.

Source Voltage (E) _____			
R1 E(R1) R2 E(R2) R(total) I(A) I(B) I(C)	1000 ohms ‒‒‒‒ 100 ohms ‒‒‒‒ ‒‒‒‒ ‒‒‒‒ ‒‒‒‒ ‒‒‒‒	1000 ohms ‒‒‒‒ 1000 ohms ‒‒‒‒ ‒‒‒‒ ‒‒‒‒ ‒‒‒‒ ‒‒‒‒	1000 ohms ‒‒‒‒ 10,000 ohms ‒‒‒‒ ‒‒‒‒ ‒‒‒‒ ‒‒‒‒ ‒‒‒‒

Starting with a 100 ohm resistor (brown-black-brown) for R2, measure the voltage drop across each resistor. Then measure the current at points A, B, and C. Finally, disconnect the voltage source and measure the combined value of the paralleled resistors with an ohmmeter. Enter all of these values in Table 16-5.

Now replace R2 with a second 1000 ohm resistor (brown-black-red) and repeat all of the measurements. Then, using a 10,000 ohm resistor (brown-black-orange) repeat the experiment one more time.

Notice that the voltage across each resistor is equal to the source voltage, but the currents flowing through each resistor (measured at points B and C) are different. The current measured at point A (total circuit current) should equal I(B) + I(C).

Also notice that I(B)(the current through R1) should be a constant value. This is because neither the voltage drop across the resistor, nor the resistance are changed. Since I = E/R, the value of I should also remain stable.

When R1 = R2, I(B) should be equal to I(C). There might be some slight variation because of the tolerances in the resistor values. R1 might, for instance, be 4% above its nominal value, while R2 might be 6% below.

The component tolerances also explain why the measured value of R(total) probably won't be precisely equal to the calculated value (i.e., $1/R_t = 1/R1 + 1/R2$).

According to the formula, the nominal total resistance should be about 91 ohms when R2 is 100 ohms, 500 ohms when R2 is 1000 ohms, and 909 ohms when R2 is 10,000 ohms. Your measured values should be reasonably close to these figures.

EXPERIMENT #4   A SERIES/PARALLEL CIRCUIT

The circuit for this experiment is shown in Fig. 16-6. R1 is in series with the parallel combination of R2 and R3. R1 is 100 ohms, R2 is 1000 ohms, and R3 is 10,000 ohms. R(P) is the combined resistance of R2 and R3 in parallel. R(total) is the combined resistance of all three resistors.

Calculate the nominal values of R(P) and R(total) and enter your results in the appropriate spaces in Table 16-6. Then, with the voltage source disconnected, measure the actual resistance of R(P) and R(total). The measured and the calculated results should be similar.

Connect the voltage source and measure the voltage drop across R1 and across R(P). Then measure the current at points A through E. Enter all of these readings in the table.

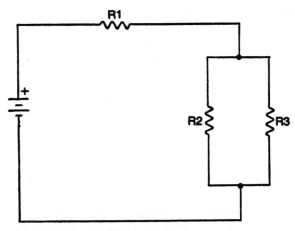

Fig. 16-6. Circuit for Experiment #4—series/parallel circuit.

Table 16-6. Worksheet for Experiment #4—Series/Parallel Circuit.

	Calculated	Measured
R1	100 ohms	
R2	1000 ohms	
R3	10,000 ohms	
R(P)	————	————
R (total)	————	————
E (source)		————
E (R1)		————
E (RP)		————
I(A)		————
I (B)		————
I (C)		————
I (D)		————
I (E)		————

Notice that E(R1) + E(RP) = the source voltage. Also currents I(A), I(B), and I(E) should all be equal. This is because all of these currents are in series.

Finally, the currents in the parallel section of the circuit should add up to equal the total circuit current. That is, I(C) + I(D) = I(A).

You might want to repeat this experiment with a different source voltage. For example, if you are using a battery, you could add a third cell for a source voltage of about 4.5 volts. All of the resistances will remain the same, but the voltage and current values will change. But the relationships between the values discussed above will remain the same.

EXPERIMENT #5 CHECKING A CAPACITOR WITH A DC OHMMETER

Take a 0.1 μF (microfarad) capacitor and short the leads together, as in Fig. 16-7. This will discharge the capacitor if it contains any residual charge.

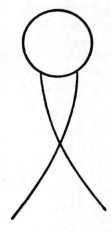

Fig. 16-7. Capacitor with shorted leads to discharge it.

Set your ohmmeter to its highest range and connect it across the leads of the capacitor. See Fig. 16-8. Watch the meter's pointer very carefully while you're doing this. The pointer should jump down the scale, indicating some finite dc resistance, then it will slowly creep back up the scale towards the infinity (open circuit) mark. Discharge the capacitor by shorting the leads and repeat the procedure several times, until you get a good feel for what is happening.

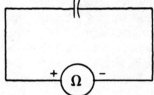

Fig. 16-8. Connecting an ohmmeter across a capacitor.

The equivalent circuit for this experiment is shown in Fig. 16-9. Since an ohmmeter consists of a voltage source, a meter, and a range resistor, the capacitor is charged by the voltage source through the resistor. That is, it is an RC circuit.

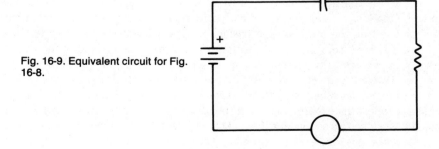

Fig. 16-9. Equivalent circuit for Fig. 16-8.

When the capacitor has no charge (i.e., when the ohmmeter leads are first applied) quite a bit of current flows through the circuit. As the capacitor reaches its charged condition, less and less current can flow. Finally, the capacitor is fully charged, and no further current can flow in the circuit at all.

If you draw a graph of the resistance shown on the meter against the elapsed time, it would look like Fig. 16-10. The same graph shows the charge on the capacitor plates against time.

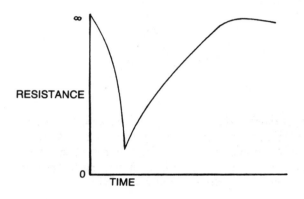

Fig. 16-10. Graphing resistance reading vs charging time—Experiment #5.

If the pointer doesn't move at all, the capacitor could be open, or the meter's range might be too low for a readable indication. A ×1 megohm range should give a readable movement when charging a 0.1 μF capacitor. If the pointer jumps to 0, or a very low resistance value, and stays there, it means the capacitor is shorted.

With many capacitors the pointer won't return completely to the infinity position. This is because dc leakage through the dielectric is providing a limited current path. As long as the final value is very high (50 to 100 megohms, or more), the capacitor can be considered good.

Now, perform the experiment again and carefully time the motion of the meter from its minimum to its maximum resistance position.

Next, try the same procedure with a 0.01 μF capacitor. Notice that the pointer doesn't move as far down the scale as with the 0.1 μF unit, and it returns to the infinity position much faster. This indicates that the capacitance is less. Capacitors smaller than about 0.01 μF usually won't give a readable indication on most ohmmeters.

Now, try the same procedure with a 10 μF electrolytic capacitor. Be sure to observe the polarity markings. Some electrolytic capacitors have their positive lead identified by a "+" or a dot on the body of their casing. Others indicate the negative lead with a "−". The red lead from the ohmmeter must go to the positive lead of the capacitor, and the black lead from the ohmmeter must go to the capacitor's negative lead.

Notice that the pointer jumps further down the scale than the smaller capacitors did. Also, it takes much longer to move back up the scale to the fully charged condition.

Finally, repeat the experiment with a 100 μF electrolytic capacitor (observe polarity). It should take close to a second for the pointer to return to the infinity position. Actually, with a capacitor this size you will probably see some leakage resistance in the meter indication, but the pointer should be fairly close to the infinity mark when it stops moving.

EXPERIMENT #6   A DC RC CIRCUIT

Place a resistor and a capacitor in series, as shown in Fig. 16-11. Short the leads of the capacitor together with a metal bladed screwdriver, or a piece of bare wire. Then connect the ohmmeter leads across the RC series combination.

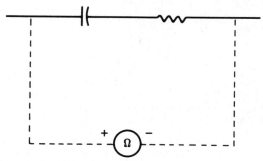

Fig. 16-11. Basic circuit for Experiment #6—a dc circuit.

Table 16-7. Worksheet for Experiment #6—A DC RC Circuit.

	100 ohms	1000 ohms	10,000 ohms
0.01 μF 0.1 μ 10 μ 100 μ			

Start with a 100 ohm resistor (brown-black-brown) and a 0.1 μF capacitor. Notice that this combination takes longer to charge than just the capacitor alone. Repeat this procedure several times, until you can determine exactly how long it takes the capacitor to charge. Enter this value in Table 16-7.

Do this experiment with each of the resistor-capacitor combinations indicated in the table. Don't forget to observe the polarity markings on the electrolytic capacitors.

Notice that increasing either the resistance or capacitor proportionately increases the time required for charging the capacitor (that is, the time constant of the combination). The 10,000 ohm resistor (brown-black-orange)/100 μF capacitor combination should take the longest time (nominally a full second—disregarding the internal resistance of the meter).

EXPERIMENT #7   AN AC RC CIRCUIT

Again we'll use a resistor in series with a capacitor, but in this experiment we will examine what happens when the combination is powered by an ac voltage source. You will need a variable frequency oscillator for this experiment.

Set your variable frequency oscillator to a fairly low frequency and measure the ac voltage coming from the oscillator itself (inexpensive oscillator circuits don't put out the same voltage at all frequencies).

192

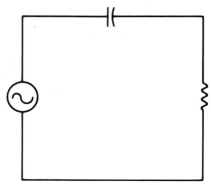

Fig. 16-12. Basic circuit for Experiment #7—an ac RC circuit.

Table 16-8. Worksheet for Experiment #7—An AC RC Circuit.

Frequency	Source Voltage	0.1 μF	0.01 μF
LOW MEDIUM HIGH	—— —— —— —— ——	—— —— —— —— ——	—— —— —— —— ——

Connect the oscillator across the resistor/capacitor combination, as shown in Fig. 16-12, and measure the voltage drop across the resistor.

Perform this experiment—once with the 0.1 μF capacitor, and once with the 0.01 μF unit. Use the 1000 ohm resistor (brown-black-red) in each case.

Do not try to perform this experiment with the electrolytic capacitors! They cannot hold up under an ac voltage.

Enter your voltage readings in Table 16-8.

Disconnect the circuit and set the oscillator to a medium frequency and measure the ac voltage. Connect the oscillator to the RC combination again and repeat the experiment. Finally, repeat the experiment once more with the oscillator set to generate a high frequency.

Notice that with my oscillator, the source voltage decreased as the frequency increased (this isn't always true). Even so, the voltage passed by the capacitor increased noticeably with each increase in frequency. This is because a capacitor has a lower reactance at higher frequencies.

Also notice that the 0.1 μF capacitor passed more voltage than the smaller, 0.01 μF unit. This is because the larger capacitor has a lower impedance.

EXPERIMENT #8   COILS

Measure your source voltage carefully, then build the circuit shown in Fig. 16-13 with a 100 ohm resistor (brown-black-brown) and a 0.1 mH coil. Measure the voltage drop across the resistor. It should be just about equal to the source voltage.

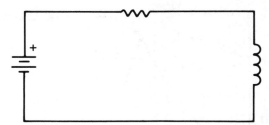

Fig. 16-13. A simple RL circuit for Experiment #8.

Table 16-9. Worksheet for Experiment #8—Coils.

Frequency	Source Voltage	Voltage Drop
LOW MEDIUM HIGH	—— —— ——	—— —— ——

Now disconnect the dc power source and measure the voltage drop across the resistor at three different ac frequencies. Be sure to measure the source voltage at each frequency. Enter your results in Table 16-9.

Notice that the results of this experiment should display the exact opposite of the pattern found in the experiment with capacitors. As the source frequency increases, the voltage passed by the coil decreases. In other words, the voltage dropped across the coil increases with frequency.

Remember, an inductance exhibits more reactance to high frequencies, very low frequencies, and at dc a coil acts like a simple length of wire.

EXPERIMENT #9   RESONANCE

Connect the series circuit shown in Fig. 16-14. Vary the frequency of the oscillator while carefully watching the pointer of the ammeter. As the frequency is increased, the current drawn through the circuit should also increase, until some maximum value is reached. Increasing the frequency beyond this point should cause the current to fall off again. The maximum current occurs at the resonant frequency of the coil/capacitor combination.

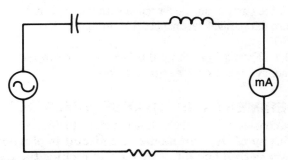

Fig. 16-14. Simple RL series circuit.

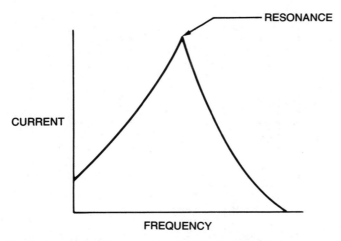

Fig. 16-15. Graph of current vs frequency for the circuit of Fig. 16-14.

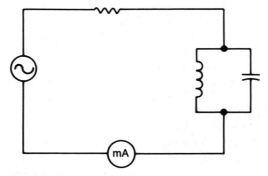

Fig. 16-16. Simple RL parallel circuit.

If you graphed the results of this experiment, it would look something like the chart shown in Fig. 16-15.

Again, set the oscillator at its minimum position, and slowly increase the frequency, while watching the ammeter. The current should start out at a fairly high value, and decrease to a minimum value at the same frequency it reached a peak in the last step. From that point on, the current should increase with the applied frequency. In this case we have passed through parallel resonance. The graph for this circuit would resemble the one in Fig. 16-17.

Replace the 0.1 μF capacitor with a 0.01 μF unit, and repeat the experiment. Is the resonant frequency higher or lower than before?

EXPERIMENT #10   TRANSFORMER ACTION

Very carefully construct the circuit shown in Fig. 16-18.

Do not touch any of the wires once the circuit is plugged into the wall socket! Carelessness could be fatal! Be sure to include the half amp fuse—it is for your protection. The fuse will limit the current in case of an accident.

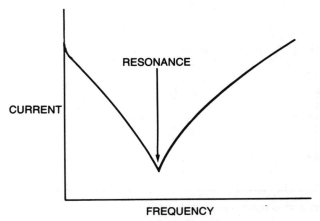

Fig. 16-17. Graph of current vs frequency for the circuit of Fig. 16-16.

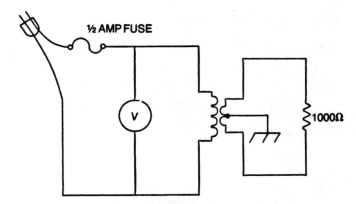

Fig. 16-18. Measuring a transformer's primary voltage.

Wrap all connections and bare wires with several layers of electrical tape for insulation!

Be sure the voltmeter is connected to the circuit before ac power is applied!

Plug in the line cord, read the voltage on the meter and immediately unplug the cord. (It probably wouldn't hurt anything to leave it plugged in for a few minutes, but why take chances?) You should have gotten somewhere between 100 to 120 volts ac.

Now, with the circuit unplugged, move the voltmeter to the position shown in Fig. 16-19. Again, carefully plug in the line cord, read the meter and unplug the circuit. This time you should have gotten about 6.3 volts. This is a step down transformer. It transforms one ac voltage to a lower ac voltage.

This same transformer could be reversed and used as a step up transformer, but **do not try it!** The very high voltage produced could be extremely dangerous in an unshielded circuit like this.

Now, connect the voltmeter between one of the secondary end leads and the center tap as shown in Fig. 16-20A. Of course, the circuit should be unplugged while you are working on it. Plug the circuit in, read the voltage, and unplug the circuit. Move the

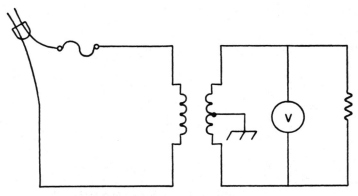

Fig. 16-19. Measuring a transformer's secondary voltage.

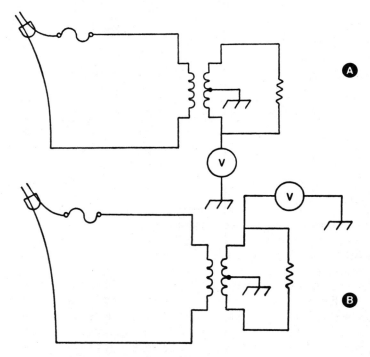

Fig. 16-20. Measuring the output of a center-tapped transformer.

voltmeter so it is connected between the center tap and the other end lead of the secondary, as illustrated in Fig. 16-20B. Plug in the line cord, read the voltage, and unplug the cord.

In these last two steps you should have read exactly one half of the total secondary voltage. Since this transformer has a 6.3 volt ac secondary, the voltage between the center tap and either end of the secondary will be about 3.15 volts ac.

EXPERIMENT #11 RELAY ACTION

Hook up the circuit illustrated in Fig. 16-21. Watch the metal contacts inside the relay as you slowly rotate the shaft of the potentiometer. At some point along the potentiometer's rotation, you should see them move and hear a faint but distinct click. The ohmmeter will now show that this set of contacts are now shorted together. This condition will remain stable as long as the potentiometer is not turned in the opposite direction.

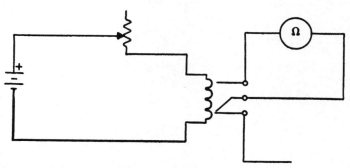

Fig. 16-21. Simple SPDT relay circuit for Experiment #11.

Slowly turning the potentiometer back in the opposite direction will produce the opposite result, of course. A point will be reached when the relay's contacts will click back to their original position, and the ohmmeter will show infinite resistance, or an open circuit.

If you connected the voltmeter shown in the dotted lines, you'd find the relay contacts remain open as long as the voltage shown on the meter is below 6 volts. Any voltage over 6 volts will energize the relay and cause the contacts to close.

Reconnect the ohmmeter as shown in Fig. 16-22 and notice that the relay works in exactly the same way, but backwards. That is, increasing the voltage above 6 volts will cause the switch contacts to open, or vice versa.

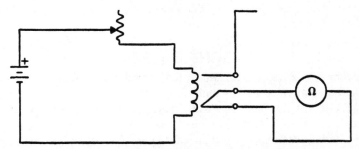

Fig. 16-22. Figure 16-21 with an ohmmeter across the normally closed contacts of the relay.

If you don't understand how the voltage is being varied in this experiment, remember that the relay coil has a certain amount of dc resistance, so you essentially have two resistors in series (as far as dc voltage is concerned). Changing the value of one series resistor (that is, turning the shaft of the potentiometer, will alter the amount of voltage dropped by the other resistor. Refer back to experiment #2 for further clarification.

17

Tubes

By themselves, the circuits described so far are of relatively little usefulness. This is because all of the components discussed in the previous chapters have been *passive* devices. They can reduce (or *attenuate*) a signal, but they cannot increase (or *amplify*) it.

For practical electronic circuits we also need *active* devices, that is, components that can amplify, or in some other way actively alter a signal.

The first practical active device, and probably the simplest, was the *triode vacuum tube*. But before we can examine how this device works, we need to look at a couple of related passive devices.

LIGHT BULBS

The vacuum tube is closely related to the common light bulb. In fact, a light bulb could be called a single element vacuum tube.

Figure 17-1 shows the construction of a typical light bulb. A thin, specially prepared wire is enclosed in a glass bulb and all of the air is pumped out, creating a vacuum within the bulb. Electrical connections to the wire (called the *filament*) can be made from outside the bulb via a metal base.

When an electric current passes through the filament, its resistance causes it to heat up. The special type of wire used for the filament will glow when heated, producing light. Some of the filament material is inevitably destroyed by this process, which is why light bulbs eventually burn out. If you looked inside a burnt out light bulb, you'd see that the filament wire is broken.

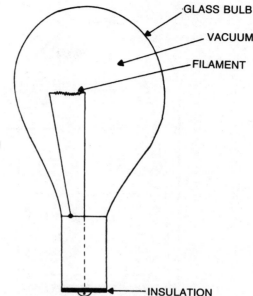

GLASS BULB

VACUUM

FILAMENT

INSULATION

Fig. 17-1. Construction of a typical
light bulb.

The resistance of the filament determines the wattage consumed by the bulb (and thus the brightness of the emitted light). For example, if the filament is 144 ohms, and works off of standard house current (nominally 120 volts), the current drawn by the bulb will be equal to the voltage divided by the resistance. (Ohm's law—$I = E/R$). In this example, $I = 120/144$, or about 0.83 ampere. This means the power consumed by this particular light bulb ($P = EI$) is approximately 100 watts.

It takes more energy to heat up the filament to the glowing point, than to maintain its temperature once it is heated. In other words, the resistance of the filament is higher when it is cold. This means when power is first applied to a light bulb, the current drawn will flow in a large surge before settling down to its nominal value. This surge current can be several times larger than the nominal current flow. For this reason, no power is saved by turning out a light if it will be turned back on within a few minutes.

THE DIODE

Besides emitting light, the heated filament in a light bulb also emits a stream of electrons. If a second element is placed within the vacuum tube envelope, and given a positive charge, it will attract these electrons. That is, a current can be made to flow between the elements within the bulb, or tube. Because this type of tube has two elements, it is called a *diode*. Actually, most practical diodes have three elements, as shown in Fig. 17-2.

The positively charged element is called the *plate*, or *anode*. The stream of electrons is emitted from the *cathode*, which is given a negative charge by the external circuit. The filament, or *heater* is generally not considered an active element in the tube. It simply heats up the cathode so it can emit electrons easily. Heating the cathode directly would result in less efficient operation and a tube with a shorter life expectancy.

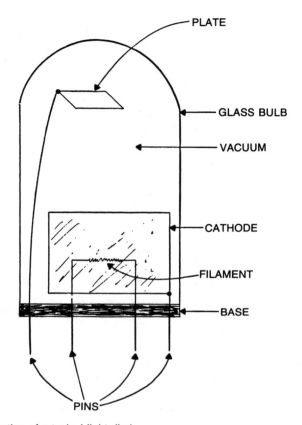

Fig. 17-2. Construction of a typical light diode.

Usually the heater circuit is electrically isolated from the main circuit. In most tube equipment, there is a separate power source (or transformer winding) just for powering the filaments of the tubes. For the longest possible life, the filaments should be heated with an ac voltage, rather than dc.

The most common schematic symbols for diodes are shown in Fig. 17-3. Sometimes the filament is not shown in the schematic diagram at all, as in Fig. 17-3B and 17-3C. The symbol in Fig. 17-3C isn't often used for vacuum tube diodes (see the next chapter), but it occasionally shows up in certain schematics.

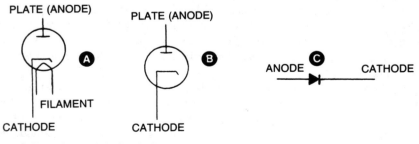

Fig. 17-3. Schematic symbols for diodes.

Figure 17-4 shows a simple circuit for testing the action of a diode. When the power source is connected as shown in Fig. 17-4A, a current flows through the ammeter. The value of this current will be determined primarily by the resistor. The diode electrically looks like a very small resistance—almost a short circuit. We say the diode is *forward-biased*.

However, if the polarity of the dc voltage source is reversed, as in Fig. 17-4B, no current will flow (or very, very little), because the plate cannot emit electrons. The diode is now *reverse-biased*, and its resistance is extremely high.

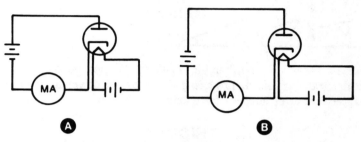

Fig. 17-4. Test circuit for a diode: A. forward bias; B. reverse bias.

This is the basic principle of a diode. Current can flow through it in one direction, but not in the other. An ideal diode would have zero resistance if measured from cathode to anode, but infinite resistance from anode to cathode. Practical diodes have some resistance when forward biased, but the value will be very low. Similarly, some current will flow through a diode when it is reverse-biased, but the resistance will be so high the current will be of a negligible value.

Considering the way a diode behaves in a dc circuit, what would happen if it was placed in an ac signal path?

In an ac circuit, only that portion of the applied signal with the correct polarity can pass through the tube, while the rest of the signal will be blocked. Figure 17-5 shows the effect of a basic diode circuit on a simple sine wave.

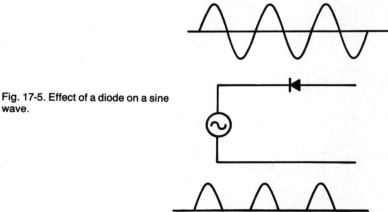

Fig. 17-5. Effect of a diode on a sine wave.

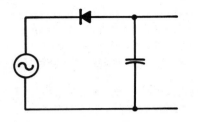

Fig. 17-6. Effect of a diode and a filter capacitor on a sine wave.

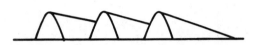

If we add a capacitor, as shown in Fig. 17-6, its charging and discharging times will tend to smooth out the waveform, producing a more or less dc voltage from an ac source. See the chapter on power supplies for additional information.

TRIODES

As useful as a diode is, it still can't amplify. To do that we need to add a third active element to our tube (ignoring the heater). This new element is called the *grid* (sometimes the *control grid*), and such a three element tube is called a *triode.*.

Figure 17-7 shows the construction of a typical triode, and Fig. 17-8 shows the most common schematic symbols for the device. As with the diode, the heater is sometimes omitted from the schematic diagram, because it is not a part of the actual circuit. The heater connections are always assumed.

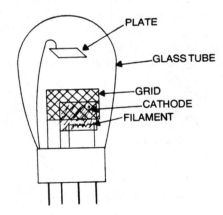

Fig. 17-7. Construction of a typical triode.

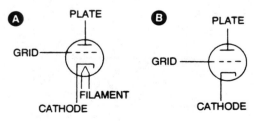

Fig. 17-8. Schematic symbols for a triode.

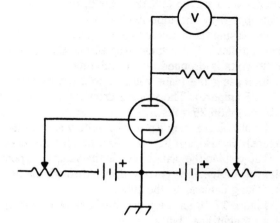

Fig. 17-9. Demonstration circuit for a triode.

Figure 17-9 shows a simple demonstration circuit for a triode tube. For simplicity and convenience the heater circuit is not shown in the diagram—the heater circuit would be identical to the one in the diode circuit discussed previously. As a matter of fact, all tube circuits use the heaters in essentially the same way—this is why they can be omitted from the diagrams. When we see a tube in a circuit, we automatically know we need to apply a voltage across the heater. The level of this voltage varies from type to type, and will be specified by the manufacturer.

The grid in a triode tube is a metallic mesh. That is, it has holes in it that allow electrons from the cathode to pass through it on their way to the plate. Just how many electrons can pass through the grid (i.e., the current) depends on its electrical charge. If the grid is made very negative with respect to the cathode, it will repel all of the electrons (which are also negatively charged), and let none of them pass through to the plate. The voltage at which all current through the tube is blocked is called the *cut-off point* of the tube.

As the voltage on the grid is made more positive (or less negative) with respect to the cathode, more and more electrons can pass through the mesh and get to the plate. At some specific point all of the electrons emitted by the cathode will reach the plate. This is called the *saturation point* of the tube.

If the grid is made even more positive past the saturation point, it will start to attract the electrons itself, once again preventing them from reaching the plate. Usually the grid is slightly negative with respect to the cathode in practical circuits.

Using a hypothetical tube, let's examine some of the effects that take place in this kind of circuit.

For this example, we'll use the cathode as our reference point. That is, the cathode is grounded, which means its voltage is, by definition zero volts.

Let's assume the grid voltage (E_g) is 0 volts. If the plate voltage (E_p) is also 0, obviously no current will flow through the tube.

If we increase the plate voltage to 25 volts, about 1.8 mA (0.0018 ampere) of current will flow through the plate circuit. If the load resistor (R_L) is 5000 ohms, the voltage drop across it will be 0.0018 × 5000, or 9 volts. The rest of the plate voltage is used up (dropped) by the tube itself.

Increasing the plate voltage further, to 50 volts, will cause a 4 mA (0.004 ampere) current to flow. The voltage drop across R_L is now equal to 0.004 × 5000, or 20 volts, and 30 volts is dropped by the tube itself.

Increasing the plate voltage, to 75 volts increases the current flow to 7.25 mA (0.00725 ampere). The voltage drop across the load resistor is now equal to 0.00725 × 5000, or 36.25 volts.

Finally, increasing Ep to 100 volts will increase the current flow to 11 mA (0.011 ampere). R_L drops 0.011 × 5000, or 55 volts under these conditions.

100 volts in the plate circuit is the saturation point of this particular tube with a zero volt grid voltage. The current drawn through the tube cannot be increased further without risking damage to the tube.

Figure 17-10 shows a graph of this plate voltage to current ratio. Notice that it is not a straight line, but a curve.

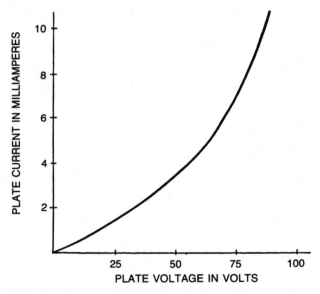

Fig. 17-10. Characteristic plate current curve for a typical tube—Eg = 0 volts.

Figures 17-11 and 17-12 show similar graphs for the same tube, but with the grid voltage at −2 volts and −4 volts, respectively. These graphs are collectively called a *family of plate characteristic curves* for this specific tube. Other tubes will have somewhat different curves, but they will always exhibit basically the same shape.

Obviously we could eliminate the tube altogether and just vary the resistance through R_L directly. In actual practice, the plate voltage is usually held at a constant level, and the grid voltage (Eg) is varied.

Let's assume a plate voltage of 100 volts. We already know that if Eg equals 0 and Ep equals 100, then the current will be 11 mA, and the voltage drop across the 5000 ohm load resistor will be 55 volts.

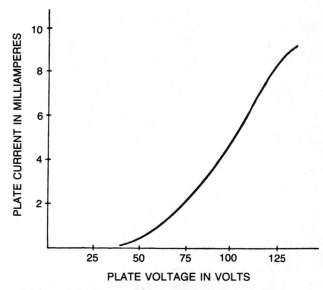

Fig. 17-11. Characteristic plate current curve for a typical tube—Eg = −2 volts.

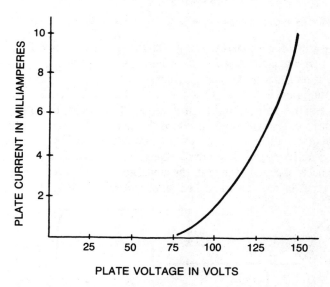

Fig. 17-12. Characteristic plate current curve for a typical tube—Eg = −4 volts.

If the grid voltage is changed to −1 volt (remember, the grid should be negative with respect to the cathode), the current through the plate circuit will be 8 mA (0.008 ampere). The negative charge on the grid is repelling some of the electrons from the cathode. The voltage drop across the load resistor will be equal to 0.008 × 5000, or 40 volts.

At a grid voltage of −2 volts only 5 mA (0.005 ampere) will flow through the plate circuit. The load resistor will drop 0.005 × 5000, or 25 volts.

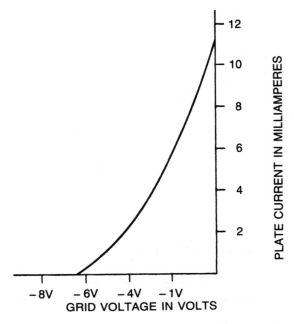

Fig. 17-13. Characteristic grid voltage curve for a typical tube—Ep = 100 volts.

The entire graph for a constant Ep of 100 volts, and a variable Eg is shown in Fig. 17-13. Notice that when Eg is −6 volts or less, no current will flow through the plate circuit at all. Plainly, this is the cut-off point.

Compare the graph in Fig. 17-13 to the one in Fig. 17-12. Notice that while it takes a 100 volt range in the plate voltage to produce an 11 mA range of plate current, it takes only a 6 volt range of grid voltage to produce the same plate current range. A relatively small change in grid voltage produces a relatively large change in the plate current, and this produces a fairly large voltage drop change across the load resistor.

If an ac signal is applied to the grid, the signal across the load resistor will be a larger replica of the input signal. This is called amplification. See Fig. 17-14.

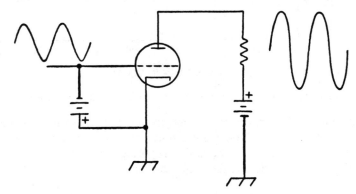

Fig. 17-14. Amplification.

Of course, the energy across the load has to be provided by the plate voltage source—you can't get something for nothing. The voltage drop of the load resistor will always be less than the voltage applied to the plate circuit.

The amount of amplification in any given circuit is called the *gain*. How much gain a specific tube is capable of is called the *amplification factor*, which is usually represented by the Greek letter μ (pronounced Mu). The amplification factor is determined by the ratio of the change in grid voltage needed to produce a given change of current and the change in plate voltage required for the same amount of current change. That is:

$$\mu = \frac{\Delta\ Ep}{\Delta\ Eg}$$

Equation 17-1

where μ is the amplification factor, Δ Ep is the change in plate voltage (Δ is *delta*, and is used to represent a changing value). Δ Eg is the change in grid voltage.

In our sample tube, increasing the plate voltage 20 volts will increase the output current about 2.5 mA, while a change of about 1 volt in the grid will produce the same change in current. Therefore, μ equals 20/1, or an amplification factor of 20.

As in the diode, current can flow through a triode in only one direction—from cathode ($-$) to plate ($+$). Reversing the polarity of the plate voltage will automatically result in zero current flow, regardless of the value of either Ep or Eg. This is true of all tubes.

TETRODES

A major problem with triodes is due to *interelectrode capacitance*. That is, the electrodes within the tube act like the plates of a capacitor. See Fig. 17-15.

Fig. 17-15. Interelectrode capacitances in a triode.

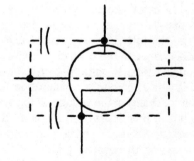

The capacitance between the plate and the grid is particularly significant, because it can allow ac current from the plate circuit to leak back into the grid circuit, putting a severe limitation on how much gain the tube can put out.

This effect can be greatly reduced by adding a second meshed element called a screen grid, which is placed between the original control grid and the plate. Figure 17-16 shows the schematic symbol for this type of four element tube, which is called a *tetrode*.

The screen grid is connected so that it is positive with respect to the cathode, but somewhat negative with respect to the plate.

A capacitor is usually connected from the screen grid to the cathode. This will have no effect on the dc voltage levels, but any ac signal that manages to get into the screen

208

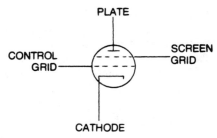

Fig. 17-16. Schematic symbol for a tetrode.

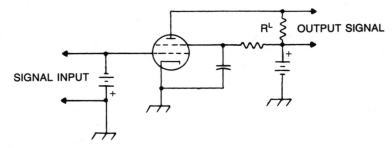

Fig. 17-17. Basic tetrode circuit.

grid circuit will be shorted to the cathode, which is generally at ground potential (0 volts). The diagram for the basic tetrode circuit is illustrated in Fig. 17-17.

Because the screen grid is physically closer to the cathode than the plate is, its positive charge has a greater effect on pulling the electrons through the holes in the control grid than does the plate. This means the plate voltage has very little effect on the current flow through the tube. A very large change in plate voltage would be needed to equal a very small change in the control grid voltage. Of course, this means the amplification factor of such a tube is quite high. A typical triode might have an amplification factor of 20 to 25, but a tetrode's amplification figure is often more than 600.

Of course, changing the voltage on the screen grid could alter the current flow through the tube, but in practical tetrode circuits the screen grid is virtually always held at a constant voltage. The current flow through a tetrode is determined almost exclusively by the voltage on the control grid.

Because the screen grid is an open mesh, most of the electrons pass right through the large holes in it and go on to strike the even more positively charged plate. A few electrons do strike the screen grid, however, causing a small current to flow through the screen grid circuit.

Passing through the positively charged screen grid tends to speed up the electrons in their path, causing them to strike the plate with considerable force. If this force is large enough, many of the electrons can ricochet off the plate and return to the screen grid. Obviously this is undesirable, because it represents a loss of current flow through the plate circuit. This problem is referred to as *secondary emission*.

PENTODES

The problem of secondary emission can be greatly reduced by the addition of yet another grid element. This one is called a *suppressor grid*.

The suppressor grid is placed between the screen grid and the plate, and it is usually connected directly to the cathode, so it is quite negative with respect to the plate.

The main electron stream is speeded up by the screen grid. The electrons pass through the holes in the suppressor grid so fast the negative charge doesn't have a chance to repel them, but it does slow them down a bit. Any secondary electrons that bounce off of the plate are repelled by the suppressor grid's negative charge, so they are forced to return to the positively charged plate.

The plate voltage in a *pentode* (5 element tube) can vary over an extremely large range without appreciably changing the current in the plate circuit. As a matter of fact, the plate voltage can even drop slightly below the screen grid voltage without a serious drop in the output current.

The schematic symbol for a pentode is shown in Fig. 17-18. As with all other tubes, the heater circuit is often omitted from schematic diagrams.

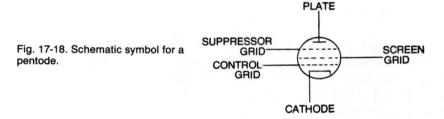

Fig. 17-18. Schematic symbol for a pentode.

In most pentodes, the suppressor grid is brought out to its own terminal pin and is connected to the cathode via the external circuit. In some pentodes, however, the suppressor grid is internally connected to the cathode. This type of tube is usually shown schematically as in Fig. 17-19.

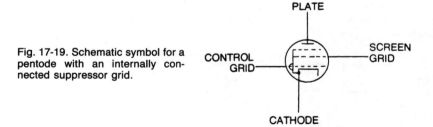

Fig. 17-19. Schematic symbol for a pentode with an internally connected suppressor grid.

The amplification factor of a pentode can be extremely high. Some tubes have an amplification factor of 1,500, or even more. Compare this to the amplification factor of a simple triode!

Of course, since pentodes have more elements, and are more complicated to manufacture, they are more expensive. Triodes and tetrodes are usually used whenever possible to achieve the desired results.

MULTIUNIT TUBES

Some tubes actually contain more than one set of electrodes in a single bulb. In other words, more than one tube is contained in a single glass envelope. The most common combinations are dual diodes, dual triodes, and diode/triode combinations. Tetrodes and pentodes are rarely found in multiunit tubes.

Some dual tubes have a common cathode, and many share a common heater filament. This means the element is used in both tubes.

Figure 17-20 shows the schematic symbol for a dual triode. In many circuits the two sections of the tube can be used in entirely different circuits.

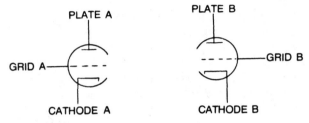

Fig. 17-20. Schematic symbol for a dual triode.

CATHODE RAY TUBES

There are a number of special tube types available for specific, unique applications. One that merits special discussion here is the *cathode-ray tube*, or *CRT*. The key principle in a cathode ray tube is that certain special materials, called *phosphors*, will glow when struck by an electron beam.

The basic structure of a cathode ray tube is illustrated in Fig. 17-21. The elements that make up the section called the *electron gun* are shown in more detail in Fig. 17-22.

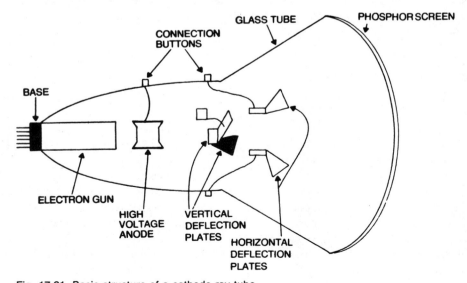

Fig. 17-21. Basic structure of a cathode ray tube.

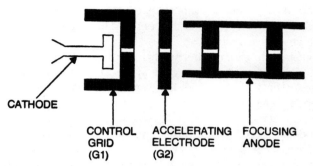

Fig. 17-22. Basic structure of an electron gun from a CRT.

The cathode is indirectly heated (i.e., there is a separate heater filament), and emits a stream of electrons, as in any tube. There is one difference, however. In ordinary tubes, the cathode generally emits electrons from its sides, but the cathode in a CRT is designed so that it emits electrons primarily from the end facing the *phosphor screen*.

The cathode is enclosed in a metal cylinder that acts as the control grid. There is a minute opening at the end of this grid, facing the screen. This hole is for the electrons to pass through. Because it is so small, it forces the electrons to travel in a narrow beam.

By making the control grid negative with respect to the cathode, some of the electrons are repelled, and thus, aren't allowed to pass through the opening. If the control grid is made negative enough, it will cut off the electron beam to the rest of the tube entirely.

In other words, changing the voltage to the control grid with respect to the cathode controls (or *modulates*) the intensity of the electron beam. Holding the voltage on the control grid constant and varying the cathode voltage would have exactly the same effect. The intensity of the electron beam is determined by the difference between these two voltages. Both methods are commonly used in practical circuits.

The more intense the beam (i.e., the greater the number of electrons) striking the phosphors, the brighter they will glow. So obviously, modulating the cathode-control grid voltages will control the amount of light emitted.

Once the electron beam has passed through the control grid, it moves through a second grid element, called the *accelerating electrode*, or *grid 2*. This electrode is a metal cylinder or disk with a small opening for the electron stream to pass through. A high positive voltage is applied to the accelerating electrode. This voltage is held constant—that is, it is not modulated.

As the name implies, the purpose of this element is to accelerate, or speed up the electrons as they pass through. In this respect, it is somewhat similar to the screen grid in a regular tetrode.

Because the accelerating electrode is highly positive it drains off some of the electrons from the passing stream. But the electrons are moving too fast for the positive voltage to deflect them from the narrow beam created by the narrow opening in the end of the control grid.

Next, the electron beam passes through the *focusing anode*. Again, the name suggests the function—this element focuses, or tightens the stream of electrons into a still finer beam.

The focusing anode is a metal cylinder that is open at both ends. Inside the cylinder are two metal plates with tiny holes in the center. The element acts similarly to a glass focusing lens in an optical system.

Besides focusing the electron beam, this electrode also speeds it up still further. A rather large, constant positive voltage is applied to the focusing anode.

These four elements (the cathode, the control grid, the accelerating electrode, and the focusing anode) comprise the electron gun. The electron gun is so named because it "shoots" a narrow beam of electrons at the phosphor screen. Electrical connections to these electrodes are brought out through metal pins in the base of the tube, just as with ordinary tubes. Once the electron beam leaves the electron gun, it passes through a second anode. Because an extremely high (several thousand volts) positive potential is applied to this element, it is called the *high voltage anode*. The electrical connection for this element is brought out to a metallic button on the body of the tube.

Within the electron gun, the accelerating electrode and focusing anode (sometimes called *anode #1*) are both held at a positive voltage, and might tend to attract a large number of electrons out of the beam if the higher positive voltage of the high voltage anode (*anode #2*)didn't have such a strong attraction that it pulls the electrons on through. Despite this high attraction, even the high voltage anode doesn't drain many electrons out of the beam. Because the electron beam is very tightly focused, and moving at an extremely high speed, and since the high voltage anode is an open cylinder, almost all of the electrons pass through it to strike the phosphor screen.

If the tube consisted only of the elements described so far, the electron beam would always strike the exact center of the screen. Obviously, this wouldn't be particularly useful. We need a way to deflect the beam so that it can strike any portion of the screen we choose. There are two basic ways of accomplishing this—*electrostatic deflection* and *electromagnetic deflection*.

The cathode ray tube shown in Fig. 17-21 is of the electrostatic deflection type. In this kind of tube there are four *deflection plates*, with electrical connections made to metal knobs on the outside of the glass envelope.

The plates at the top and bottom of the tube are called the *vertical deflection plates*. The other set, at the sides, are called the *horizontal deflection plates*. The electron beam passes between all four plates.

For simplicity, we'll ignore the horizontal deflection plates (the ones on the sides) for the time being. If both vertical deflection plates have the same voltage applied to them, they will have no effect on the path of the electron beam, and it will strike the center of the screen. See Fig. 17-23.

Now, if the lower plate is made more negative than the upper plate, the lower plate will repel the stream of electrons, and the upper plate will attract it. This means the electron beam will move at an upward angle. It will strike the screen near the top—see Fig. 17-24.

The exact location of the lighted spot on the screen will depend on the voltage difference between the deflection plates. The greater the difference between the plate voltages, the further the spot will be displaced from the center of the screen.

It is very important to realize that the displacement is dependent on the difference of voltage on the plates—not necessarily their absolute values. When we say the lower

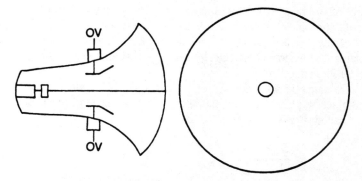

Fig. 17-23. A CRT with equal voltages applied to the vertical plates.

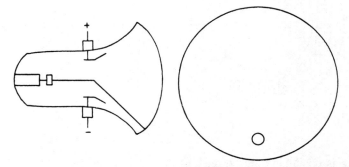

Fig. 17-24. A CRT with the lower vertical plate negative with respect to the upper vertical plate.

plate is negative, we are speaking of its relation to its partner—not necessarily with respect to ground. For instance, if the lower plate has an applied voltage of −25 volts (with respect to ground), and the upper plate is at +25 volts, the voltage difference is 50 volts. The exact same effect on the electron beam can be achieved if the lower plate is at +100 volts over ground and the upper plate is at +150 volts.

Of course, if the relative polarities of the deflection plates are reversed, as in Fig. 17-25, the effect on the electron beam will also be reversed. A negative upper plate

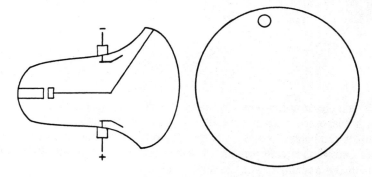

Fig. 17-25. A CRT with the lower vertical plate positive with respect to the upper vertical plate.

and a positive lower plate will cause the electron beam to move down the screen. The horizontal deflection plates work in the same way, moving the electron beam from side to side. By combining the effects of the horizontal deflection plates and the vertical deflection plates, the electron beam can be aimed so that any desired spot on the phosphor screen can be illuminated.

The electromagnetic deflection system works basically in a similar manner, but instead of internal deflection plates, electromagnets are placed around the neck of the tube in an assembly called a *yoke*. See Fig. 17-26.

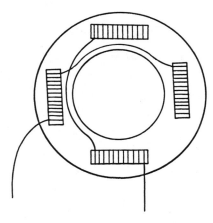

Fig. 17-26. Construction of a CRT yoke.

The yoke is positioned on the neck of the tube so the electromagnets are placed in the places shown in Fig. 17-27. Notice that these positions correspond directly to the positions of the deflection plates in an electrostatic deflection cathode ray tube.

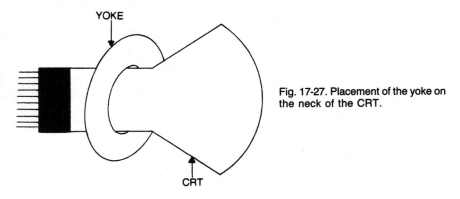

Fig. 17-27. Placement of the yoke on the neck of the CRT.

Because an electron can be attracted or repelled by a magnetic field (it can be considered as a microscopic magnet itself), the relative strength of the electromagnet's magnetic fields can control the angle of the electron beam, and thus, the position of the lighted spot on the phosphor screen. Of course, the strength of each magnetic field is dependent on the amount of voltage applied to the appropriate electromagnet.

These two deflection systems are very similar. Generally, the electromagnetic deflection type CRT is more complex to manufacture, and is, therefore, more expensive, as a rule, than the electrostatic deflection type CRT. However, the electromagnetic deflection system allows for more precise control of the electron beam's angle. This means the image formed on the screen is sharper, or has more *resolution* (i.e., finer detail).

In *oscilloscopes* and *radar monitors*, high resolution isn't particularly critical, so the less expensive electrostatic deflection type CRT's are usually used. A *television picture tube*, on the other hand, demands a very high degree of resolution, so an electromagnetic deflection type CRT is usually employed for that application.

If we apply a repeating ac waveshape to the horizontal deflection plates (or electromagnets) the electron beam will move back and forth across the screen in step with the ac frequency. The same voltage is applied to each of a pair of deflection plates (magnets), but one is inverted 180°, so as one voltage increases, the other decreases, so the difference between the two plate voltages will vary in the same manner as the applied signal. See Fig. 17-28.

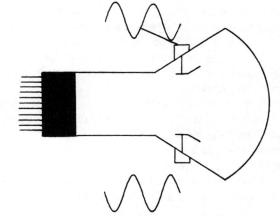

Fig. 17-28. Applying a repeating ac wave to the horizontal plates.

Usually the best waveform for moving the lighted dot across the screen is the *sawtooth*, or *ramp wave*. This waveshape is illustrated in Fig. 17-29. Notice that the voltage starts at some specific minimum value and gradually builds up to a maximum level. Then it quickly drops back to the original minimum value, and the entire cycle is repeated.

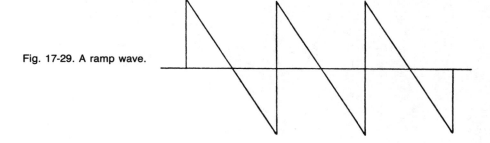

Fig. 17-29. A ramp wave.

At the minimum voltage point of the cycle, the electron beam is angled to strike the far left edge of the screen (facing the screen from the front of the tube). The left deflection plate is exhibiting maximum attraction, and the right deflection plate is exhibiting maximum deflection. As the voltage increases, the left deflection plate gradually loses some of its attraction, and the right deflection plate loses some of its repulsion. The lighted dot moves across the screen from left to right. When it is in the center of the top, both deflection plates are at an equal voltage. From this point on, the right deflection plate starts to attract the electron beam, and the left deflection plate starts to repel it. The lighted dot continues to move across the screen, until, at the maximum applied voltage, it is at the far right edge of the screen. This part of the cycle is called the *sweep*. The line drawn by the electron beam across the screen is called the *trace*.

During the next part of the cycle, the applied voltage drops quickly back to the original minimum level, causing the electron beam to snap back to its original far left position. This is called the *retrace*, or *flyback*.

In most practical circuits, the electron gun is cut off (no electron beam at all) during the flyback time. It is impossible to produce a sawtooth wave with an instantaneous flyback. It takes a certain finite amount of time to go from the maximum voltage to the minimum voltage. If the beam was allowed to strike the screen during the retrace time, it could produce a confusing trace image. So the screen is only illuminated by the left to right movement of the electron beam. During the retrace it is dark.

The frequency of this sawtooth waveform is called the sweep frequency, since it determines how rapidly (and how many times per second) the electron beam will sweep across the screen.

If, at the same time the horizontal plates are being fed by the sweep signal, we apply another waveform to the vertical deflection plates, something quite interesting (and useful) takes place. Between any two given instants, the electron beam will be moved a small amount, so each instantaneous value of the vertical deflection voltage will be displayed in a different horizontal position on the screen. In other words, if a sine wave of the same frequency as the sweep signal is applied to the vertical deflection plates, the electron beam will draw the pattern shown in Fig. 17-30 on the phosphor screen.

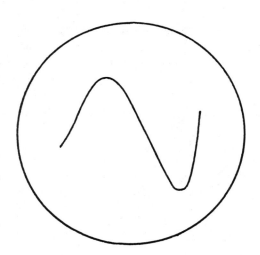

Fig. 17-30. CRT screen—vertical frequency equals horizontal frequency.

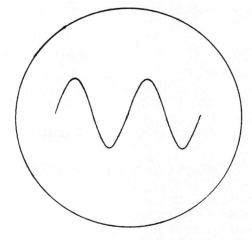

Fig. 17-31. CRT screen—vertical frequency equals two times the horizontal frequency.

If the frequency of the sine wave is doubled, two complete vertical cycles will take place in the time required for a single horizontal cycle, so two complete waveforms will be displayed on the screen, as in Fig. 17-31.

The sweep frequency is selected to be fast enough so that the trace will appear to be a solid, continuous line. Actually, at any given instant, the electron beam is striking only one tiny spot on the screen. The phosphors glow due to a property called *fluorescence*. Another property of these materials, which is known as *phosphorescence* allows them to continue glowing for a brief time after the electron beam stops striking the spot. This property, coupled with the persistence of vision (the eye continues to see a light source for a brief moment after it is removed) gives the illusion of a solid image.

The exact chemical properties of the phosphors used determine the phosphorescence time. Different applications require different amounts of after-glow. A typical oscilloscope generally uses a phosphor that produces a green trace with a moderate after-glow time. If the oscilloscope is intended to display non-cyclic voltage patterns of very short duration, a greater degree of phosphorescence is necessary. For television pictures, on the other hand, a relatively short after-glow time is preferable. In a black and white picture tube, the phosphors glow white. In a color picture tube, three types of phosphors are used together. These phosphors glow red, green and blue. This will be explained in the chapter on color television.

The screen of a cathode ray tube can either be round (as in most oscilloscopes and radar monitors), or rectangular (as in most television picture tubes). With the round type, the size is specified by the diameter of the screen, while with the rectangular shape, the size is defined by the diagonal. See Fig. 17-32.

Because the electron beam strikes the phosphor screen at an extremely high speed, secondary emission could be a problem, producing reflections at undesired portions of the screen. This problem is generally prevented by lining the interior surface of the glass tube with a conductive graphite coating called the *Aquadag*. This Aquadag is tied electrically to the high voltage anode. Because it has a high positive potential, any electrons bouncing off of the screen will be attracted to the Aquadag coating, rather than striking the screen a second time.

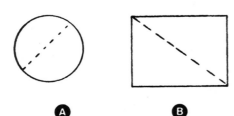

Fig. 17-32. Measuring CRT screen sizes.

A   B

Self-Test

1. What are the active parts of a diode tube?

A *Anode, grid, and cathode*
B *Anode and plate*
C *Anode and cathode*
D *Cathode and grid*
E *None of the above*

2. What is the simplest type of tube capable of amplification?

A *Triode*
B *Anode*
C *Diode*
D *Tetrode*
E *None of the above*

3. Under what conditions will a diode conduct?

A *At all times*
B *When it is forward biased*
C *When it is reverse biased*
D *When it is amplifying*
E *None of the above*

4. Which of the following best describes the grid in a tube?

A *A large flat plate*
B *A metallic mesh*
C *A cone-like shape*
D *A filament*
E *None of the above*

5. What happens if the grid is made more positive than the saturation point?

A *Electrons are drawn to the grid and do not reach the plate*
B *No further amplification takes place*

C *The tube elements may be damaged*
D *The tube stops conducting*
E *None of the above*

6. What is the term specifying the maximum gain a tube is capable of?

A μ—*characteristic curve*
B Ω—*amplification factor*
C β—*attenuation factor*
D μ—*amplification factor*
E *None of the above*

7. What is the purpose of the screen grid?

A *To allow greater amplification*
B *To reduce the effect of interelectrode capacitances*
C *To reduce impedance of the tube*
D *To make the tube more durable*
E *None of the above*

8. How many electrodes does a pentode have?

A *Two*
B *Three*
C *Four*
D *Five*
E *Six*

9. What is the name of an electrode found in a pentode, but not in a tetrode?

A *Control grid*
B *Screen grid*
C *Suppressor grid*
D *Signal grid*
E *None of the above*

10. What type of tube is used to display signals on an oscilloscope?

A *Tetrode*
B *Cathode-ray tube*
C *Filament tube*
D *Pentode*
E *None of the above*

18

Semiconductors

Way back in the first chapter we learned that certain substances allow electrons to flow through them fairly easily. Such substances are called conductors. Other substances, called insulators, tend to oppose the flow of current.

There is a third important class of substances with properties somewhere between conductors and insulators. These substances are called *semiconductors*. As you will soon see, semiconductors are extremely important to modern electronics.

SEMICONDUCTOR PROPERTIES

All substances, whether they are conductors, insulators, or in between will offer some resistance to current flow. Conductors present a very small resistance, while insulators present a very large resistance. As might be expected, a semiconductor offers a moderate amount of resistance to the flow of electrons through it.

Copper is an excellent conductor. A cubic centimeter of this substance has a resistance of about 1.7×10^{-6} (0.0000017) ohm. This is clearly a very minute amount of resistance.

On the other hand, a cubic centimeter of slate (a good insulator) has a resistance of about 100 megohms (100,000,000 ohms). Compared to copper, virtually no current can flow through slate.

Now compare both of these substances to germanium. A cubic centimeter of this material has a resistance of approximately 60 ohms. Germanium is a semiconductor. Another common semiconductor is silicon.

A germanium atom has four electrons in its outermost ring. The electrons in the outermost ring of any atom are called *valence electrons*.

The valence electrons in germanium pair up with the electrons of other germanium atoms in a crystaline structure. The pattern of these interlinked atoms is illustrated in Fig. 18-1. The atoms within the crystal are held together by a force called the *covalent bond*. As this term suggests, the atoms share their valence electrons.

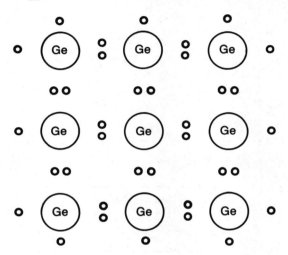

Fig. 18-1. Germanium crystal structure.

Pure germanium has no particularly unique electrical properties, beyond being a fair material for making small resistors. But if selected impurities are added to a germanium crystal, a number of interesting effects can be achieved. The process of adding impurities to a piece of semiconductor material is called *doping*.

First, let's look at what happens if a pure germanium crystal is doped with a small amount of arsenic. We'll assume a single arsenic atom has been added.

The arsenic atom will try to act like a germanium atom, but, since arsenic has five valence electrons, there will be an extra electron left over. See Fig. 18-2.

This extra electron can drift freely from atom to atom throughout the crystal. The crystal as a whole is electrically neutral, because the total protons equal the total electrons.

If the germanium crystal is doped with a number of arsenic atoms, there will be an equal number of surplus electrons drifting through the crystal. The crystal itself will still be electrically neutral, of course.

If a voltage source is connected to the crystal, as shown in Fig. 18-3, the extra electrons will be drawn to the positive terminal of the voltage supply, and removed from the crystal.

Because the crystal now has fewer electrons than it has protons, it possesses a positive electrical charge, and draws electrons out of the negative terminal of the voltage source. These electrons will move through the crystal and out to the positive terminal of the voltage source. In other words, current will flow through the crystal. So far we still don't have anything special, but wait.

222

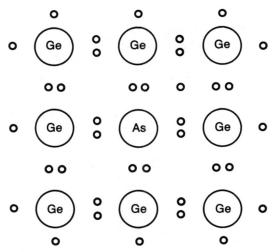

Fig. 18-2. Germanium crystal doped with arsenic.

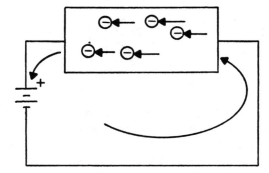

Fig. 18-3. Voltage source applied to a Germanium crystal doped with arsenic.

There are two types of doped semiconductors. The type we have just discussed is called an *N-type semiconductor*, because negatively charged electrons move through it. The other kind of doped semiconductor is called a *P-type semiconductor*. It is quite similar to the N-type, except an impurity with just three valence electrons is used to dope the crystal. Indium is frequently used. In this case, some of the covalent bonds are incomplete. That is, there are *holes* where electrons belong. See Fig. 18-4.

The various covalent bonds will steal electrons from each other to fill their holes, causing the positions of the holes to apparently drift. We can say we have a flow of holes. Actually electrons are being moved about, as in any electric circuit, but in this situation it is simpler to think of the holes as moving. Remember, a hole is simply the absence of an electron. By thinking of the holes as positively charged particles (since subtracting an electron will leave a positive charge) we can greatly simplify discussion of semiconductor action.

If the impurity adds extra electrons (like arsenic), it is called a *donor impurity*. If it adds extra holes (fewer electrons) (like indium) it is an *acceptor impurity*.

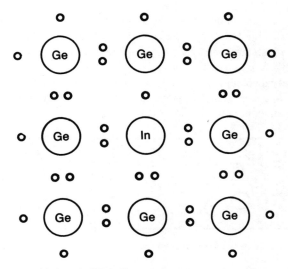

Fig. 18-4. Germanium crystal doped with Indium.

Electrical current will flow through either an N-type or a P-type semiconductor. In a P-type semiconductor we speak of a flow of holes from positive to negative, instead of the usual flow of electrons from negative to positive, but it really amounts to exactly the same thing. See Fig. 18-5.

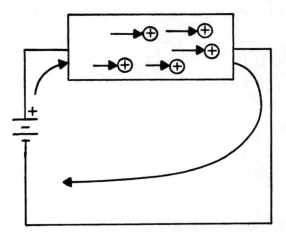

Fig. 18-5. Voltage source applied to a Germanium crystal doped with Indium.

Electrons and holes are referred to as *current carriers,* or simply *carriers.* Both types of semiconductors contain both types of carriers, but one kind of carrier will be much more plentiful than the other. In an N-type semiconductor, electrons are the *majority carriers,* and holes are the *minority carriers.* That is, there are more electrons than holes. In a P-type semiconductor, the situation is reversed. Holes are the majority carriers and electrons are the minority carriers.

Neither type of semiconductor exhibits any special electrical properties when used separately, but when the two types are welded together we find a very unique situation.

The point at which different types of semiconductors are joined is called a *junction*, or, more precisely, a *pn junction*.

When no external voltage is applied to a pn junction, the carriers are randomly placed, as in Fig. 18-6. Remember that despite the extra electrons or holes, the net charge of each type of semiconductor is electrically neutral.

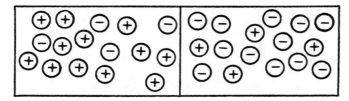

Fig. 18-6. A pn junction with randomly placed carriers.

Now, suppose we hook up a voltage source with its positive terminal connected to the N-type semiconductor, and its negative terminal connected to the P-type semiconductor. This is shown in Fig. 18-7. The holes in the P-type semiconductor will be drawn towards the end of the crystal with the negative charge, while the excess electrons in the N-type semiconductor will be drawn towards the positive charge. Virtually no majority carriers will be found near the junction. This means virtually no electrons can cross from one type of semiconductor to the other. Almost no current will flow through the crystal. We call a crystal under these conditions *reverse-biased*.

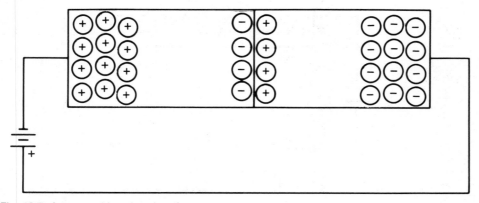

Fig. 18-7. A reverse biased pn junction.

If we now reverse the polarity of the voltage source, as in Fig. 18-8, we will find a completely different situation. The positive charge on the P-type material will attract its minority carriers (i.e., electrons). Some of these electrons will leave the semiconductor and flow towards the voltage sources's positive terminal. Since some electrons have been removed from the P-type material, and it still has the same number of protons, it now has an overall positive charge.

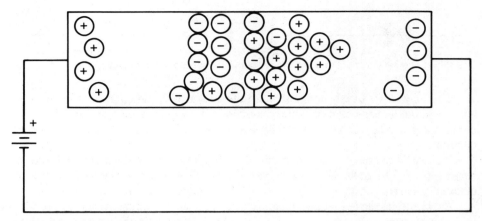

Fig. 18-8. A forward biased pn junction.

Meanwhile, the negative terminal of the voltage source is connected to the slab of N-type material, repelling its majority carriers (electrons) towards the junction. Since there are many electrons being pushed towards the junction, and a positive charge pulling them from the other side, they are forced through the narrow junction area to neutralize the positively charged P-type side. Meanwhile, the voltage source is drawing away more electrons from the P-type material, so it retains its positive charge. Clearly, current flows through the semiconductor under these conditions.

You could also look at the whole procedure from the point of view of a flow of holes. The voltage supply's negative terminal on the N-type material adds electrons to it. These added electrons fill the holes in the N-type material (minority carriers). Since there are now more electrons than in the neutral state, this material acquires a negative charge which pulls holes from the P-type side. These holes (majority carriers in the P-type section) are also forced towards the junction by the positive terminal of the voltage source.

In both cases we are describing exactly the same phenomenon, only our point of reference has changed.

SEMICONDUCTOR DIODES

No doubt you've recognized the similarity between a pn semiconductor junction and the diode tube discussed in the last chapter. Both devices will pass current in only one direction, and block current if the polarity is reversed. In fact, the device we have been describing in the last section is a *semiconductor diode*.

The schematic symbol for a semiconductor diode is shown in Fig. 18-9.Notice that this is identical to one of the symbols used to represent tube diodes.

Current flows in the direction indicated by the small arrow. This arrow is *not* part of the schematic symbol.

Semiconductor diodes can be used for most of the same applications as tube diodes. There are a number of advantages to using semiconductor devices instead of vacuum tubes. Semiconductor diodes tend to be less expensive and smaller than their tube counterparts. Their operation produces less heat, and no separate filament circuit is needed.

Fig. 18-9. Schematic symbol and two typical cases for semiconductor diodes.

The only major disadvantage is that tubes, as a rule, can operate at higher power levels without damage than can semiconductors. However, modern semiconductors will comfortably handle virtually all power levels you're likely to encounter in electronic circuits.

It should also be mentioned that tube diodes generally have a higher resistance when reverse biased, because the minority carriers in a semiconductor diode will allow a small amount of current to flow. This rarely is of any significance in practical circuits.

As with tube diodes, semiconductor diodes are most commonly used in rectifying (see the chapter on power supplies), and demodulation (see the chapter on radio).

The most important specification for a semiconductor diode is the *PIV*, or *peak inverse voltage*. Sometimes this specification is referred to as *PRV*, or *peak reverse voltage*. This title is pretty much self explanatory. It is the maximum voltage that can be applied to a diode with a reverse bias without the diode breaking down. With ordinary diodes this voltage must never be exceeded.

You will learn more about how diodes work in the next set of experiments.

ZENER DIODES

A specialized variation of the semiconductor diode is the *zener diode*. This type of diode responds to a reverse polarity voltage in a unique way. The schematic symbol for a zener diode is shown in Fig. 18-10.

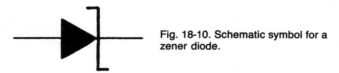

Fig. 18-10. Schematic symbol for a zener diode.

In the circuit illustrated in Fig. 18-11, the zener diode is reverse biased, and the input voltage is variable via the potentiometer. The zener diode is a 6.8 volt unit. The meaning of this specification will soon be clear.

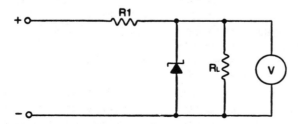

Fig. 18-11. Basic zener diode circuit.

When the applied voltage is zero, the voltmeter will read 0 volts, of course. When the input voltage is increased to 1 volt, the meter reads just under 1 volt. Resistor R1, which is included to limit the current through the diode, drops a small amount of the source voltage. With one volt reverse bias, the diode does not conduct—the circuit acts essentially as if the diode wasn't there at all.

This will hold true up until the point when the source voltage exceeds the voltage rating of the zener diode (6.8 volts in our example). This is the voltage at which the diode begins to conduct when reverse biased. It is often called the *avalanche point* of the diode, because the current through the diode abruptly rises from practically zero to a very high value, limited only by the low internal resistance of the diode. This is why it is necessary to include R1 in the circuit. R1 increases the series resistance, and therefore lowers the circuit current.

Since the zener diode sinks any voltage greater than its reverse bias avalanche point to ground, the voltmeter will read 6.8 volts, even if the source voltage is raised to 7 volts, 8 volts, or even higher.

This basic zener diode circuit also serves to *regulate* the voltage to the load. That is, the voltage remains fairly constant, regardless of the amount of current drawn by the load.

To understand this, let's see what happened when we vary the load resistance (R_L), and leave the zener diode out of the circuit. This experiment circuit is illustrated in Fig. 18-12.

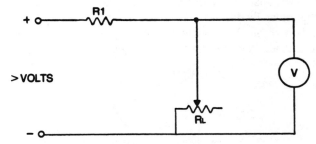

Fig. 18-12. Varying the load resistance to a zener diode circuit.

If R1 is a constant 500 ohms, Table 18-1 shows the current and voltage drop across the load resistor for various values of R_L.

Notice that the voltage across the load varies a great deal as the load resistance changes. The load resistance has to be close to 10,000 ohms before E_L gets close to the desired 6.8 volts (with a 7-volt supply).

When the zener diode is in the circuit, however, it will hold the output voltage to a fairly constant 6.8 volts. Since the load resistance is in parallel with the zener diode, E_L is also a constant 6.8 volts. Remember, when two components are in parallel, the voltage dropped across them is equal. In other words, Fig. 18-11 is a simple *voltage regulation* circuit.

If the effective load resistance drops for any reason (that is, if the load circuit starts to draw more current), this will cause an increase in the voltage drop across R1 (corresponding to the decreasing voltage drop across R_L). This decreases the voltage to the

**Table 18-1. Effects of Varying the
Load Resistance in an Unregulated Circuit.**

E (source) = 7 volts R1 = 500 ohms			
R_L	R (total)	I (mA)	E (R_L)
100	600	11.7	1.17
200	700	10.0	2.00
300	800	8.8	2.63
400	900	7.8	3.11
500	1000	6.4	3.82
600	1100	5.8	4.08
800	1200	5.4	4.31
900	1400	5.0	4.50
1000	1500	4.7	4.67
1100	1600	4.4	4.81
1200	1700	4.1	4.94
1300	1800	3.9	5.06
1400	1900	3.7	5.16
1500	2000	3.5	5.25
10,000	10500	0.7	6.67

diode. As less voltage is applied to the diode, it draws less current. That means the voltage drop across R1 must decrease, forcing the output voltage to stabilize at the level determined by the zener diode.

Figure 18-13 is a graph showing the relationship of current and voltage through a zener diode. Zener diodes are available for voltages up to about 200 volts.

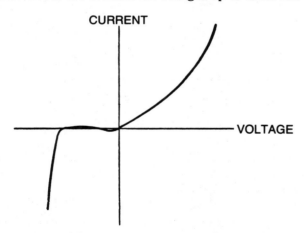

Fig. 18-13. Current vs voltage graph through a zener diode.

VARACTOR DIODES

The pn junction of any diode has a certain amount of internal capacitance, along with its semiconductor properties. The value of this internal capacitance is dependent on the width of the junction itself.

A *varactor diode* is a special type of diode that takes advantage of this concept. Varying the reverse bias voltage to this kind of diode will vary the effective size of the pn junction. In other words, the diode acts like a voltage variable capacitor.

When a varactor diode is used as the capacitor in a resonant circuit (either series or parallel), the resonant frequency can be electrically controlled. Obviously, this device is ideal for automatic tuning systems. The specific applications will be discussed in later chapters.

The schematic symbol for a varactor diode is shown in Fig. 18-14. Sometimes a varactor is called a *voltage controlled capacitor,* for obvious reasons.

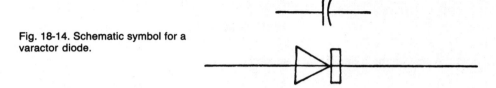

Fig. 18-14. Schematic symbol for a varactor diode.

SHOCKLEY DIODES

Another special purpose diode is the *Shockley diode*. Unlike most other diodes, the Shockley diode is made up of more than a single pn junction. Its construction includes two of each type of semiconductor—npnp. For this reason this component is also known as a *four-layer diode*.

Like most other diodes, the Shockley diode has two terminals—an anode and a cathode. The schematic symbol for this device is shown in Fig. 18-15.

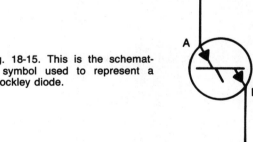

Fig. 18-15. This is the schematic symbol used to represent a Shockley diode.

The Shockley diode has switching properties similar to those of a neon glow lamp. It has an inherent *trigger voltage*. Below the trigger voltage, the device is in its ''off'' state, and exhibits a very high resistance. If the applied voltage exceeds the trigger value, the diode will be switched on and the resistance will drop to an extremely low value of just a few ohms.

The trigger voltage is also known by other names, including *''threshold voltage,''* *''firing voltage,''* and *''avalanche voltage.''*

If a third electrode, called a *gate* is added to a Shockley diode, the result will be a *SCR*. This more complex device will be discussed in Chapter 21.

FAST RECOVERY DIODES

Ordinary diodes are designed to operate on fairly low frequencies. One of the most common applications is to rectify 60 Hz sine waves. A certain finite amount of time is required for the diode to recover. This is the time it takes for the diode to turn back off when the polarity of the applied voltage is reversed. In low-frequency applications, the recovery time is not particularly critical.

In high-frequency applications, such as in a television flyback circuit, the recovery time becomes critical, since the diode must respond to very short duration spikes. An ordinary diode could cause erratic or incorrect operation of the circuit. For high-frequency applications a special purpose diode called a *fast recovery diode* is used.

Self-Test

1. Which of the following is a semiconductor?

A *Rubber*
B *Copper*
C *Germanium*
D *Carbon*
E *None of the above*

2. A P-type semiconductor has a shortage of which of the following?

A *Neutrons*
B *Holes*
C *Electrons*
D *Doping*
E *None of the above*

3. How many junctions are there in a semiconductor diode?

A *None*
B *One*
C *Two*
D *Four*
E *It varies*

4. The arrow in the schematic symbol for a diode points which way?

A *Toward the cathode*
B *Toward the anode*
C *In the direction of current flow*
D *Toward magnetic north*
E *None of the above*

5. What is the most important specification for a semiconductor diode?

A *Forward resistance*
B *Reverse resistance*
C *Peak inverse voltage*
D *Current capacity*
E *None of the above*

6. What happens when the voltage applied to a Zener diode exceeds its avalanche point?

A *The current through the diode abruptly rises to a very high value*
B *The current through the diode abruptly drops to zero*
C *The diode is destroyed*
D *The capacitance of the diode increases*
E *None of the above*

7. What type of circuit would a Zener diode be most likely used in?

A *Amplifier*
B *Rectifier*
C *Oscillator*
D *Voltage regulator*
E *None of the above*

8. What is another name for a varactor diode?

A *Zener Diode*
B *Voltage controlled capacitor*
C *Four-layer diode*
D *Fast recovery diode*
E *None of the above*

9. Which of the following is *not* another name for "trigger voltage?"

A *Threshold voltage*
B *Avalanche voltage*
C *Firing voltage*
D *Peak inverse voltage*
E *None of the above*

10. What is another name for a four-layer diode?

A *Varactor diode*
B *Fast recovery diode*
C *Shockley diode*
D *Voltage controlled inductor*
E *None of the above*

19

LEDs

A special type of diode is the *LED*, or *light-emitting diode*. This device is shown in Fig. 19-1. Like any other diode, an LED will pass current in only one direction. But, as the name implies, it will glow, or emit light when forward biased. When reverse biased an LED will remain dark.

While some clear (white light) LEDs have been developed, most of these devices emit colored light. Red is, by far, the most common color for LEDs, but green and yellow are also frequently used. In addition to these visible color types, some LEDs are designed to emit light in the infrared region, which is outside the visible spectrum.

Figure 19-2 shows the most commonly used schematic symbols for LEDs. As you can see, the symbol in Fig. 19-2B simply omits the circle shown in Fig. 19-2A, otherwise the two symbols are identical.

LEDs are used primarily as indicator devices. That is, an operator can tell whether or not a specific voltage is present in a circuit by whether the LED is lit up, or dark. For example, in the circuit shown in Fig. 19-3, the LED lights whenever the circuit is activated (switch closed) and serves as a reminder to turn off the equipment when it is not in use. When this LED is lit it also indicates that the circuit is getting power and is presumably operating properly.

Within certain limits, the higher the voltage applied to an LED, the brighter it will glow. And, of course, lowering the applied voltage will dim the LED. This cannot be used to measure exact values, but it can be used for simple relative comparisons.

LEDs are relatively durable, but they are intended for use in low-power circuits only. Typically no more than 3 to 6 volts should be applied to an LED. Excessively high

232

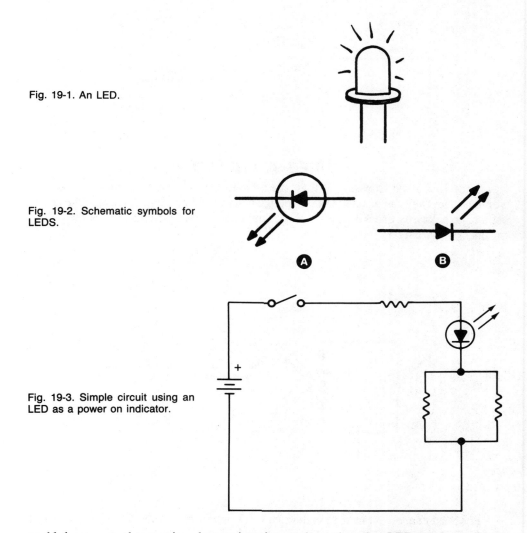

Fig. 19-1. An LED.

Fig. 19-2. Schematic symbols for LEDS.

Fig. 19-3. Simple circuit using an LED as a power on indicator.

could burn out the semiconductor junction and render the LED useless. Some consideration should also be given to the amount of current flowing through an LED. Excessive current can damage or destroy the component very quickly.

The LED, being a diode, exhibits a very high resistance when reverse biased. According to Ohm's law (I = E/R) this means the diode will draw very little current. When forward biased, on the other hand, the LED's resistance drops to a very low value, allowing the current to climb to a relatively high level. Depending on the rest of the circuit, the LED may attempt to pass more current than it can safely handle.

To limit the current through the LED to a safe value, a series resistor is often added to the circuit, as illustrated in Fig. 19-4. This resistor should have a relatively low value, usually between about 100 and 600 ohms. 330 ohms is a commonly used value.

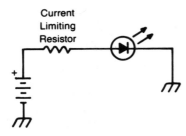

Fig. 19-4. A series resistor is used to reduce the current flow through a LED.

THREE-STATE LED

Since LEDs will glow only if they are forward biased, and not when they are reverse biased, they can be used to test voltage polarity. Figure 19-5 illustrates the circuit for a simple polarity checker. The resistor limits the current through both of the LEDs, so only one is needed.

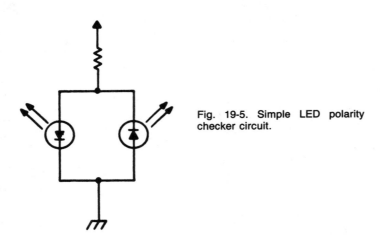

Fig. 19-5. Simple LED polarity checker circuit.

For most convenient indication the two LEDs in this circuit should be of contrasting colors. For instance, LED 1 might be red, and LED 2 could be green.

If the lead is connected to an unknown voltage source (both the tester and the circuit being tested should have a common ground) we can easily tell the polarity of the unknown signal. If the voltage is positive with respect to ground, red LED 1 will light up, and green LED 2 will remain dark. If the polarity is reversed, green LED 2 will glow instead of red LED 1. If neither LED lights up, the applied voltage must be zero, or very close to it.

Single unit dual LEDs are available. These are simply two differently colored LEDs within a single package. These LEDs are internally connected like in the circuit of Fig. 19-5, without the current-limiting resistor.

The dual LED will glow one color when a voltage of one polarity is applied to it, and the other color will glow when the polarity is reversed. The device is known as a three-state LED, because there are three ways it can light up. We have already covered two. Before we get to the third, let's consider what happens when an ac voltage is applied to an LED.

By definition the polarity of the ac signal keeps reversing itself. For half of each cycle the LED is forward biased and lit. For the other half of each cycle the polarity is reversed so the LED is reverse biased and dark. In other words, the LED blinks on and off.

If the applied frequency is low enough, you would actually be able to see the LED blink on and off in step with the applied signal. When the frequency is increased, the LED will still blink on and off, but it will do so too rapidly for the human eye to follow each separate blink. The LED will appear to be continuously lit, although it may seem somewhat dimmer than what a similar dc voltage would produce.

Now, let's return to the three-state LED and examine what happens when an ac voltage is applied. For our discussion we will assume the two internal LEDs are red and green. For half of each ac cycle the red LED will be lit, and the green LED will be dark. For the other half of each cycle the green LED will be lit and the red LED will be dark.

If the ac frequency is very low, we will be able to see the alternation between red and green. Because the two LEDs are so closely placed within a single package, the device will appear to be changing color.

When the applied frequency is raised, the two colors will both appear to be continuously on. They will tend to blend together producing a yellow glow.

The three-state LED has three different color states;

RED—dc polarity A
GREEN—dc polarity B
YELLOW—ac

The three-state LED is an extremely useful indicator device.

MULTIPLE SEGMENT DISPLAYS

LEDs are useful indicators, but they can be even more useful if a number of them are used together to indicate a wider range of circuit conditions.

Figure 19-6 shows a simple circuit in which three LEDs indicate the position of a rotary switch. The switch could have additional poles that simultaneously perform other functions. If the switch is in position A, only LED 1 will be lit. LED 2 and LED 3 will remain dark. Advancing the switch to position B will extinguish LED 1's glow and light LED 2. LED 3 will remain dark. If the switch is moved to position C, both LED 2 and LED 3 will be lit, while LED 1 will be dark. If only LED 3 is lit and the other two LEDs are dark, we know the switch must be in position D. Position E illuminates all three LEDs, while in position F, all three LEDs are off.

Study this circuit diagram carefully to make sure that you understand how the three LEDs are being controlled in each of the switch positions. This kind of multiple indication system can be extremely helpful in operating complex circuits.

Notice that the cathodes of all three LEDs are electrically tied together. Effectively, all three LEDs share a single cathode. Such a system is called a *common-cathode display*.

236

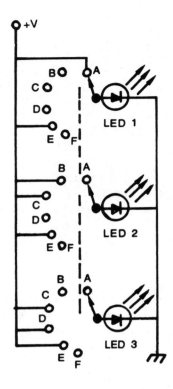

Fig. 19-6. LEDs to indicate the position of a 3P6T rotary switch.

Alternatively, the LEDs could be connected as a *common-anode display*. This is illustrated in Fig. 19-7.

Notice that in these cases the word "common" does not necessarily refer to the circuit's common ground point. It simply refers to a shared element which is common to each of the component LEDs.

Bargraphs

Figure 19-8 shows another multiple LED display. In this circuit, the higher the voltage, the more LEDs will light. For an example, we'll assume that each LED requires at least 1.5 volts to light, and that each resistor has a value of 1000 ohms. For simplicity we will ignore the internal resistances of the LEDs.

The resistors act as a voltage divider network. Since all four resistors are equal, one quarter of the applied voltage will be dropped across each resistor.

For instance, if 2 volts is applied to the circuit, we would be able to read the full 2 volts at point A. R1 would drop 0.5 volt (one quarter of the source voltage), leaving 1.5 volts at point B. This is enough to illuminate LED 1. Another 0.5 volt is dropped by R2 so only 1 volt can be read at point C, so LED 2 is off. LED 3 also remains dark because after the 0.5 volt drop across R3, there is only 0.5 volt at point D. Point E, of course, is grounded, so it is always at zero potential.

Now, what happens if we increase the source voltage to 4 volts? One volt will be dropped across each resistor in this case, so we'll have 3 volts at point B, 2 volts at

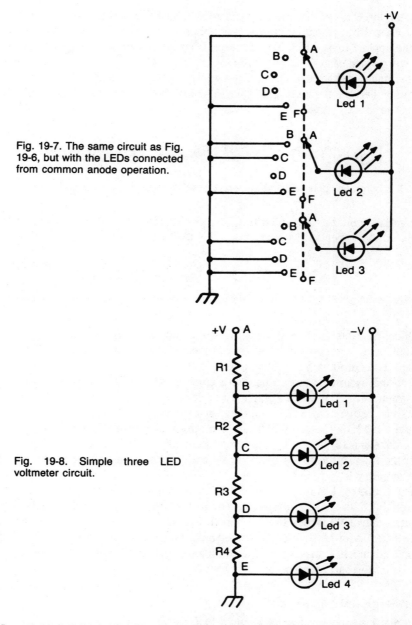

Fig. 19-7. The same circuit as Fig. 19-6, but with the LEDs connected from common anode operation.

Fig. 19-8. Simple three LED voltmeter circuit.

point C, and 1 volt at point D. Since points B and C are greater than the minimum turn-on voltage (1.5 volts), LED 1 and LED 2 will light up, while LED 3 will still stay dark.

Raising the input voltage to 6 volts will result in a 1.5 volt drop across each resistor. Point B will be at 4.5 volts, point C will be at 3 volts, and point D will be at 1.5 volts. Of course, this means all three LEDs will be lit.

Naturally, this type of circuitry can readily be expanded to include more than just three LEDs. But remember, there is an inherent limitation on how much voltage can safely be applied to an LED. To measure higher level signals, some sort of attenuation stage will be necessary.

This kind of display is often called a *bargraph*, because the LEDs are usually arranged as a line or bar, as shown in Fig. 19-9. The longer the lighted portion of the bar, the greater the input voltage.

Fig. 19-9. A bargraph lights up all of the indicator LEDs below the measured level.

A variation on the bargraph is the *dot graph*. In this case only the highest appropriate LED is lit. All lower LEDs stay dark. This is illustrated in Fig. 19-10.

Fig. 19-10. A dot-graph lights up only the single LED that represents the measured level.

Of course, this method of measurement is not as precise as a meter, but it is quite sufficient, and very convenient in certain pieces of equipment. For example, a dot or bargraph is ideal for a VU meter in a tape recorder.

Dot and bargraphs are so useful, a number of manufacturers sell strips of LEDs in bargraph form. Dot and bargraph driver circuits are also available in IC form. These circuits use active comparators, rather than the passive resistances of Fig. 19-8. For example the TL490C and TL491C are ten-step analog level detectors designed for use in bargraphs in a 16-pin IC. The user merely needs to add an external reference voltage (to set the range), the LEDs themselves, and their current-limiting resistors. The rest of the circuitry is included within the IC chip (see Chapter 24).

The LM3914, LM3915, and LM3916 take things a step further. The driver IC and a row of 10 LEDs are sold premounted on a small PC board (printed circuit). The board measures only 1.99 inches by 0.850 inch. The chip is protected by an opaque plastic cover. The row of LEDs are also under a plastic cover strip with a square window for each individual LED, making the bargraph look more like a bar. These devices may be used for either bargraphs or dot graphs.

Seven-Segment Displays

Perhaps the most widely useful multi-LED display arrangement is the seven-segment display. This consists of seven LEDs shaped like narrow rectangles and arranged in a figure-8 pattern, as shown in Fig. 19-11.

A seven-segment display may be of either the common-cathode or the common-anode type. There is only a single lead for the common element, and the other leads are brought out individually.

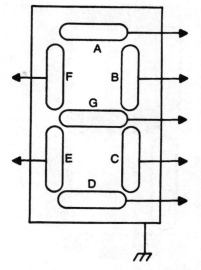

Fig. 19-11. Seven segment common cathode LED display.

If all seven LEDs are lit, the display will, of course, look like the digit 8. But, by lighting only selected LEDs, or segments, a single digit from 0 to 9 can be formed. For example, if segments A, B, D, E, and G are lit, but segments C and F are dark, the number 2 will be formed.

Figures 19-12 through 19-21 show how each digit can be formed. If a number higher than 9 must be displayed, more than one seven-segment display is used.

Certain letters of the alphabet can also be displayed on a seven-segment display, although some require a little imagination to read. Can you determine which segments

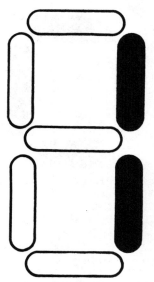

Fig. 19-12. Displaying 1 on a seven segment display.

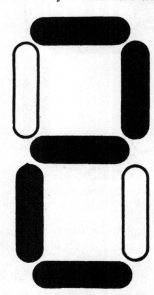

Fig. 19-13. Displaying 2 on a seven segment display.

240

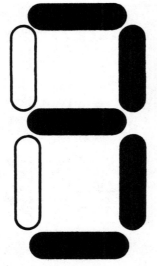

Fig. 19-14. Displaying 3 on a seven segment display.

Fig. 19-15. Displaying 4 on a seven segment display.

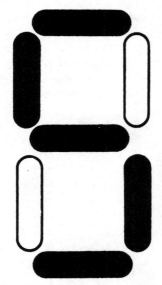

Fig. 19-16. Displaying 5 on a seven segment display.

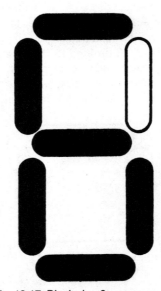

Fig. 19-17. Displaying 6 on a seven segment display.

must be lit to display the following letters? A, B, C, E, F, G, H, I, J, L, O, P, S, and U. Can you find a way to display any other letters?

LEDS AS LIGHT DETECTORS

LEDs are ordinarily used to produce light, but they can also be used to detect the presence and approximate level of an external light source. If a light-emitting diode is

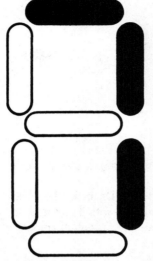

Fig. 19-18. Displaying 7 on a seven segment display.

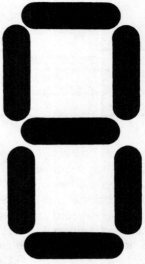

Fig. 19-19. Displaying 8 on a seven segment display.

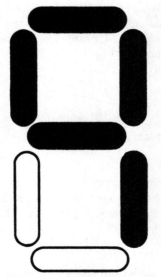

Fig. 19-20. Displaying 9 on a seven segment display.

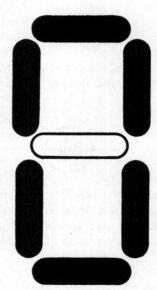

Fig. 19-21. Displaying 0 on a seven segment display.

exposed to a strong light source, a small voltage will be generated between its leads. The magnitude of this voltage will be determined by the brightness of the light source and the actual structure of the LED itself.

LED light detectors are most sensitive to the type of light they were designed to emit. For instance, a red LED will respond best to red light, a green LED will be more sensitive to green light, and so forth.

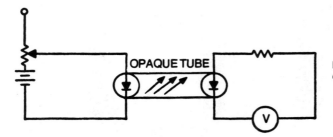

Fig. 19-22. An LED optoisolator circuit.

If two LEDs are connected as shown in Fig. 19-22, the signal applied to circuit A, will be transferred to circuit B (at a reduced amplitude) without a direct electrical connection between the two circuits.

To avoid interference from outside light, the two LEDs should be enclosed in some kind of tight opaque structure, such as a cardboard tube with its interior painted black. They should also be positioned so that LED 1 sheds the maximum amount of light onto LED 2. This arrangement is called an *optoisolator*.

FLASHER LEDS

An interesting variation on the basic LED may be difficult to recognize at first glance. If you look very carefully you will see a small black speck within the clear epoxy case of this LED. The speck is a tiny oscillator IC. Just applying a voltage to the two leads of this device will cause the LED to blink on and off at a 3-Hz rate. The device is known as a flasher IC, and the schematic symbol is shown in Fig. 19-23. Flasher ICs make very eye-catching displays and indicator devices.

Fig. 19-23. The LED schematic symbol is modified slightly to indicate a flasher LED.

The flasher LED includes its own internal current dropping resistor. It uses a +5 volt power supply and draws about 200 mA. Higher voltages may be applied if a series dropping resistor is used, as shown in Fig. 19-24. For a 9-volt power supply, the dropping resistor should have a value of 1000 ohms, ½ watt.

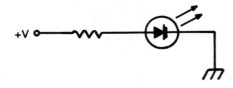

Fig. 19-24. This simple flasher LED circuit requires only a voltage supply and a current limiting resistor.

The flasher LED can even use an ac power source by placing a diode in parallel across the LED. This is illustrated in Fig. 19-25. We can speed up the flash rate by adding a capacitor in parallel across the dropping resistor, as shown in Fig. 19-26.

Experiment with various component values to obtain the desired flash rate. Capacitance values should be kept in the 500 to 3000 μF range. Flash rates above about 10 to 12 Hz tend to blend together to the human eye. The LED will appear to be continuously lit.

Fig. 19-25. A flasher LED can be powered by ac if a diode with reverse polarity is placed in parallel with the flasher LED.

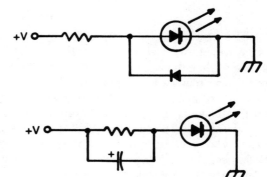

Fig. 19-26. The blink rate of a flasher LED can be speeded up by adding a capacitor in parallel with the current-limiting resistor.

Figure 19-27 shows a dual flasher circuit. LED 1 is a flasher LED, and LED 2 is a standard LED, perhaps of a contrasting color. When the components are wired as shown here, the second LED will be under the control of the flasher IC. When LED 1 is lit, LED 2 will be dark, and vice versa. This can make for an extremely eye-catching display. Typical component values for this circuit are as follows;

$$R1 = 680 \text{ ohms}$$
$$R2 = 680 \text{ ohms}$$
$$C1 = 47 \ \mu F$$
$$C2 = 250 \ \mu F$$

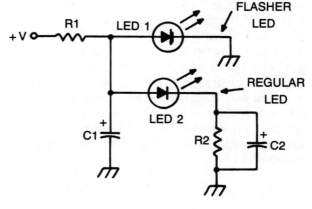

Fig. 19-27. Here is an alternate flasher circuit. When the flasher LED goes off, the other (standard type) LED is lit up.

The flash rate will be determined by the supply voltage. The useful range runs from about +3 to +6 volts. For a +6 volt supply, the LEDs will flash at the flasher's nominal 3-Hz rate. Raising the voltage above +6 volts will cause the flasher LED to stop flashing and stay on. If this condition lasts too long, the IC can be damaged.

Decreasing the supply voltage will cause the flash rate to increase. Around +3 volts the flash rate will exceed the 10 to 12 Hz limit of perception. Below this voltage both of the LEDs will appear to be continuously lit, but dim.

LCDS

LEDs are handy for many indicator and display applications. They are small, inexpensive, sturdy, and easy to use. However, there are disadvantages. They are difficult to see when the ambient light is bright. Also, they tend to eat up a lot of current, especially in circuits with a number of LEDs. This is often a problem in battery-operated circuits where multiple-digit seven-segment displays are used. A more recently developed alternative is the *LCD*, or *Liquid Crystal Display*.

An LCD display panel is an optically transparent "sandwich," usually with an opaque backing. The inner faces of the two panels that make up the sandwich contain a thin metallic film. On one of the panels, the film has been deposited in the form of the desired display, such as the standard seven-segment display, illustrated in Figs. 19-11 through 19-21. Figure 19-28 shows a 3½-digit display with colon for use in digital clocks.

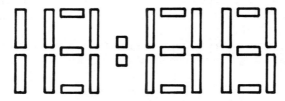

Fig. 19-28. A multiple-digit display is an expansion of the basic seven-segment display system.

Between the two panels is a special fluid called a *nematic liquid*. Ordinarily this fluid is transparent. But when an electrical field is passed between the back metalized panel and one of the metalized segments on the front panel, the liquid between these portions of the panels, will darken and become opaque. The segment will appear as a black mark. When the electrical field is removed, the liquid will become transparent again.

A very small amount of current is required to darken an LCD segment. The segments are easily visible in most ambient lighting conditions, except in dim lighting conditions. LCDs can be designed to display almost anything.

In modern commercial electronic products, LCDs are increasingly replacing LEDs. There are even miniature TV sets using LCD screens to form the images. Early LCD devices were relatively slow, and moving pictures would be extremely blurred. Recent improvements in faster LCDs have made applications such as LCD screen TV sets possible.

Self-Test

1. What voltage can safely be applied to a typical LED?

A *+12 to + 18 volts*
B *−6 to +3 volts*

C *+3 to +6 volts*
D *120 volts*
E *None of the above*

2. When will an LED glow?

A *When power is applied*
B *When forward biased*
C *When reverse biased*
D *When the PIV is exceeded*
E *None of the above*

3. How is current through an LED limited to a safe value?

A *A small value series resistor*
B *A parallel diode*
C *A large value parallel resistor*
D *A capacitor to ground*
E *None of the above*

4. How many segments are needed to display any digit?

A *6*
B *8*
C *7*
D *10*
E *None of the above*

5. What is the purpose of an optoisolator?

A *Display indication*
B *Measurement*
C *Isolation between a controlling and a controlled circuit*
D *Isolation between unrelated circuits*
E *None of the above*

6. What is the normal flash rate of a flasher LED?

A *Once per second*
B *10 to 12 times per second*
C *60 times per second*
D *3 times per second*
E *None of the above*

7. What digit is displayed when all 7 segments of a seven-segment display are lit?

A *7*
B *0*
C *8*
D *9*
E *None of the above*

8. What does "LCD" stand for?

A *Liquid crystal display*
B *Liquid crystal diode*
C *Light crystal display*
D *Light crystal diode*
E *None of the above*

9. Which of the following is *not* an advantage of LEDs?

A *Small size*
B *Visible in dark environments*
C *Sturdy*
D *Visibility not affected by bright ambient light*
E *None of the above*

10. Which segments in a seven-segment display need to be lit to create the digit "4"?

A *A, B, F, and G*
B *B, C, F, and G*
C *A, B, C, and D*
D *B, C, E, F, and G*
E *None of the above*

20

Transistors

Since there is a semiconductor equivalent to the vacuum tube diode, it would be reasonable to ask, is there a semiconductor equivalent to the vacuum tube triode? The answer is—yes and no.

A *transistor* is a three terminal semiconductor device that can perform most of the functions of a triode tube, but has some very different properties of its own.

There are a number of different types of transistor. The simplest, and most common type is the *bipolar transistor*. As the name implies, this is a device with two pn junctions.

NPN TRANSISTORS

Figure 20-1 shows the basic structure of one kind of bipolar transistor. You can see that this device consists of a thin slice of P-type semiconductor material sandwiched between two thicker slabs of N-type semiconductor material. Leads are brought out from each of these semiconductor sections.

One of the N-type sections is identified as the *emitter*, and the other N-type section is called the *collector*. The center P-type section is called the *base*. These terms will be explained shortly.

For obvious reasons, a transistor built according to this model is called a *npn transistor*. Later in this chapter we will also discuss its mirror image, the *pnp transistor*.

The entire semiconductor sandwich is enclosed in a protective plastic or metal case. When the case is metal, one of the leads is often electrically connected to the case. Most frequently, this is the collector, but there are exceptions. When in doubt, check the

247

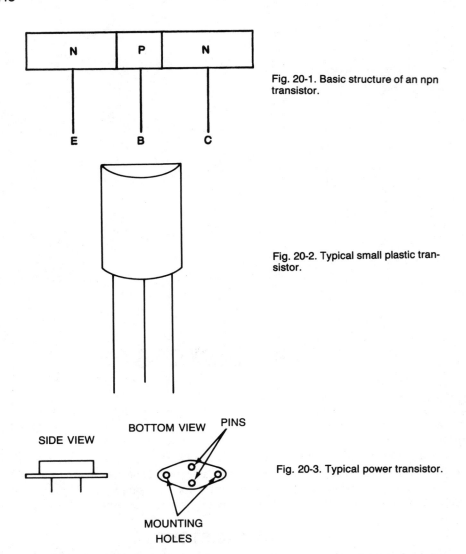

Fig. 20-1. Basic structure of an npn transistor.

Fig. 20-2. Typical small plastic transistor.

Fig. 20-3. Typical power transistor.

manufacturer's data sheet or use an ohmmeter to test for continuity (zero resistance) between each of the leads and the case.

Figures 20-2 and 20-3 show some typical transistors. The device shown in Fig. 20-2 is intended for use under fairly low wattage conditions. Figure 20-3 shows a *power transistor*, which is designed to safely handle a moderately large wattage. Notice that one of the power transistors has only two leads. The third connection (the collector) is made directly to the case.

Power transistors can get very hot in operation, and this self-generated heat can damage the semiconductor material. To prevent this, power transistors are usually mounted on *heatsinks*. A heatsink is simply a piece of metal that conducts heat away from the component and dissipates it into the air. Many heatsinks are finned for maximum surface to air contact area. If the transistor is in a metal case, it is usually necessary to insulate it with a sheet of mica to prevent a short circuit. For maximum heat transfer, the transistor is often smeared with *silicon grease*.

The schematic symbol for a npn transistor is shown in Fig. 20-4. The lead marked "E" is the emitter, "B" is the base, and "C" is the collector. (The emitter and the collector are usually doped somewhat differently, so they are rarely electrically interchangeable.) Some schematics will have the leads marked in this manner, but usually it is assumed that you can tell from the symbol which lead is which. The lead with the arrow is always the emitter.

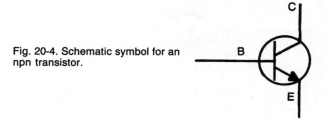

Fig. 20-4. Schematic symbol for an npn transistor.

Unfortunately, there isn't much standardization of lead positions on the actual transistors. See Fig. 20-5. For this reason, every technician and hobbyist should have a good transistor specification book that identifies the leads on various transistors.

BOTTOM VIEWS

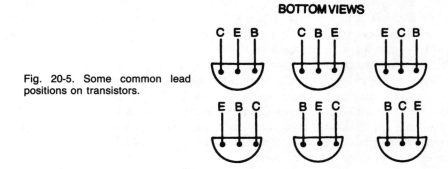

Fig. 20-5. Some common lead positions on transistors.

Such a specification book is also usually a substitution guide for transistors. Literally thousands of different transistors have been manufactured all over the world over the years, and many are very difficult, if not impossible, to locate. Fortunately, many transistor types are interchangeable (at least in most circuits), so you can often substitute one type for another. However, you should bear in mind that some circuits are extremely fussy and will work with only one specific type of transistor. Such special requirements will usually be noted on the schematic or parts list. A good transistor substitution guide is an absolute necessity for anyone working in electronics, whether professionally, or as a hobby.

Returning to the schematic symbol of the npn transistor in Fig. 20-4. Notice that the arrow on the emitter points outward. This is what identifies the transistor as a npn unit. To help you remember this, you can think, "NPN Never Points iN."

How a NPN Transistor Works

Figure 20-6 shows the basic electrical connections for normal operation of a npn transistor. Notice that there are two voltage sources. This is for convenience in our discussion. Later, you'll learn how these two voltages can be obtained from a single power source.

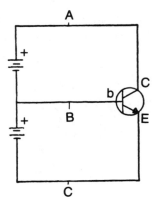

Fig. 20-6. Biasing an npn transistor.

Take careful notice of the polarities within this circuit. The base is more positive than the emitter, but more negative than the collector. The actual voltage applied to the base (measured from the common ground) may be either positive or negative, but the polarity relationships between the transistor's leads always follow this pattern.

You'll recall that a N-type semiconductor has extra electrons, and a P-type semiconductor has extra holes (spaces for electrons). Since the P-type section in a npn transistor is much smaller than either of the N-type sections, it has fewer holes than they have spare electrons.

The negative charge from the terminal connected to the emitter, forces the spare electrons in the emitter section towards the base region. The base-emitter pn junction is forward biased, so the electrons can cross into the base, filling the holes. But there are too many electrons and not enough holes. Because the base section now has more electrons than in its normal state, it acquires an over-all negative charge that forces the extra electrons out of the base region.

Some electrons will leave through the base lead to the positive terminal of the base-emitter battery. The base lead is kept positive with respect to the emitter. But the collector lead is even more positive, drawing the extra electrons out of the collector section, leaving it with a strong positive charge. This will pull most of the electrons out of the base section, and into the collector section, where they are drained off into the positive terminal of the base-collector battery. In other words, the emitter emits electrons, and the collector collects them.

About 95% of the current flow will pass through the collector, while only about 5% will leave the transistor via the base lead. Imagine that there are milliammeters at the points labeled A, B, and C in Fig. 20-6. If the current drawn by the emitter (meter C) is 10 mA (0.01 ampere), then meter A (collector current) will read 9.5 mA (0.0095 ampere), and meter B (base current) will show only 0.5 mA (0.0005 ampere) passing through it.

Just how much current is drawn by the emitter is determined by the characteristics of the specific transistor being used, and the level of the voltage applied to the base terminal.

We can adapt the basic circuit of Fig. 20-6 to the circuit shown in Fig. 20-7. The setting of potentiometer R1 will determine the voltage to the base, which will, in turn, determine the current drawn by the transistor. Regardless of the amount of current drawn, only about 5% will flow through the base lead, and the remaining 95% will flow out of the collector lead, and through the load resistance, R2.

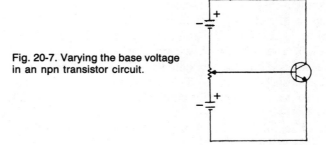

Fig. 20-7. Varying the base voltage in an npn transistor circuit.

It can be seen that a very small change in the base current will result in a very large change in the collector current. For this reason, transistors are sometimes called *current amplifiers*. They amplify current rather than power. Of course, thanks to Ohm's law, the net effect is basically the same, since varying the current through the load resistor (R2) will vary the voltage dropped across it.

Figure 20-8 shows a more advanced version of this circuit. Notice that there is only one battery in this version. The combination of resistors R1, R2 and R3 is called a *volt-*

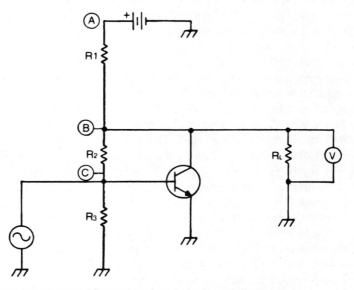

Fig. 20-8. Common-emitter amplifier powered by a single voltage source.

age divider. To see how this works, let's assume the battery generates 9 volts, and all three resistors are of identical value. That is, each resistor drops one third of the voltage, or three volts (ignoring, of course, the effects of the other components in the circuit, such as the internal resistance of the transistor itself).

Measuring between point A and ground, we naturally have the full 9 volts. At point B (the collector lead) R1 is within the section being measured, so the three volt drop across it is subtracted. At point B we would measure 6 volts above the ground. This is the voltage seen by the collector. The base, however, is connected to point C. There are two resistors between this point and the source voltage, so, subtracting the two 3 volt drops, we find that the voltage applied to the base is only 3 volts.

R3 subtracts another 3 volts from the voltage allowed to reach the emitter. In other words, the emitter is at 0 volts (ground), and the base is positive with respect to the emitter (+3 volts). The collector, however, is even more positive— +6 volts.

If we connected a voltmeter between point C (negative lead) and point B (positive lead), we'd get a reading of +3 volts. This means the base, while it has a positive voltage with respect to ground, is negative with respect to the collector. This is the correct bias for a npn transistor.

Thus, you can see that we can obtain all of the required polarities in the transistor circuit with a single voltage source.

Now, if a very small ac voltage source is also applied to the base (as shown in Fig. 20-8) the voltage on the base will vary above and below its nominal dc value. This causes the collector current, and thus the voltage dropped across the load resistance (R_L) to vary in step with the varying voltage applied to the base. Because of the current gain through the transistor, the ac voltage across R_L will be much larger than the original ac voltage applied to the base. That is, the signal is amplified.

BASIC TRANSISTOR AMPLIFIER CONFIGURATIONS

Because the emitter in Fig. 20-8 is at common ground potential, this type of amplifier circuit is called a *common-emitter amplifier*. The emitter is used as the common reference point for both the input and output signals.

In most practical common-emitter circuits there will be a resistor between the emitter and the actual common ground point. The resistor is included to improve stability of the circuit. The emitter is still considered to be grounded.

The common emitter amplifier configuration exhibits a low input impedance and a high output impedance. Current, voltage, and power gain are all high. The output will always be 180° out of phase with the input. That is, when the input signal goes positive (above the dc bias level), the output signal will go negative, and vice versa. The common-emitter amplifier is probably the most commonly used configuration, but there are others.

Figure 20-9 shows a *common-base amplifier* circuit. Notice that the polarity relationships between the transistor leads remain the same. The base is positive with respect to the emitter, but negative with respect to the collector. In other words, the emitter is at a negative voltage (below common ground), and the collector is at a positive voltage (above common ground). The base is grounded, so its nominal value is 0 volts. If you are having trouble visualizing what is happening here, the voltage drop across R2 causes the voltage on the emitter to be below ground potential. That is negative.

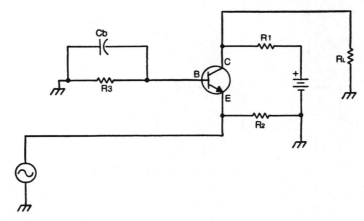

Fig. 20-9. Common-base amplifier.

R3 and C_b which are between the base and the actual ground point, are for stability. Their values are quite small to keep the voltage drop across them negligible. For all intents and purposes, the voltage applied to the base is zero.

The power gain (current gain times voltage gain) of a common-emitter amplifier is slightly lower than that of a common-base amplifier using the same transistor, but its voltage gain is much higher.

Another important difference between these circuit configurations is their input and output impedances. Remember, power is transferred between circuits most efficiently if their impedances match.

As already mentioned, the input impedance of a common-emitter amplifier is fairly low (typically between about 200 and 1000 ohms) and the output impedance is fairly high (typically between about 10,000 to 100,000 ohms).

The impedances of a common-base amplifier are similar, but the difference between the input and the output are much more dramatic. The input impedance of a common-base amplifier is generally below 100 ohms, and the output impedance can be up to several hundred kilohms (1 kilohm is 1000 ohms).

Another difference between these circuit configurations is that the output signal of a common-base amplifier is in phase with its input signal. Remember, a common-emitter inverts the signal (throws it 180° out of phase).

The third transistor amplifier configuration is rather unique. As you may have guessed, this is the *common-collector amplifier.* See Fig. 20-10. Notice that this circuit employs a positive ground point—the operating voltages within the circuit are all negative.

The emitter is the most negative, and R1 drops some of the negative voltage so that the base is less negative (more positive) than the emitter, but it is still more negative than the collector. Thus, the relative polarity requirements are still met.

One of the unique features of the common-collector amplifier configuration is that the voltage gain is always negative. That is, the output voltage is less than the input voltage. Another way of saying this is that the voltage gain is less than *unity*. There is, however, some positive power gain (voltage gain × current gain) in this type of circuit, but it is relatively small compared to the power gains of the common-base and common-emitter configurations.

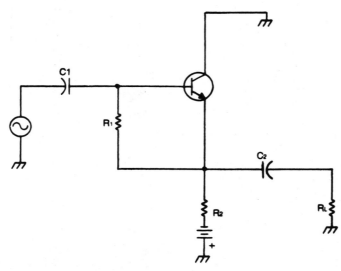

Fig. 20-10. Common-collector amplifier.

**Table 20-1. Comparing the three
basic transistor amplifier configurations.**

COMMON ELEMENT	BASE	EMITTER	COLLECTOR
INPUT IMPEDANCE	VERY LOW	LOW	MEDIUM-HIGH
OUTPUT IMPEDANCE	VERY HIGH	HIGH	LOW
CURRENT GAIN	NEGATIVE	HIGH	HIGH
VOLTAGE GAIN	MEDIUM	HIGH	NEGATIVE
POWER GAIN	HIGH	HIGH	LOW
OUTPUT IN PHASE WITH INPUT?	YES	NO	YES

For obvious reasons, the common-collector circuit doesn't make a very good amplifier, as such. But it is quite a useful circuit for impedance matching. With the other configurations the input impedance is always lower than the output impedance. In the common-collector circuit, this is reversed. The output of a common-collector amplifier is in phase with its input signal.

The differences between these three basic circuit configurations are summarized in Table 20-1.

ALPHA AND BETA

There are literally thousands of different types of bipolar transistors available. They differ in a number of factors. For example, two transistors might differ in the maximum amount of power they can safely dissipate, their internal impedances, and most importantly, how much current gain they can produce.

The current gain of any specific transistor is defined by two interrelated specifications. These are α*(alpha)*, and β*(beta)*.

Alpha is the current gain between the emitter and the collector. That is, for any given change in the emitter current (with the supply voltage held constant), the collector will change with a fixed relationship to the emitter. The basic equation for determining alpha is:

$$\alpha = \frac{\Delta I_c}{\Delta I_e} \qquad \qquad \textbf{Equation 20-1}$$

The symbol Δ is read as *delta*. It is used to identify a changing value. I_c is the collector current, and I_e is the emitter current.

For a typical transistor a 2.6 mA (0.0026 ampere) change in the emitter current would result in 2.4 mA (0.0024 ampere) change in the collector current. Notice that the collector current changes less than the emitter current does. This is because there is always a negative current gain from the emitter to the collector in a bipolar transistor. Remember, only about 95% of the emitter current gets through to the collector.

For our example, $\alpha = \Delta I_c / \Delta I_e = 2.4/2.6 =$ approximately 0.92.

Alpha will always be less than unity (one). A small change in the base current, however, results in a large change in the collector current. 5% of the total current through a transistor flows through the base lead, while the remaining 95% flows through the collector.

For our sample transistor, the same 2.4 mA (0.0024 ampere) current change in the collector can be achieved with a mere 0.2 mA (0.0002 ampere) current change in the base circuit. The ratio between base current and collector current is β, or beta. The formula for beta is:

$$\beta = \frac{\Delta I_c}{\Delta I_b} \qquad \qquad \textbf{Equation 20-2}$$

So beta for our sample transistor equals 2.4/0.2, or about 12. Beta is always greater than one. Alpha and beta are closely interrelated. If you know one, you can calculate the other with the following formulas:

$$\alpha = \frac{\beta}{1 + \beta} \qquad \qquad \textbf{Equation 20-3}$$

$$\beta = \frac{\alpha}{1 - \alpha} \qquad \qquad \textbf{Equation 20-4}$$

For example, we just found the beta for our sample transistor was 12, so we could calculate the alpha as 12/(12 + 1) = 12/13 = about 0.92. Of course, this is the same value we found when we figured the alpha directly from the emitter and collector currents.

Let's suppose we have another transistor with an alpha of 0.88. The beta would be equal to $\alpha/(1 - \alpha) = 0.88/(1 - 0.88) = 0.88/0.12$ or just slightly over 7.

On the other hand, a transistor with an alpha of 0.97 would have a beta of $0.97/(1 - 0.97) = 0.97/0.03 =$ approximately 32.

You can see that alpha and beta increase together. If one is increased, the other will also be increased.

PNP TRANSISTORS

The other basic type of bipolar transistor is the *pnp transistor*. This device is a direct mirror image of the npn type.

The basic structure of a pnp transistor is shown in Fig. 20-11. Its schematic symbol is shown in Fig. 20-12. Notice that the arrow on the emitter points in for a pnp transistor. A handy memory aid is "Points iN Perpetually."

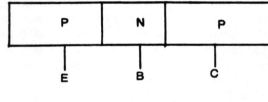

Fig. 20-11. Basic structure of a pnp transistor.

Fig. 20-12. Schematic symbol for a pnp transistor.

If we think in terms of the flow of holes rather than the flow of electrons, a pnp transistor works exactly like a npn transistor, except all of the polarities are reversed. That is, the base must be negative with respect to the emitter, but positive with respect to the collector.

Holes are forced out of the emitter, through the base, and into the collector.

Pnp transistors can be used in any of the basic amplifier configurations (common-base, common-emitter, or common-collector), and alpha and beta are calculated in the same manner as with npn transistors. The input/output phase relationships also remain the same.

Self-Test

1. What is the most basic type of transistor?

A *Npn*
B *Bipolar*
C *Triode*
D *Emitter*

2. What are the three leads to a standard transistor called?

A *Emitter, base, and collector*
B *Emitter, anode, and cathode*
C *Emitter, anode, and base*
D *Base, collector, and source*
E *None of the above*

3. What can be done to ensure maximum heat transfer from a transistor?

A *Ground the transistor's case*
B *Ground the collector*
C *Keep the transistor moist*
D *Smear the transistor's surface with silicon grease*
E *None of the above*

4. Which of the following is *not* a standard transistor circuit type?

A *Common-emitter*
B *Common-cathode*
C *Common-collector*
D *Common-base*
E *None of the above*

5. How much of the current through a transistor will flow through the collector?

A *5%*
B *63%*
C *95%*
D *100%*
E *None of the above*

6. What is the phase relationship between the input and output of a common-emitter amplifier?

A *In-phase*
B *90° Out-of-phase*
C *180° Out-of-phase*
D *360° Out-of-phase*
E *None of the above*

7. What is the input impedance in a common-base amplifier?

A *Very low*
B *Medium*

C *High*
D *Very high*

8. Which is larger, alpha or beta?

A *Alpha*
B *Beta*
C *Either*
D *Neither—the values are equal*

9. Which of the following formulas is correct for determining the beta of a transistor from its alpha?

A $\beta = \dfrac{1}{\alpha}$

B $\beta = 1 - \alpha$

C $\beta = \dfrac{\alpha}{1 + \alpha}$

D $\beta = \dfrac{\alpha}{1 - \alpha}$

E *None of the above*

10. What is the biggest difference between a npn transistor circuit and a pnp transistor circuit?

A *The emitter and collector are reversed*
B *All polarities are reversed*
C *Input and output impedances are reversed*
D *Resistance values must be recalculated*
E *None of the above*

21

Special-Purpose Transistors

While the bipolar transistor is probably the most commonly used type, there are a number of other kinds of transistors that are designed for various special purposes.

DARLINGTON TRANSISTORS

Bipolar transistors can often become quite unstable if a high output current is required of them. A more stable higher gain can be achieved by connecting two bipolar transistors in series, as shown in Fig. 21-1. The emitter of transistor Q1 (transistors are often identified as "Q" in schematic diagrams) is connected to the base of transistor Q2. Both collectors are tied together. When transistors are connected in this manner, they are called a *Darlington pair*.

A Darlington pair can be used in most circuits almost as if they were a single "super" transistor. The current at the emitter of Q2 is virtually the same as the current at the collectors. This allows for excellent balance. The transistors in a Darlington pair must be very closely matched for the best performance.

Often a specific *Darlington transistor* is used. This is essentially two identical transistors within a single housing. Externally, such a device looks just like a regular transistor. To indicate that it is a single unit rather than two *discrete* (separate) transistors, a ring is usually drawn around the schematic symbol. See Fig. 21-2.

UNIJUNCTION TRANSISTORS

So far we have been discussing bipolar transistors. That is, transistors with two complete diode junctions.

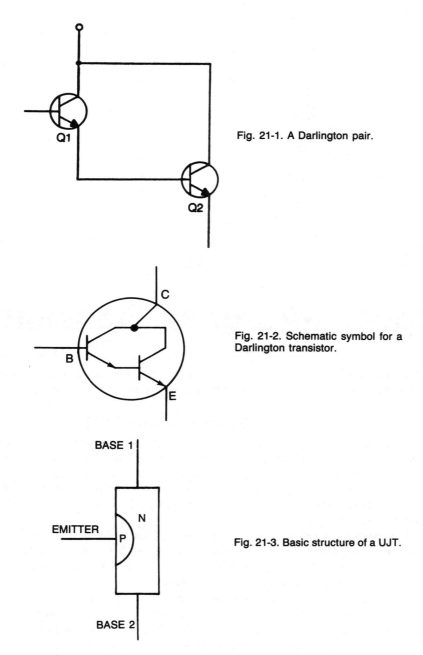

Fig. 21-1. A Darlington pair.

Fig. 21-2. Schematic symbol for a Darlington transistor.

Fig. 21-3. Basic structure of a UJT.

A *unijunction transistor* (UJT), as the name implies, has only a single pn junction. The internal structure of this device is shown in Fig. 21-3. The schematic symbol is given in Fig. 21-4.

Notice that there are three leads—an emitter (the P-type section) and two base connections on either end of the length of N-type semiconductor material.

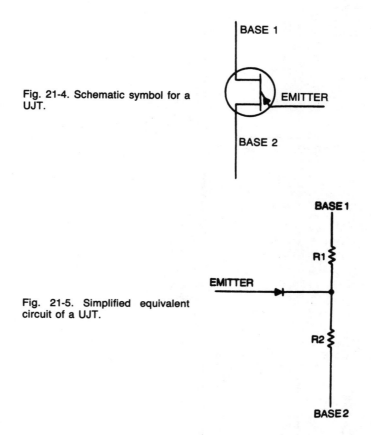

Fig. 21-4. Schematic symbol for a UJT.

Fig. 21-5. Simplified equivalent circuit of a UJT.

Electrically, the N-type section acts like a voltage divider, with a diode (the pn junction) connected to the common ends of the two resistances. Figure 21-5 shows a roughly equivalent circuit for a unijunction transistor.

A voltage applied between Base 1 and Base 2 reverse biases the diode. Of course, this means no current will flow from the emitter to either base.

Now, let's suppose there is an additional variable voltage source connected between the emitter and Base 1 (see Fig. 21-6). As this emitter-Base 1 voltage is increased, a point will be reached when the diode becomes forward biased. Beyond this point current can flow between the emitter and the bases.

Unijunction transistors are most often used in *oscillator* circuits. These will be discussed in a later chapter.

FETS

It was mentioned in the last chapter that the operation of a bipolar transistor doesn't quite directly correspond to that of a triode vacuum tube, so they are not suitable for some circuits.

The *field-effect transistor* (FET) is a semiconductor device that more closely resembles a vacuum tube in operation.

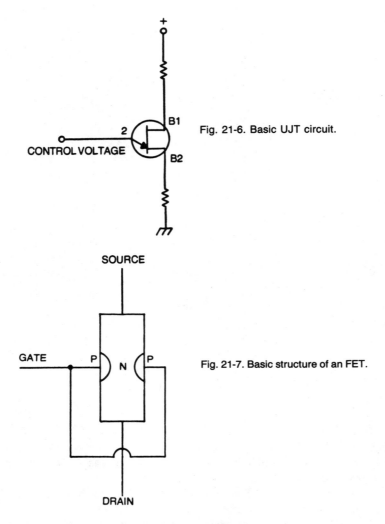

Fig. 21-6. Basic UJT circuit.

Fig. 21-7. Basic structure of an FET.

The basic structure of an FET is illustrated in Fig. 21-7. Like the unijunction transistor, its body is a single, continuous length of N-type semiconductor material. But in an FET there is a small section of P-type material placed on either side of the N-type section. Both P-type sections are electrically tied together internally.

The lead connected to the two P-type sections is called the *gate*. Additional leads, called the *source* and the *drain*, are connected to either end of the N-type material. The schematic symbol for an FET is shown in Fig. 21-8.

To get a general idea of how an FET works, consider the mechanical system illustrated in Fig. 21-9. When the valve (gate) is opened, as in Fig. 21-9A, water can flow through the pipe (from source to drain). If, on the other hand, the valve is partially closed, as in Fig. 21-9B, the amount of water that can flow through the pipe is limited. Less water comes out of the drain.

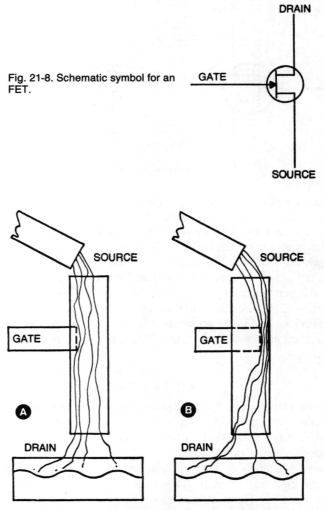

Fig. 21-8. Schematic symbol for an FET.

Fig. 21-9. Mechanical equivalent of an FET: A. open valve; B. partially closed valve.

Quite similarly, the gate terminal of a field-effect transistor controls the amount of electric current that can flow from the source to the drain. See Fig. 21-10.

A negative voltage applied to the gate lead reverse biases the pn junction, producing an *electrostatic field* (electrically charged region) within the N-type material. This electrostatic field opposes the flow of electrons through the N-type section, acting somewhat like the partially closed valve in our mechanical model. The higher the negative voltage applied to the gate, the less current that is allowed to pass through the device from source to drain.

Notice how this is directly analogous to the action of a vacuum tube. The gate, of course, corresponds to the grid, controlling the amount of current flow. The source is the equivalent to the cathode (it acts as the source of the electron stream) and the drain

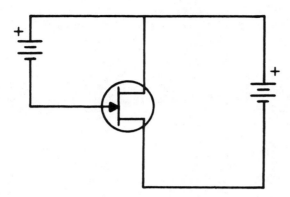

Fig. 21-10. Biasing an FET.

serves the same function as the plate (it drains off the electrons from the device). The path from the source to the drain is sometimes called the *channel*.

FETs can also be made with a P-type channel and an N-type gate. These devices work in the same way if we think of a positive voltage on the gate opposing the flow of holes through the channel. The schematic symbol for a P-type FET is the same as for an N-type unit, but the direction of the arrow on the gate lead is reversed.

FETs have a very high input impedance, so they draw very little current (since I = E/Z, increasing Z will decrease I). This means they can be used in highly sensitive measuring applications, and in circuits where it is important to avoid loading down (drawing heavy currents from) previous circuit stages. This is another way in which FETs resemble vacuum tubes.

MOSFETS

The field-effect transistors discussed in the last section are sometimes called *junction field-effect transistors*, or *JFETs*. Another type of FET does not have an actual pn junction. These devices are called *insulated gate field-effect transistors* (*IGFETs*) because the gate is insulated from the channel.

Most commonly, this is done by using a thin slice of metal as the gate (rather than a piece of semiconductor crystal). This metal is oxidized on the side against the semiconductor channel. This insulates the gate because metal oxide is a very poor conductor. When a metal oxide gate is used, the device is often called a *metal-oxide-silicon field-effect transistor*, or *MOSFET*.

The semiconductor channel is backed by a *substrate* of the opposite type of semiconductor. If, for example, the device is built around an N-type channel, it will be backed by a P-type substrate.

The basic structure of an N-type MOSFET is shown in Fig. 21-11, and the schematic symbols are given in Fig. 21-12.

Even though the gate is physically insulated from the channel, it can still induce an electrostatic field into the semiconductor to limit current flow. The substrate is generally kept at the same voltage as the source. The basic MOSFET biasing circuit is illustrated in Fig. 21-13.

Fig. 21-11. Basic structure of an N-type MOSFET.

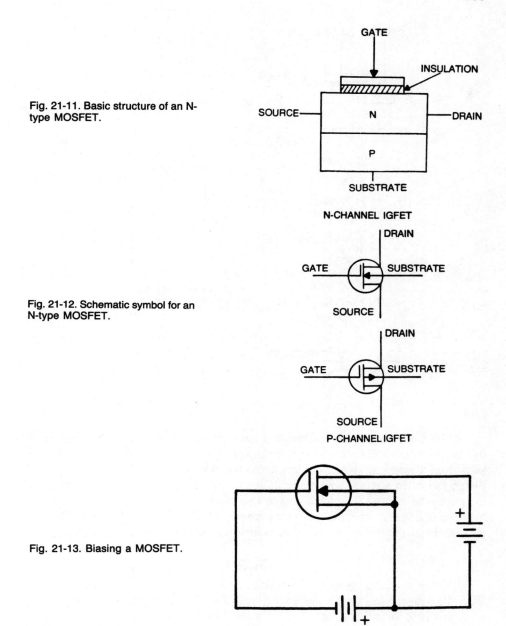

Fig. 21-12. Schematic symbol for an N-type MOSFET.

Fig. 21-13. Biasing a MOSFET.

ENHANCEMENT MODE FETS

So far we have been talking about *depletion mode FETs*. These devices reduce current flow by increasing the negative voltage on the gate (assuming an N-type channel).

Another type of FET operates in the *enhancement mode*. In this type of FET the source and the drain are not parts of a continuous piece of semiconductor material. The basic structure of this component is shown in Fig. 21-14.

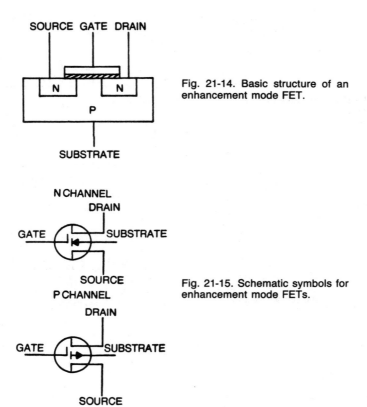

SOURCE GATE DRAIN

N   N

P

SUBSTRATE

Fig. 21-14. Basic structure of an enhancement mode FET.

N CHANNEL
DRAIN
GATE   SUBSTRATE
SOURCE
P CHANNEL
DRAIN
GATE   SUBSTRATE
SOURCE

Fig. 21-15. Schematic symbols for enhancement mode FETs.

A positive voltage is applied between the gate and the source. The higher this voltage is, the greater the number of holes drawn from the N-type source into the P-type substrate. These holes are then drawn into the drain region by the voltage applied between the drain and the source. In other words, increasing the voltage on the gate increases the current flow from source to drain.

The schematic symbols for enhancement mode FETs are shown in Fig. 21-15. Enhancement mode devices are always insulated gate field-effect transistors.

SCRS

Another special purpose semiconductor device is the *silicon controlled rectifier*, or *SCR*. Its schematic symbol is shown in Fig. 21-16. Notice that this is basically the same as the symbol for a diode, but there is a third lead, called the *gate*.

If a voltage is applied between the cathode and the anode (see Fig. 21-17A), but the gate is at zero volts (grounded), no current will flow through the SCR.

Now, if a voltage greater than some specific value (which varies from unit to unit) is applied to the gate (see Fig. 21-17B), current will start to flow from cathode to anode against only a small internal resistance (as with an ordinary diode).

This current will continue to flow—even if the voltage on the gate is removed. The only way to stop the current flow through an SCR once it's started is to decrease the

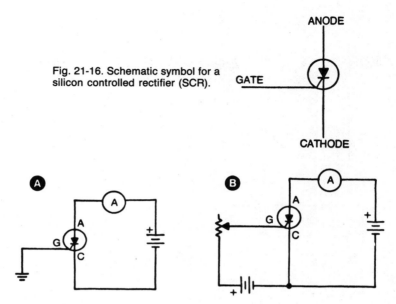

Fig. 21-16. Schematic symbol for a silicon controlled rectifier (SCR).

ANODE

GATE

CATHODE

Fig. 21-17. Basic SCR circuits.

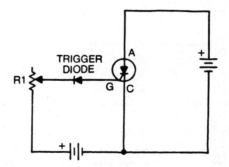

Fig. 21-18. Basic SCR circuit with a trigger diode.

positive voltage on the anode (or remove it altogether). When the anode voltage drops below a pre-determined level, current flow will be blocked, even if the anode voltage returns to its original value, unless the gate receives the required *triggering* voltage.

A special type of diode, called a *trigger diode* is usually connected in series with the gate (see Fig. 21-18). This diode is ordinarily reverse biased, and exhibits a sharp pulse when its trigger voltage is exceeded. This is shown graphically in Fig. 21-19. The trigger diode can provide cleaner, and more reliable triggering of the SCR.

Silicon controlled rectifiers are typically used to control ac voltages. Figure 21-20 shows a simplified circuit of this type.

When the voltage applied to the trigger diode reaches a specific point in the cycle, it turns the SCR on with a sharp voltage pulse. When the applied voltage to the anode drops below the holding level of the SCR, the current is cut off until the cycle is repeated.

268

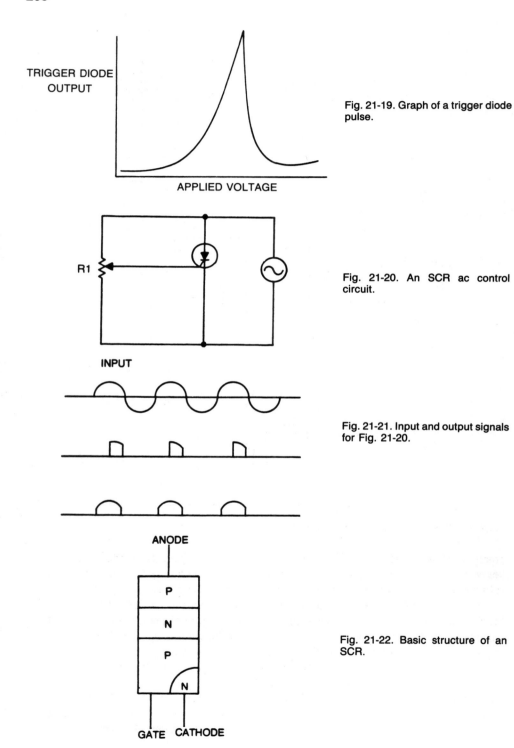

TRIGGER DIODE
OUTPUT

APPLIED VOLTAGE

Fig. 21-19. Graph of a trigger diode
pulse.

R1

Fig. 21-20. An SCR ac control
circuit.

INPUT

Fig. 21-21. Input and output signals
for Fig. 21-20.

ANODE

| P |
| N |
| P |
| N |

Fig. 21-22. Basic structure of an
SCR.

GATE CATHODE

Figure 21-21 illustrates the input signal and several possible output signals. Varying R1 in Fig. 21-20 will alter the voltage to the trigger diode, and thus, the point in the cycle when the SCR is turned on.

Because part of the ac cycle is cut out, and the voltage stays at zero during those times, the average value of the output signal must be lower than that of the input signal. The less time the current is allowed to flow through the SCR during each cycle, the lower the effective value of the output voltage. Figure 21-22 shows the basic structure of a silicon controlled rectifier.

DIACS AND TRIACS

Trigger diodes and SCRs are single polarity devices. They can work only during one half of each ac cycle. Of course, this means the energy in the other half cycle is wasted.

A *diac* is a dual trigger diode that will produce an output pulse on each half cycle. It is effectively two separate diodes that are internally connected as shown in Fig. 21-23. The schematic symbol for a diac is shown in Fig. 21-24.

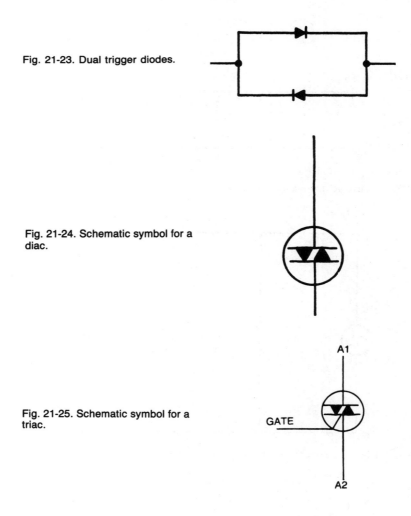

Fig. 21-23. Dual trigger diodes.

Fig. 21-24. Schematic symbol for a diac.

Fig. 21-25. Schematic symbol for a triac.

A1

GATE

A2

Similarly, a *triac* is a dual SCR. It acts like two separate SCRs that are connected in parallel and in opposite directions, so current of either polarity can flow through the device. When one of the SCRs is conducting, the other one is cut off. The schematic symbol for a triac is shown in Fig. 21-25.

Figure 21-26 illustrates the operation of a typical diac/triac circuit. The circuit itself is shown in Fig. 21-27. Notice that this is essentially the same circuit that is used with the regular, single polarity trigger diode/SCR combination, except, of course, this circuit is operative during both half cycles. This means less potential power is wasted. This circuit is much more efficient.

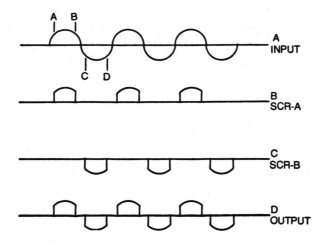

Fig. 21-26. Triac/diac operation.

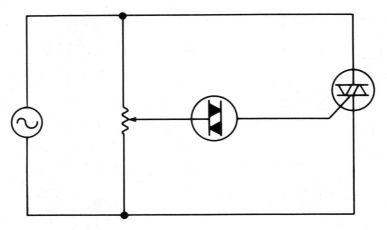

Fig. 21-27. Triac/diac control circuit.

When point A is reached in the cycle, diode A conducts, triggering SCR A. At point B, SCR A is turned back off. See Fig. 21-26.

During the second half cycle, diode A and SCR A are inactive, while diode B and SCR B are employed. At point C in the cycle, SCR B is turned on by the pulse from trigger diode B. Finally, at point D, SCR B is cut off again. Then the entire cycle is repeated.

Figure 21-26D is a graph of the combined output of SCR A and SCR B. This is the signal that will be applied as the power source to the load.

Self-Test

1. What is a Darlington pair?

A *Two bipolar transistors in parallel*
B *Two bipolar transistors in series*
C *An npn transistor and a pnp transistor in series*
D *A symmetrical circuit*
E *None of the above*

2. How are the three leads on a UJT labeled?

A *Emitter, base 1, base 2*
B *Emitter, base, collector*
C *Gate, drain, source*
D *Emitter 1, emitter 2, base*
E *None of the above*

3. What type of circuitry most commonly uses UJTs?

A *Amplifiers*
B *Metering circuits*
C *Oscillators*
D *Filters*
E *None of the above*

4. How many pn junctions are there in an FET?

A *One*
B *Two*
C *Three*
D *It varies*
E *None of the above*

5. Which type of transistor most closely resembles a vacuum tube?

A *Bipolar*
B *UJT*
C *Darlington*
D *FET*
E *None of the above*

6. Which of the following is *not* a variation on the basic FET?

A *JFET*
B *MOSFET*
C *IGFET*
D *EFET*
E *None of the above*

7. What is an SCR?

A *A diode with an electrical switch*
B *A modified FET*
C *Two triacs in a single housing*
D *An enhancement mode UJT*
E *None of the above*

8. What is the primary disadvantage of an ordinary SCR?

A *Slow reaction time*
B *Easily damaged*
C *Wastes half of each ac cycle*
D *Tends to overheat*
E *None of the above*

9. What is a diac?

A *A dual SCR*
B *A dual-polarity trigger diode*
C *A simplified triac*
D *A Darlington pair*
E *None of the above*

10. What are the three connections made to an SCR?

A *Anode, cathode, gate*
B *Emitter, base 1, base 2*
C *Drain, gate, source*
D *Anode 1, anode 2, gate*
E *None of the above*

22

Light Sensitive Devices

Semiconductors have an unusual property that has not been taken advantage of until fairly recently. Semiconductor materials are light sensitive. That is, the amount of light striking them affects their electrical characteristics.

Ordinarily, this photosensitivity is very undesirable in most applications, because the light level is generally uncontrolled. For this reason, most semiconductor components (transistors, diodes, ICs, etc.) are normally enclosed in a light-tight housing made of plastic or metal. This housing also protects the delicate semiconductor crystal from moisture, and stray particles (dust, etc.).

In certain applications, photosensitivity can be highly desirable, or even essential. An obvious example is a light meter. An electronic circuit cannot measure the light level unless it can "see" it.

A number of special semiconductor photosensitive devices (or light sensors) are now available. Some feature PN junctions (like ordinary diodes and transistors), while others are made from a single, junctionless slab of semiconductor crystal. The specific reaction to detected light depends on the construction of the photosensitive component and the specific semiconductor material used.

A junctionless semiconductor light sensor is generally known as a *photocell*. There are two basic types of photocells. They are the *photovoltaic cell* and the *photoresistor*.

PHOTOVOLTAIC CELLS

A photovoltaic cell is a simple slab of semiconductor crystal, usually primarily made up of silicon. The semiconductor material is spread out onto a relatively large, thin plate for the largest possible contact area. The more of the semiconductor material that is exposed to light, the stronger the response to the light will be.

In essence, a photovoltaic cell functions something like a dry cell (battery). This is indicated in the schematic symbol for the device, illustrated in Fig. 22-1. Notice that this symbol is similar to the one used to represent an ordinary voltage cell or battery.

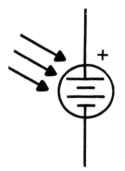

Fig. 22-1. Schematic symbol for a photovoltaic cell.

When the silicon surface is shielded from light, no current will flow through the cell. But when it is exposed to a bright light a small voltage is generated, due to what is known as the *photoelectric effect*.

If an illuminated photovoltaic cell is hooked up to a load, a current will flow through the circuit. See Fig. 22-2. Just how much current will flow is dependent on the amount of light striking the surface of the cell. The brighter the light, the higher the current available.

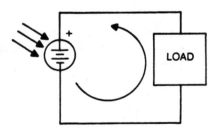

LOAD

Fig. 22-2. Current flowing through a photovoltaic cell.

The cell's output voltage, on the other hand, is relatively independent of the light level. The voltage produced by most photovoltaic cells is in the range of about half a volt.

The most obvious use for a photovoltaic cell is as a substitute for an ordinary dry cell. Of course, the half volt output of a single cell is too low for most applications, so a number of photovoltaic cells can be added in series to form a battery, just as with ordinary dry cells.

If the circuit you want to power from photovoltaic cells requires more power than your photocells can provide, more cells can be added in a parallel battery.

You can power almost any dc circuit with a combination of series and/or parallel connected photocells. This type of grouping of photocells is often called a *solar battery*, but it will work just as well under artificial light.

There's one factor that should be kept in mind—the more cells there are in a solar battery, the larger the total surface area will be, and the harder it will be to arrange the cells so they will all be lighted evenly. This means that usually solar batteries are best suited for fairly low power circuits.

A related use for photovoltaic cells is in a recharger for nickel-cadmium batteries. Figure 22-3 shows a typical arrangement. The diode can be almost any standard type, such as a 1N914. It is needed to protect the photocells. If the diode wasn't included in the circuit, and the cells were inadvertently darkened, the nickel-cadmium battery could start to discharge through them. This reverse polarity could quickly ruin the photocells.

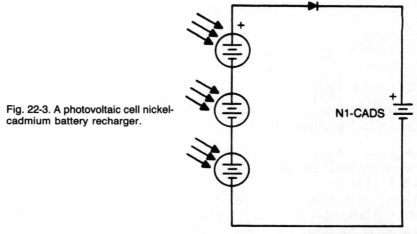

Fig. 22-3. A photovoltaic cell nickel-cadmium battery recharger.

N1-CADS

Bear in mind that photovoltaic cells, like any other dc voltage source have definite polarity. That is, one lead is always positive and the other is always negative. These should never be reversed.

Another frequency application for photovoltaic cells is in light metering circuits. A very simple light meter is shown in Fig. 22-4. Since the current flowing through a photocell is proportional to the amount of light striking its surface, measuring the current with a milliammeter will give a direct indication of the degree of illumination. The meter scale can be calibrated in whatever units are convenient.

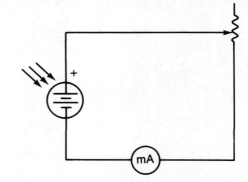

Fig. 22-4. A photovoltaic light meter.

The measurements made with this type of simple circuit won't be very precise, but the circuit is perfectly functional for comparative measurements.

The potentiometer adjusts the sensitivity of the meter. Generally this should be a trimpot that is set once for calibration, then left alone for the measurements. For best results the meter should be as sensitive a unit as possible, but care should be taken that the meter is not so sensitive that a strong light source can slam the pointer off the scale.

A photovoltaic cell can be used to trigger a relay. The basic set-up for this is shown in Fig. 22-5. Notice that the controlled circuit requires a separate power supply. The voltage generated by the photocell only opens or closes the relay contacts.

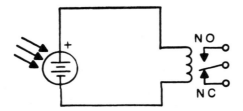

Fig. 22-5. A light operated relay.

Because the output of the photocell is fairly small, it can only drive a relatively light duty relay. If the circuit you want to control requires a heavier relay, there are two ways to solve the problem.

One method is to use the light duty relay to control the heavy duty unit as shown in Fig. 22-6. Alternatively, the output of the photocell can be amplified with a transistor, as illustrated in Fig. 22-7.

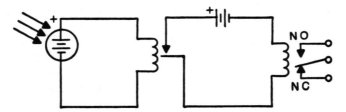

Fig. 22-6. Using a light-duty relay to drive a heavy-duty relay.

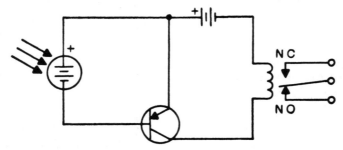

Fig. 22-7. Using a photovoltaic cell to drive a heavy-duty relay with the aid of a transistor amplifier.

The potentiometer is again used to adjust sensitivity. Notice that both of these methods require an extra voltage source in addition to the photocell and the controlled circuit's power supply.

All of these relay circuits respond only to the presence or absence of light. They are not adjustable to trigger on a specific amount of lighting, because a photocell's output voltage is essentially constant.

PHOTORESISTORS

Another popular light sensitive device is the photoresistor, or *light dependent resistor*. As the name implies, a photoresistor changes its resistance value in step with the level of illumination on its surface. Photoresistors generate no voltage themselves. Photoresistors are usually made of cadmium-sulfide, or cadmium-selenide.

Functionally, a photoresistor is a light controlled potentiometer. The light intensity corresponds to the position of the potentiometer shaft.

These devices generally cover quite a broad resistance range—often on the order of 10,000 to 1. Maximum resistance—typically about 1 megohm (1,000,000 ohms) is achieved when the cell is completely darkened. As the light level increases, the resistance decreases.

Since photoresistors are junctionless devices, like regular resistors, they have no fixed polarity. In other words, they can be hooked up in either direction without affecting circuit operation in any way. Figure 22-8 shows the schematic symbol for a photoresistor.

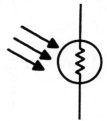

Fig. 22-8. Schematic symbol for a photoresistor.

Photoresistors are perfect for a wide range of electronic control applications. They can easily be used to replace almost any variable resistor in virtually any circuit.

Photoresistors can also be used in many of the same applications as photovoltaic cells, usually with just the addition of a battery and another resistor. They offer the advantage of being sensitive to different light levels. For example Fig. 22-9 shows a light controlled relay circuit. In this circuit, R1 can be adjusted so that the relay switches at any desired light intensity level.

Figure 22-10 shows a simple light meter built around a photoresistor. Photoresistors can also be used in many applications where a photovoltaic cell would not be used. Since a photoresistor is a variable resistance, it can be used in place of almost any standard potentiometer in almost any circuit. If appropriate for the application at hand, a photoresistor can be substituted for a fixed resistor.

OTHER PHOTOSENSITIVE COMPONENTS

Any standard semiconductor device (with one or more pn junctions) can be made photosensitive, simply by placing a transparent lens in the component's housing so the

278

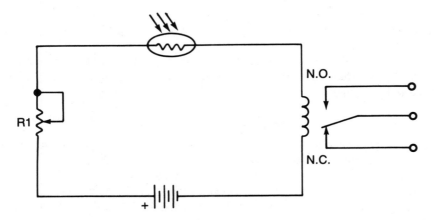

Fig. 22-9. Light activated relay using a photoresistor.

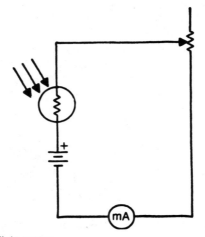

Fig. 22-10. A photoresistor light meter.

semiconductor material is exposed to light. A great variety of light sensitive components are now available. Most of these are light controlled versions of more familiar semiconductor devices.

Figure 22-11 shows the schematic symbol for a photodiode. A phototransistor is illustrated in Fig. 22-12. Figure 22-13 shows a *Light Activated SCR* (Silicon Controlled Rectifier). This last component is commonly known by its acronym, *LASCR*.

Phototransistors are especially useful in a large number of applications because they can be used as amplifiers whose effective gain is controlled by light intensity. Usually (though not always) the base lead is left unconnected in the circuit, the base-collector current being internally generated by the photoelectric effect. Many phototransistors don't even have an external base lead.

Many circuits that use standard bipolar transistors could be adapted and rebuilt with phototransistors. This could result in some very unique effects.

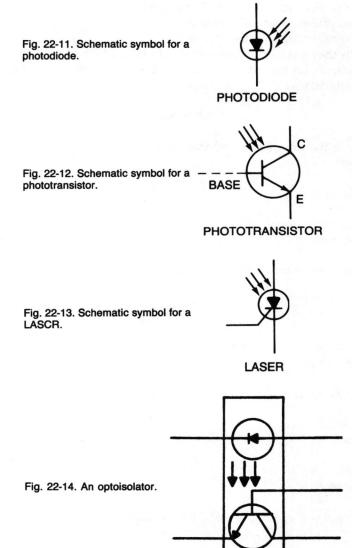

Fig. 22-11. Schematic symbol for a photodiode.

PHOTODIODE

Fig. 22-12. Schematic symbol for a phototransistor.

PHOTOTRANSISTOR

Fig. 22-13. Schematic symbol for a LASCR.

LASER

Fig. 22-14. An optoisolator.

Of course, the base lead could be used too, so the output would depend on both the signal on the base lead, and the light intensity.

An *optoisolator* is another extremely useful device. As the name implies, it isolates two interconnected circuits so that their only connection is optical.

In an earlier chapter we learned how a rudimentary optoisolator can be made from two LEDs. Most practical optoisolators, however, consist of a LED and a phototransistor, or phototransistor encapsulated in a single, light-tight package. The schematic symbols are shown in Fig. 22-14.

The LED is wired into the controlling circuit, and the photocell is wired into the circuit to be controlled. This provides a convenient means of control with virtually no undesirable crosstalk between the two circuits.

Essentially the same effects can be achieved with a separate LED and photoresistor (or phototransistor), but they must be carefully shielded from all external light to prevent uncontrolled interference.

Self-Test

1. What is the name for a component that changes its resistance in response to light intensity?

A *Photovoltaic cell*
B *Photoresistor*
C *Potentiometer*
D *Photopotentiometer*

2. What is the nominal voltage of a photovoltaic cell?

A *0.5 volt*
B *1.0 volt*
C *1.5 volt*
D *Varies in proportion to the size of the cell*
E *None of the above*

3. What is the name for the property used in light sensitive devices?

A *Piezoelectric effect*
B *Solar Energy*
C *LASCR*
D *Photoelectric effect*
E *Photoresistance*

4. Which of the following is *not* a light sensitive transducer?

A *Photodiode*
B *LASCR*
C *Photoresistor*
D *SCR*
E *None of the above*

5. Which type of material is most likely to be photosensitive?

A *Conductors*
B *Semiconductors*
C *Insulators*

6. Which lead is often omitted in a phototransistor?

A *Collector*
B *Emitter*
C *Base*
D *Gate*
E *None of the above*

7. Why are ordinary transistors not sensitive to light?

A *Different materials are used in their construction*
B *They only have 2 pn junctions*
C *They are enclosed in light-tight housings*
D *Different biasing is used*
E *None of the above*

8. How many pn junctions does a photoresistor have?

A *One*
B *Two*
C *Three*
D *None*
E *More than three*

9. What semiconductor material is most often used in photovoltaic cells?

A *Silicon*
B *Cadmium-sulfide*
C *Gallium-arsenic*
D *Cadmium-selenide*
E *None of the above*

10. Which of the following performs a similar function to that of a photovoltaic cell?

A *Capacitor*
B *DC battery*
C *Inductor*
D *LASCR*
E *None of the above*

23

Experiments 2

Table 23-1 lists the equipment and parts you will need to perform the experiments in this chapter.

EXPERIMENT #1 DC RESISTANCE OF A DIODE

Connect a 1N914 diode to an ohmmeter as shown in Fig. 23-1A. The negative ohmmeter lead (usually black) is attached to the end of the diode marked by a colored band. The ohmmeter's positive lead (usually red) is connected to the other end of the diode.

Now read the resistance on the meter. I got about 4500 ohms (4.5 kilohms). You may get a somewhat different resistance reading, but it should be fairly low (under 10,000 ohms). This indicates that the diode is conducting. It is forward biased.

Now, reverse the ohmmeter leads on the diode, as shown in Fig. 23-1B. Now the banded end of the diode should be connected to the positive lead (red) of the ohmmeter, and the unbanded end is connected to the negative lead (black). In this case the meter's pointer should scarcely budge from the full scale, infinity position. The diode is reverse biased.

Remember, a diode has a low internal resistance when it is forward biased, and an extremely high internal resistance when it is reverse biased.

If you got a fairly low resistance reading in both directions, the diode is *shorted*. If a very large, or infinity reading is found in both directions, the diode is *open*. In either case, the component is defective, and therefore, useless. It should be discarded.

Table 23-1. Equipment and Parts Needed for the Experiments.

VOM
(DC voltmeter, AC voltmeter, ohmmeter)

BREADBOARDING SYSTEM
(solderless socket, DC power supply (6 volts and 9 volts), 6.3 volt AC power supply)

EXTRA 3 VOLT POWER SUPPLY
(2,1.5 volt dry cells connected as a series battery)

1   100 ohm resistor (brown-black-brown)
2   1000 ohm resistors (brown-black-red)
2   10,000 ohm resistors (brown-black-orange)
1   100,000 ohm resistor (brown-black-yellow)
1   470,000 ohm resistor (yellow-violet-yellow)
1   10,000 ohm potentiometer
1   1N914 diode (or 1N4148)
1   3.6 volt zener diode (such as the 1N747)
1   Red LED
1   Seven Segment, Common Cathode LED Display
1   2N3904 npn transistor (or Radio Shack RS-2016)
1   2N306 pnp transistor (or Radio Shack RS-2034)
1   2N4350 FET (or Radio Shack RS-2035)

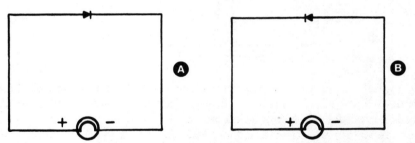

Fig. 23-1. Testing the dc resistance of a diode. Experiment #1: A. forward biased; B. reverse biased.

You can see that this method can be used to check diodes for defects with an ohmmeter. This test can also be used to determine the polarity of an unmarked diode.

EXPERIMENT #2 A DIODE IN A DC CIRCUIT

Breadboard the circuit illustrated in Fig. 23-2A with a 1N914 diode and a 1000 ohm (1 kilohm) resistor (brown-black-red).

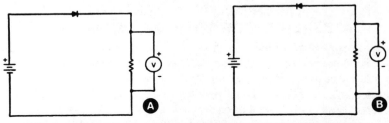

Fig. 23-2. An ac diode circuit. Experiment #2: A. forward biased; B. reverse biased.

The voltmeter should indicate a little over 4 volts (about 0.7 volts is dropped by a typical diode). Clearly, current is flowing through the circuit. The diode is forward biased.

The exact voltage reading you get is not terribly important. The key point is that current is flowing through the circuit. Now, reverse the polarity of the diode so that the circuit resembles Fig. 23-2B. Since the diode is now reverse biased, you should read very little, or no voltage across the resistor.

Virtually no current will flow through a circuit with a reverse biased diode. Some might leak through, giving you a small voltage drop across the diode, but this value should be negligible.

EXPERIMENT #3 A DIODE IN AN AC CIRCUIT

Notice that the circuit in Fig. 23-3 is similar to the one used in the last experiment, except a 6.3 volt ac source is used to power the circuit. You should read about a 3 volt voltage drop across the diode. Reversing the polarity of the diode should give you approximately the same voltage.

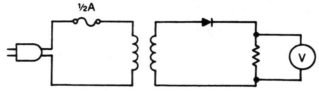

Fig. 23-3. An ac diode circuit. Experiment #3.

No matter which way you connect the diode, current will flow through the circuit for half of each ac cycle. This is because, regardless of the direction of the diode, the polarity is correct for one half cycle of each wave.

Now, repeat the experiment, and measure the *dc* voltage across the 1000 ohm (1 kilohm) resistor. When the diode is connected as shown in Fig. 23-3 you should read about +3 volts dc. This is assuming the negative lead of the ohmmeter (black) is attached to the connection of the diode and the resistor. This means the diode conducts during the positive half of each ac cycle.

If the diode is reversed, however, current will flow only during the negative half cycles. The voltage should be about the same as before, but with reversed polarity. That is, there is about −3 volts dropped across the resistor. With most voltmeters, you'll have to reverse your meter leads to read this voltage.

EXPERIMENT #4 TESTING
A ZENER DIODE WITH AN OHMMETER

Repeat experiment #1 with a 3.6 volt zener diode, such as a 1N747. With the banded end of the diode connected to the negative lead (black) of the ohmmeter, you should get a fairly low resistance. I read about 4,500 ohms (4.5 kilohms).

Reverse biasing the diode should result in a somewhat higher resistance reading, but the difference will probably not be as dramatic as with the 1N914 diode in Experiment #1. I got a reverse bias resistance of about 150,000 ohms (150 kilohms).

The exact reading you get will depend on the voltage source of the ohmmeter you are using.

EXPERIMENT #5 USING A ZENER DIODE

Breadboard the circuit shown in Fig. 23-4. R2 and R3 should each be 1000 ohms (1 kilohm) (brown-brown-red). R1 is a 10,000 ohm (10 kilohm) potentiometer. Observe what happens to the voltage dropped across R3 as the source voltage is varied via the potentiometer.

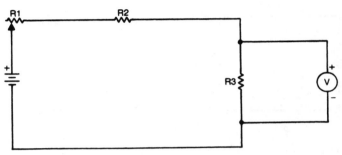

Fig. 23-4. An unregulated circuit. Experiment #5.

Now, add your 3.6 volt zener diode to the circuit, as shown in Fig. 23-5. Again vary the resistance of R1 and watch what happens to the voltage drop across R3. Notice that the voltage never goes beyond the 3.6 volt rating of the zener diode. Any voltage above this level is routed to ground through the diode.

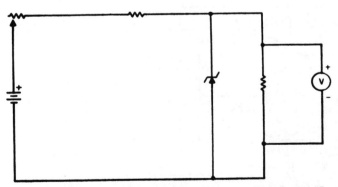

Fig. 23-5. The circuit from Fig. 23-4 with zener diode regulation. Experiment #5.

EXPERIMENT #6 LEDS

Repeat Experiment #1 and Experiment #2, but use a light emitting diode instead of the 1N914 diode.

Notice that the resistance readings for an LED resemble those of a conventional diode, although the resistance of a forward biased LED may be somewhat higher (mine read about 200,000 ohms, or 200 kilohms).

Also, depending on the voltage of your ohmmeter, the LED may or may not light up when it is forward biased. It should not be lit when reverse biased.

For the "in circuit" tests, use the circuit shown in Fig. 23-6. The resistor is essential to prevent excessive current from flowing through the LED, and quite possibly damaging it. Use a 1000 ohm (1 kilohm) resistor (brown-black-red) for this. The source voltage must be no more than six volts.

When the diode is forward biased, it should glow fairly brightly. When the polarity of the LED is reversed (that is, when it is reverse biased) as in Fig. 23-6B, the LED should remain dark.

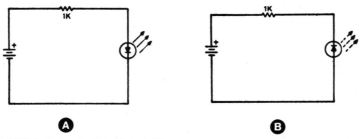

A   **B**

Fig. 23-6. An LED test circuit. Experiment #6.

EXPERIMENT #7 A SEVEN SEGMENT LED DISPLAY

Breadboard the circuit shown in Fig. 23-7 with a common cathode seven segment LED display. Connect each of the segment leads to point A, one at a time, and notice the effect.

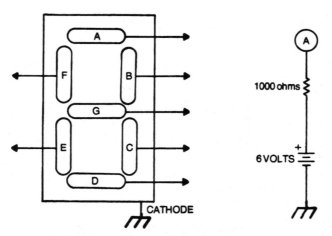

Fig. 23-7. Seven-segment LED display test circuit. Experiment #7.

Connect both segments B and C to point A and see what number is displayed. Now connect segments B, C, F, and G to point A and observe the result.

Next connect segments B, C, E, F, and G (that is, all segments, except A and D) to point A and see what is lit on the display.

In these last three steps you should have seen (in this order) a "1," a "4," and an "H."

Try connecting various other combinations of segments to point A and watch what happens to the display. See how many different numerals and letters you can form.

EXPERIMENT #8 TESTING TRANSISTORS WITH AN OHMMETER

For purposes of testing, a transistor can be thought of as two back to back diodes, as shown in Fig. 23-8.

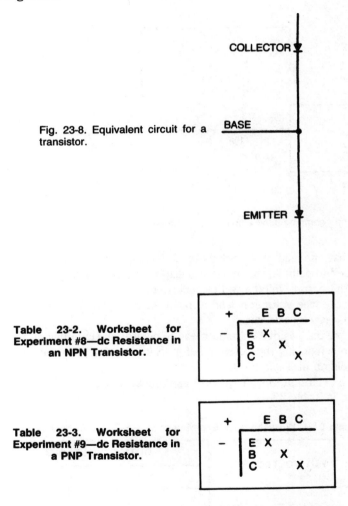

Fig. 23-8. Equivalent circuit for a transistor.

Table 23-2. Worksheet for Experiment #8—dc Resistance in an NPN Transistor.

Table 23-3. Worksheet for Experiment #9—dc Resistance in a PNP Transistor.

Measure the various internal resistances of a 2N3904 npn transistor, and enter your results into Table 23-2. You will be making a total of six measurements—remember, to measure each pair of leads with both polarities.

Now repeat the six measurements on a 2N3906 pnp transistor and enter your results into Table 23-3.

288

Now compare Table 23-2 with Table 23-3. Notice how they are essentially mirror images of each other. The measurements should all be approximately the same, but with the polarities reversed.

EXPERIMENT #9 A COMMON-BASE AMPLIFIER

Construct the circuit shown in FIg. 23-9. This is a common-base amplifier built around an npn transistor.

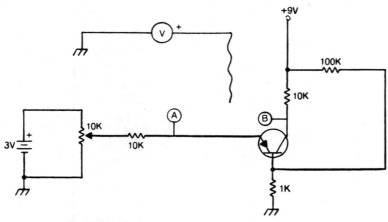

Fig. 23-9. An npn common-base amplifier. Experiment #8.

Notice that one lead of the voltmeter is left unconnected in the schematic. If this lead is attached to point A, the meter will display the input voltage. If the lead is moved to point B, the output voltage will be indicated.

Connect the free voltmeter lead to point A and adjust the potentiometer until the input signal is exactly one volt.

Now, move the positive voltmeter lead to point B and carefully measure the output voltage without touching the potentiometer. Enter this value in the appropriate space in the npn column of Table 23-4.

Return the voltmeter lead to point A and repeat the above procedure for each input voltage given in Table 23-4.

Table 23-4. Worksheet for Experiment #9—A Common-base Amplifier.

INPUT VOLTAGES	OUTPUT VOLTAGES	
	NPN	PNP
1 VOLT 1.5 VOLT 2 VOLTS 2.5 VOLTS	____ ____ ____ ____	____ ____ ____ ____

Now examine the table carefully and notice how the output voltage varies in step with the input voltage.

Now, build the circuit shown in Fig. 23-10. This is essentially the same amplifier circuit as before, but it is now built around a pnp transistor. Notice the changes in polarity.

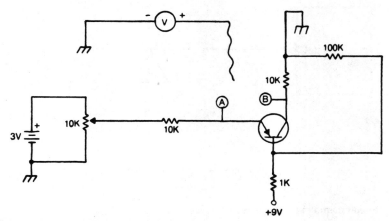

Fig. 23-10. A pnp common-base amplifier. Experiment #9.

Repeat the entire experiment with this new circuit and enter the output voltages in the appropriate spaces in the pnp column of Table 23-4.

When you are finished compare the npn column with the pnp column. The values listed in each column should be quite similar. There will perhaps be some minor variations due to component tolerances.

EXPERIMENT #10 COMMON-EMITTER AMPLIFIER

Repeat the procedure of Experiment #9 with the common-emitter amplifier circuit shown in Fig. 23-11. Enter your results into Table 23-5.

For this experiment we will just use the npn transistor, because we know the pnp transistor would give essentially the same results once all the circuit polarities are reversed.

EXPERIMENT #11 COMMON-COLLECTOR AMPLIFIER

Repeat the procedure of Experiment #9 again, this time using the circuit illustrated in Fig. 23-12. Enter your results in Table 23-6. Once again, we are using only the npn version of the circuit. A pnp common-collector amplifier would produce the same sort of results.

EXPERIMENT #12 A FIELD-EFFECT TRANSISTOR

Measure the resistance between the various leads of a 2N4350 FET (field-effect transistor) and enter your results into Table 23-7.

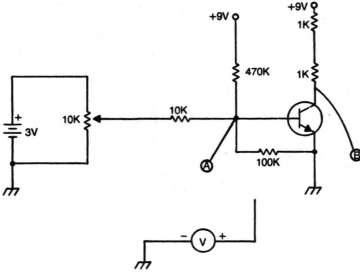

Fig. 23-11. An npn common-emitter amplifier. Experiment #10.

**Table 23-5. Worksheet for
Experiment #10—A Common-emitter Amplifier.**

INPUT VOLTAGE	OUTPUT VOLTAGE
1 VOLT 1.5 VOLT 2 VOLT 2.5 VOLT	— — — —

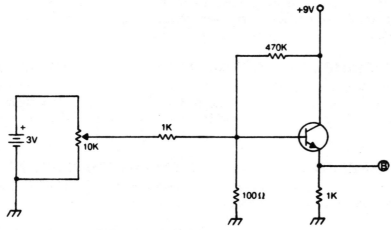

Fig. 23-12. An npn common-collector amplifier. Experiment #11.

Table 23-6. Worksheet for
Experiment #11—A Common-collector Amplifier.

INPUT VOLTAGE	OUTPUT VOLTAGE
1 VOLT 1.5 VOLT 2 VOLT 2.5 VOLT	——

Table 23-7. Worksheet for
Experiment #12—dc Resistance in
an FET.

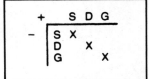

Fig. 23-13. An FET amplifier.
Experiment #12.

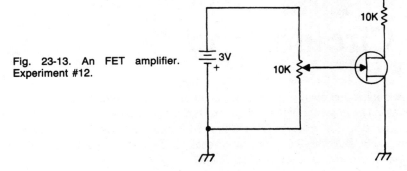

Table 23-8. Worksheet for
Experiment #12—An FET Amplifier.

INPUT VOLTAGES	OUTPUT VOLTAGES
1 VOLT 1.5 VOLT 2 VOLTS 2.5 VOLTS	——

Construct the circuit given in Fig. 23-13 and repeat the procedure of Experiment #9 one more time, entering your results into Table 23-8.

Compare the output voltages of each of the four basic amplifier types—common-base (Table 23-4), a common-emitter (Table 23-5), common-collector (Table 23-6) and FET (Table 23-8).

Which basic amplifier has the greatest amount of gain? Which has the least? Can you explain why? If not, go back and re-read Chapters 18, 19 and 20.

24

Linear Integrated Circuits and Op Amps

As electronics technology increases, circuit size has been decreasing. Many modern devices would be impractically expensive and physically unwieldy if not for recent advances in miniaturization. The switch from tubes to transistors was the first major step forward in this direction. But where miniaturization really came into its own was with the invention of *integrated circuits* (abbreviated as *ICs*).

INTEGRATED CIRCUITS

Integrated circuits are made up of a series of silicon wafers that are specially treated to simulate a number of separate transistors, diodes, resistors and capacitors. A tiny package, that is smaller than a dime can take the place of a couple dozen or more (discrete) components.

Integrated circuits are called chips, or IC chips because the internal circuitry is etched onto a chip of semiconductor material (usually silicon). There are several levels of integration possible, depending on the complexity of the circuit simulated by the IC. The standard levels of integration are as follows, from lowest to highest:

SSI	Small-Scale Integration
MSI	Medium-Scale Integration
LSI	Large-Scale Integration
VLSI	Very Large-Scale Integration

SSI (small-scale integration) devices are relatively simple circuits that could be built around a handful of discrete components. Still, a SSI IC represents a significant reduction in circuit size, and usually in cost. SSI ICs are used as basic building blocks in more complex systems. They are the most commonly used type of integrated circuits.

LSI (large-scale integration) devices include much more complicated internal circuitry. Sometimes a single integrated circuit can take the place of literally hundreds of discrete components. LSI ICs are designed for specific, specialized functions, and usually can't be used in many other applications. If you open up a typical pocket calculator, within the IC. This is an example of an LSI device.

Between these two extremes (SSI and LSI) are *MSI (medium-scale integration)* devices. An MSI IC is more complex than a SSI unit, but not as complex as a LSI device. MSI ICs are usually designed to perform some specific function within a larger system.

Recently manufacturers have been developing *VLSI (Very Large-Scale Integration)* ICs.

The first commercial IC appeared in 1961, and included four on-chip bipolar transistors. A LSI device, such as Motorola's 68000 CPU IC may contain up to 65,000 to 70,000 on-chip transistors. The VLSI chips now being designed will have as many as 250,000 transistors.

These advances are due to increasing capability for greater control, allowing finer detail. SSI ICs in the early 1970s had lines of about 20 micrometers (millionths of a meter) wide. Current LSI devices have line widths as small as 3 or 4 micrometers. Some experimental VLSI designs are shooting for 0.5-micrometer line widths.

Integrated circuits come in a number of standardized package types. Some are enclosed in round plastic, or metal cans. These devices look rather like slightly oversized transistors, except they often have ten or twelve leads, instead of just three.

Most modern integrated circuits, however, are housed in *dual-inline packages*, or *DIPs*. The integrated circuits shown in Fig. 24-1 are standard DIPs. These rectangular plastic packages have two parallel rows (or lines) of leads, hence the name. DIPs are most commonly found with 8, 14, or 16 pins. Figure 24-1 shows how these pins are

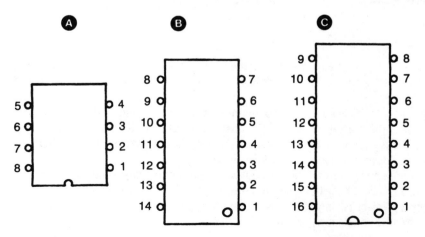

Fig. 24-1. Typical IC pin numberings: A. 8 pin DIP; B. 14 pin DIP; C. 16 pin DIP.

numbered. There will be a notch or circle etched into one end of the casing to identify pin number one. Some MSI and LSI devices may be in DIPs with 24, 28, 40, or even more pins.

Most integrated circuits are shown in schematic diagrams simply as boxes. The leads are numbered for identification, but they don't necessarily have to be drawn in numerical order. In the schematic, the leads can be arranged in any convenient pattern for the greatest clarity. See Fig. 24-2.

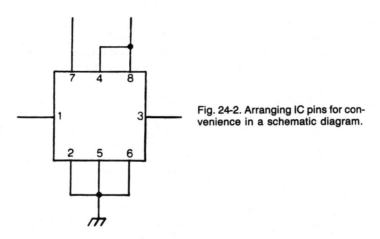

Fig. 24-2. Arranging IC pins for convenience in a schematic diagram.

Certain types of ICs have special schematic symbols that indicate their function. For example, an amplifier is often drawn as a triangle (see Fig. 24-3). These special cases will be presented as we come to them.

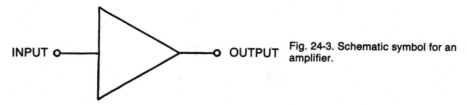

Fig. 24-3. Schematic symbol for an amplifier.

Because of the many, closely spaced pins, soldering and desoldering integrated circuits can be a problem. Like all semiconductors, too much heat can destroy an integrated circuit.

Plug-in sockets are highly recommended whenever integrated circuits are used. In some cases, the socket might actually cost more than the IC that is to be plugged into it. But the frustration and the time required to replace a defective integrated circuit makes the use of sockets a good "insurance policy."

New integrated circuits are being developed every day—especially MSI and LSI devices. So it would really be beyond the scope of this book to discuss them all. In this book we will discuss only the most common types of basic "building block" devices. These, for the most part, will be SSI units.

There are two major categories of integrated circuits. In this chapter we will be considering *linear*, or *analog* circuits. In a later chapter we'll examine *digital* ICs.

An analog circuit responds to an input signal with some output signal that varies in some (not necessarily direct) proportion to the input signal.

An amplifier is a perfect example of an analog circuit. The output signal is a larger version of the input signal.

If the relationship between the input and output signals is constant (for example, an amplifier might have an output that is 10 times the amplitude of the input signal, regardless of the frequency or waveshape of the input signal) then the circuit or device is said to be linear.

A filter is an example of an analog circuit that is not, strictly speaking, linear. Still virtually all analog integrated circuits are commonly called linear ICs.

OP AMPS

Probably the most common type of linear integrated circuit is the *operational amplifier*, or *op amp*.

Figure 24-4 shows a simplified circuit diagram of a typical operational amplifier, built around discrete components. A practical operational amplifier circuit would require many more components than the basic ones shown here, so integrated circuit op amps are obviously vastly preferable to discrete devices.

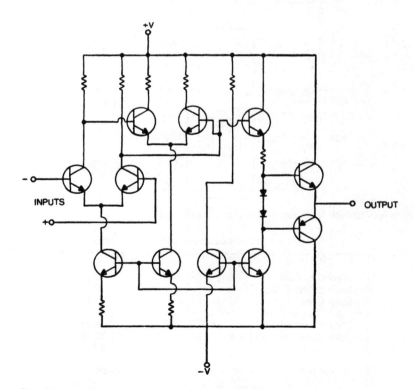

Fig. 24-4. Simplified circuit diagram for an operational amplifier.

There are a large number of IC op amps available today. One of the most popular, and least expensive is the 741. Sometimes this number is preceded or followed by key letters that identify the manufacturer, the case style, the temperature range, and so forth. For most applications, these letters can be ignored.

This integrated circuit is available in several different package configurations, but 8 pin and 14 pin DIPs predominate. The *pin-out diagrams* for these packages are shown in Figs. 24-5 and 24-6. The 741's basic specifications are listed in Table 24-1.

The standard schematic symbol for an operational amplifier is shown in Fig. 24-7. Notice that there are two voltage supply inputs— +V and −V. These voltages should be equal, but of opposite polarity with respect to ground. See Fig. 24-8. The power supply connections are often omitted from schematic diagrams, because like the filament

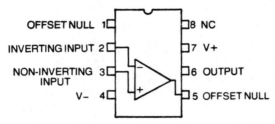

Fig. 24-5. 8 pin DIP 741 operational amplifier pin-out diagram.

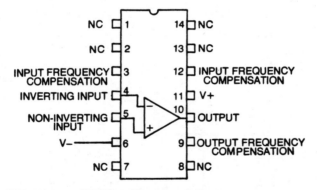

Fig. 24-6. 14 pin DIP 741 operational amplifier pin-out diagram.

Table 24-1. Specifications for a 741 Op Amp.

Maximum Gain	200,000
Input Offset Voltage	1.0 mV
Input Offset Current	20 nA
Input Bias Current	80 nA
Output Resistance	75 ohms
Common-Mode Rejection	90 dB
Slew Rate	0.5 V/μS

All Specifications Are For A ± 15 Volt Power Supply

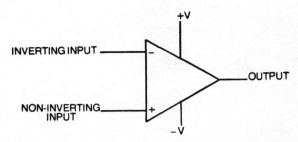

Fig. 24-7. Schematic symbol for an op amp.

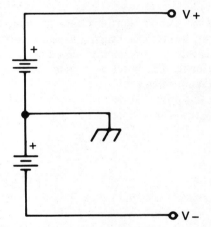

Fig. 24-8. Dual polarity power supply.

circuit of a tube, these power supply connections are automatically assumed by the presence of the IC. Remember, each and every integrated circuit package must have the appropriate power supply connections to function.

Returning to the schematic symbol (Fig. 24-7), notice that there are two inputs to an operational amplifier. One is marked with a "+". This is called the *noninverting input*. The output will be in phase with an input signal at this terminal.

The other input is identified with a "−". This input is called the *inverting input*. The output signal will be 180° out of phase with an input signal at this terminal. In other words, the signal is inverted.

There are almost countless applications for this basic device—certainly far too many to be dealt with in depth here. We will look at just a few of the many applications for an operational amplifier. The two simplest applications, of course, are inverting and noninverting amplifiers.

Inverting Amplifiers

Figure 24-9 shows the basic circuit of an inverting amplifier. The output is 180° out of phase with the input. When the input signal increases (goes more positive), the output voltage decreases (goes more negative), and vice versa.

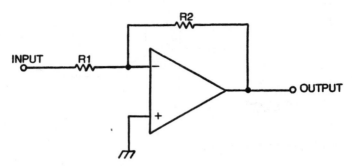

Fig. 24-9. An inverting amplifier circuit.

R2 returns some of the output signal back to the input. Since the output is 180° out of phase with the input, this returned signal will subtract from the input signal applied to the operational amplifier itself, reducing the effective gain of the amplifier. We'll see why this is necessary shortly. The process of using the output to cancel out some of the input is called *negative feedback*.

The amount of gain (G) produced by the amplifier will be determined by the ratio of R1 to R2. Specifically, the formula is;

$$G = -\frac{R_2}{R_1}$$
Equation 24-1

The negative sign is simply a mathematical indication that the signal is inverted by the circuit.

Let's assume R1 is 10,000 ohms. If R2 has a value of 100,000 ohms, the gain will be equal to −100,000/10,000 or −10.

If R2 is increased to 1 megohm (1,000,000 ohms), the gain increases to −1,000,000/10,000, or −100.

What happens if R2 is increased to infinity? That is, if the feedback path is removed altogether. The theoretical gain would be − ∞/10,000, or simply infinite gain. (∞ is a symbol used to represent an infinite quantity.)

In actual circuits, infinite gain is impossible. The upper limit of gain is determined by the internal characteristics of the op amp itself. The maximum gain is given in the specification sheet for the IC. For the 741 the maximum gain can be up to 200,000, but you'll almost never be able to use that much gain. With no feedback circuit, an inverting operational amplifier will theoretically amplify a 1 millivolt (0.001 volt) signal to −200,000 millivolts, or −200 volts at the output. But the output of an operational amplifier (or any amplifier, for that matter) is always limited to the power supply voltage. If the input signal exceeds the value that will produce an output equal to the source voltage, there will be no further change in the output. The amplifier is said to be *saturated*. This is why the feedback circuit must be used in practical circuits.

Another unique condition occurs when R1 and R2 are of equal value. For example, if both are 10,000 ohm units, the gain works out to −10,000/10,000, or simply −1. In other words, the output equals the input, except, of course, for the phase inversion. This is called *unity gain*.

An amplifier circuit with unity gain is called a *voltage follower*, because the output signal follows, or duplicates the input signal. This might not seem very useful, but sometimes it can come in quite handy for impedance matching and for *buffering*.

A *buffer amplifier* prevents a later circuit from *loading* (putting an excessive current drain on) the signal source. In other words, various sub-circuits within a system can be effectively isolated from each other via a buffer amplifier. Also, at times the phase inversion is necessary, but additional amplification is undesirable. In such a case a unity gain inverting amplifier is the obvious solution. All in all, a voltage follower is actually a useful (and frequently used) circuit.

Reducing R2's value below that of R1 will result in less than unity gain. The output level will be less than the input amplitude. For instance, if R1 was left at 10,000 ohms, and R2 was reduced to 3,000 ohms, the gain would be $-3000/10,000$, or -0.3. This would rarely be very useful. In fact, by reducing R2 to zero, we can build an amplifier with no output at all. (G = $-0/10,000$ = 0. Output = input × G = input × 0 = 0.) I seriously doubt that anybody could find a practical use for that!

Noninverting Amplifier

Figure 24-10 shows a similar circuit, except the noninverting input is used, rather than the inverting input. But the feedback is still applied to the inverting input. The output will be in phase with the input, so the circuit is called a *noninverting amplifier*.

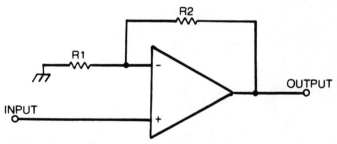

Fig. 24-10. A non-inverting amplifier circuit.

Why is the feedback applied to the inverting rather than the noninverting input? To find out why, let's assume we have a voltage gain of 200,000 with no feedback circuit. Ignoring the practical problems discussed in this last section, a 1 millivolt dc input would produce a 200 volt dc output.

Now, if we add a feedback resistor to return part of the output to the noninverting input for another pass through the amplifier (see Fig. 24-1), what will happen? We'll assume the resistance of the feedback resistor drops 195 volts, so only 5 volts is applied back to the input.

Now, we have two voltages being applied to the input of the operational amplifier. The original 1 millivolt (0.001 volt) signal, and the 5 volt feedback signal. (Of course, this is quite a large feedback signal—this is to dramatize the effects.) In other words, the effective input signal is now 5.001 volt. This is amplified by the operational amplifier's full gain, i.e., 200,000. The output jumps to 1,000,200 volts. And some of this is fed back through for additional amplification. Obviously the output level would simply continue skyrocketing towards infinity.

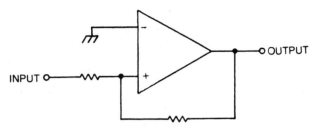

Fig. 24-11. A non-inverting amplifier with positive feedback (a nonfunctional circuit).

The feedback signal must be out of phase with the original input signal, so that it will subtract some of the input and reduce the effective gain. This can be accomplished by routing the feedback signal to the inverting input of the op amp.

If, for example, the original signal by itself produces a 10 volt output, and the feedback, by itself produces a 2 volt output, these two values would subtract because they are 180° out of phase with each other. The total effective voltate would be 8 volts, in phase with the original input signal.

Of course, you can only measure the total effective output voltage. There is no way to measure the result of the input signal and the feedback signal separately.

Notice that most of the feedback voltage is dropped by R2 and shunted to ground via R1. Obviously in practical amplification circuits the negative feedback signal should be of a lower amplitude than the input signal we want to amplify.

In all other respects, the noninverting amplifier circuit operates in the same way as the inverting amplifier. Even the formula for determining the circuit gain remains almost the same:

$$G = 1 + \frac{R_2}{R_1}$$

Equation 24-2

The absence of the negative sign, is no phase inversion.

Integration

What if the feedback component was something other than a resistor? What if R2 in a basic inverting amplifier was replaced with a capacitor, as shown in Fig. 24-12. This circuit is called an *integrator*. Let's examine how it works.

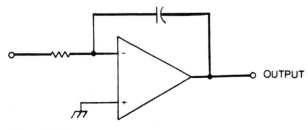

Fig. 24-12. An integrator circuit.

Obviously, the gain would vary with frequency, since the capacitor's reactance is frequency dependent. For example, let's suppose R1 is 10,000 ohms, and C is 0.01 μF (1×10^{-8} Farad). At any specific frequency the gain would be determined by the formula:

$$G = \frac{-X_c}{R1}$$

Equation 24-3

Since we know $X_c = 1/(2\pi FC)$, we can rewrite the formula as:

$$G = \frac{-1}{2\pi FC} \times \frac{1}{R1}$$

Equation 24-4

This means that for our sample circuit, the gain equals $(-1/(6.28 \times F \times 1 \times 10^{-8})) \times 1/10,000 = -1/(6.28 \times 10^{-8}) \times 1/F \times 0.001 = -1592.3567 \times 1/F$, or about $-1600/F$.

Using this formula, we find that if a 60 Hz signal is applied to this circuit, the gain will be equal to $-1600/60$, or approximately -27.

Similarly, if the input frequency is 500 Hz, the gain becomes equal to $-1600/500$, or about -3.2.

At an applied frequency of 2,000 Hz, the gain drops to $-1600/2000$, or approximately -0.8. As you can see, at this frequency (and at higher frequencies) the output is less than the input.

As the applied frequency increases, the reactance of the capacitor in the feedback circuit decreases, cancelling out more and more of the input signal at higher frequencies.

Figure 24-13 is a chart showing the frequency response of this integrator circuit. Notice that it closely resembles the frequency response graph for a simple low pass filter. In fact, an integrator is an *active* low pass filter.

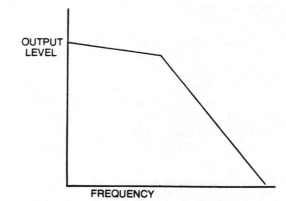

Fig. 24-13. Frequency response graph for an integrator.

The term "active" in this context means that the circuit amplifies rather than simply attenuating. A simple RC filter will subtract some signal at all frequencies (loss). An active filter will amplify the desired frequencies, as well as attenuating the undesired frequencies. This produces the effect of a sharper cut-off. You can tell from the graph

302

that the slope of this active filter is steeper than that of the simple passive filter discussed earlier.

Low pass filtering is often called integration. The terms are generally interchangeable.

Differentiator

The opposite of integration is differentation. In a differentiator circuit, such as the one shown in Fig. 24-14, R1 is replaced by a capacitor rather than R2. This means the equation for determining gain becomes:

$$G = \frac{-R2}{X_c}$$
Equation 24-5

This can be algebraically rewritten as:

$$G = -2\pi FCR_2$$
Equation 24-6

With the capacitor in the input line, low frequencies are blocked before they get a chance to reach the operational amplifier itself. The higher frequencies are passed and amplified like an ordinary inverting amplifier.

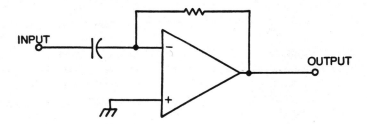

Fig. 24-14. A differentiator circuit.

Let's assume R2 is 10,000 ohms, and C is 0.01 μF(1×10^{-8} Farad). Plugging these values into the equation we find that G = $-6.28 \times F \times 1 \times 10^{-8} \times 10,000 =$ $-0.00628 \times F$.

This means the gain for a 60 Hz input would be equal to -0.00628×60, or about -0.4. There would be less than unity gain at 60 Hz.

If the input frequency is increased to 500 Hz, the gain becomes equal to -0.00628 $\times$ 500, or just over -3. Raising the input frequency to 2000 Hz, will increase the gain to $-0.00628 \times 2,000$, or about -13. Obviously a differentiator is an active high pass filter.

Difference Amplifier

If different signals are applied to each of the inputs of an operational amplifier with unity gain, the output will be equal to the difference between the input signals.

The circuit shown in Fig. 24-15 is a typical *difference amplifier*.

Suppose 5 millivolts dc (0.005 volt) is applied to the inverting input and 8 millivolts dc (0.008 volt) is applied to the noninverting input of this circuit. What will the output be?

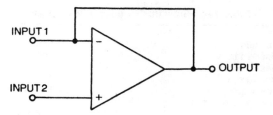

Fig. 24-15. A difference amplifier circuit.

First, ignoring the non-inverting input, the output from the inverting input alone would be − 5 millivolt (− 0.005 volt). Similarly, if we ignore the inverting input, the output from just the noninverting input would be +8 millivolts (0.008 volt).

Combining these two output signals, we have +8 millivolts, and − 5 millivolts, or an effective total of +3 millivolts—the difference between the two inputs.

It's also possible to build a difference amplifier with gain, although it isn't commonly done. If the difference amplifier had a gain of 5, the same 5 millivolt (inverting) and 8 millivolt (noninverting) inputs the output would be +15 millivolts (0.015 volt).

There are almost countless other applications for operational amplifiers. It is unquestionably the most versatile kind of integrated circuit available.

Other Op Amps

The 741 isn't the only type of integrated circuit operational amplifier, though it is the most popular. See Table 24-1.

Some additional commonly used op amp devices are the 709 (see Fig. 24-16 and Table 24-2) and the 748 (see Fig. 24-17 and Table 24-3). These devices work in the same way as the 741, but with slightly different characteristics.

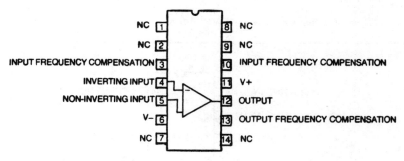

Fig. 24-16. Pin-out diagram for a 709 op amp.

Another popular integrated circuit is the 747 (see Fig. 24-18) which includes two complete 741 type operational amplifiers in a single 14 pin DIP package. Notice that the negative voltage supply terminal (pin 4) is shared by both op amps, but they each have a separate positive voltage supply terminal (pins 9 and 13). There must be the proper voltage on the appropriate pins for either op amp to function.

Table 24-2. Specifications for a 709 Op Amp.

Maximum Gain	45,000
Input Offset Voltage	1.0 mV
Input Offset Current	50 nA
Input Bias Current	200 nA
Output Resistance	150 ohms
Common-Mode Rejection	90 dB
Slew Rate	0.25 V/μS

All Specifications Are for A ± 15 Volt Power Supply

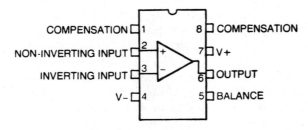

Fig. 24-17. Pin-out diagram for a 748 op amp.

Table 24-3. Specifications for a 748 Op Amp.

Maximum Gain	200,000
Input Offset Voltage	1.0 mV
Input Offset Current	20 nA
Input Bias Current	80 nA
Output Resistance	75 ohms
Common-Mode Rejection	90 dB
Slew Rate	0.5 V/μS

All Specifications Are for A ± 15 Volt Power Supply

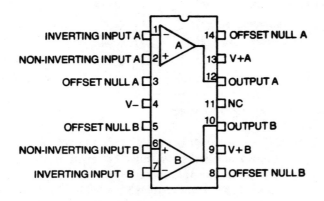

Fig. 24-18. Pin-out diagram for a 747 dual op amp.

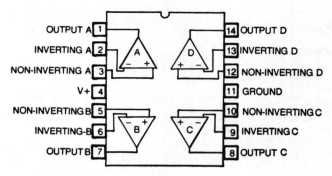

Fig. 24-19. Pin-out diagram for a 324 quad op amp.

The 324 (Fig. 24-19) goes a step further. It consists of four complete 741 type operational amplifiers in a single package. Besides saving space and costs, this particular integrated circuit can operate off a single polarity supply. Most operational amplifiers require both a positive and a negative voltage source with respect to ground.

The Norton Amplifier

Quite similar to the operational amplifiers we have been discussing, is the *Norton amplifier*. Its schematic symbol is shown in Fig. 24-20.

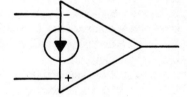

Fig. 24-20. Schematic symbol for a Norton amplifier.

Where ordinary operational amplifiers amplify the difference in the voltages applied to their inverting and noninverting inputs, the Norton amplifier amplifies the difference in applied currents. This device is also sometimes called a *current mirror*.

The Norton amplifier can be used in most op amp applications with little or no change in the basic circuitry, but this device requires only a single polarity power supply. It also has a slightly lower input impedance and a much higher output impedance than an ordinary operational amplifier. This may or may not be desirable, depending on the specific application.

The most common Norton amplifier IC is the LM3900, which contains four separate Norton amplifiers in a single package. The pin-out diagram for the LM3900 is shown in Fig. 24-21, and its specifications are listed in Table 24-4.

OTHER ANALOG ICS

Operational amplifiers are certainly the most common analog integrated circuit devices, but there are countless others. Virtually any circuit can be miniaturized on an IC chip.

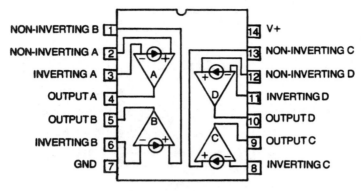

Fig. 24-21. Pin-out diagram for a 3900 quad Norton amplifier.

Input Bias Current	30 nA	Table 24-4. Specifications
Output Resistance	8,000 ohms	for a 3900 Norton Amplifier.
Input Resistance	1,000,000 ohms	
Slew Rate	0.5 V/μS	

Some other common IC devices are oscillators, audio and rf amplifiers, voltage regulators (which hold a constant voltage output regardless of current drain or source voltage fluctuations—these devices will be discussed in the chapter on power supplies) and voltage comparators (which have an output which indicates whether the input voltage is greater than, less than, or equal to a reference voltage). And, of course, there are many, many others.

It would be impossible to even list all of the various circuits that are available in IC form, much less discuss them in any detail. We will just glance at a couple of typical examples.

The XR-2206 Function Generator IC

Many types of equipment require some type of oscillator or waveform generator stage. A function generator (a circuit that can produce two or more standard waveforms) can also be used by technicians to test many types of amplifiers and other electronic equipment.

The XR-2206 is an IC function generator stage. Only a few external resistors and capacitors are required to generate sine waves, triangle waves, pulse waves, square waves, and sawtooth (or ramp) waves. A single circuit can generate two or more simultaneous or switch selectable waveforms. Figure 24-22 through 24-25 show XR-2206 function generator circuits. The circuit shown in Fig. 24-22 is a sine wave oscillator. Capacitor C sets the frequency range, with potentiometer R determining the actual output frequency. Frequency ranges for typical values of C are listed in Table 24-5.

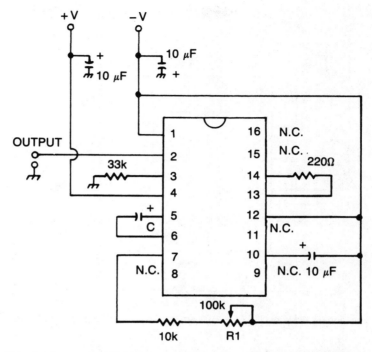

Fig. 24-22. This circuit uses an XR-2206 function generator to produce sine waves.

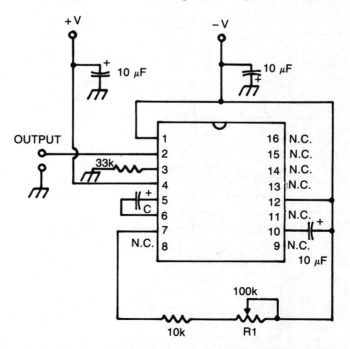

Fig. 24-23. This circuit generates triangle waves from an XR-2206 function generator.

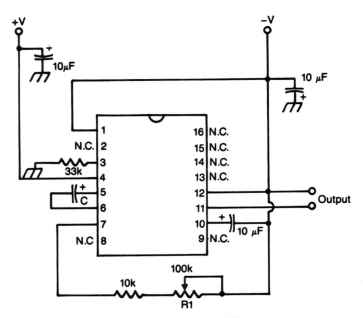

Fig. 24-24. This square wave oscillator is built around an XR-2206 function generator.

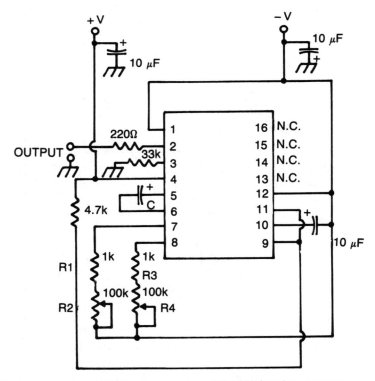

Fig. 24-25. Cx for generating sawtooth waves from an XR-2206 function generator.

Table 24-5. The Component Values to Use for the XR-2206 Oscillator.

C	F (minimum)	F (maximum)
1 μF	10 Hz	100 Hz
0.1 μF	100 Hz	1000 Hz
0.01 μF	1000 Hz	10,000 Hz
0.001 μF	10,000 Hz	100,000 Hz

The circuit illustrated in Fig. 24-23 will generate triangle waves instead of sine waves. Notice how similar this circuit is to the sine-wave oscillator circuit of Fig. 24-22. The only difference is the omission of the 220-ohm resistor between pins 13 and 14. These pins are left unconnected for the triangle wave generator.

At further modification of the same circuit is shown in Fig. 24-24. In this circuit the output consists of square waves. Notice that this time the output is taken off of pin 11 and V– instead of between pin 2 and ground.

All three of these circuits have their frequency determined in the same way. The frequency ranges as defined by the value of C in Table 24-5 are valid for all three circuits.

A somewhat different generator circuit is shown in Fig. 24-25. This circuit generates sawtooth, or ramp waves. The output frequency is determined by the values of five passive components—capacitor C, and resistors R1 through R4. Two potentiometers allow fine tuning of the output frequency and waveshape. The output frequency can be determined according to the following formula:

$$F = \frac{2}{C}\left(\frac{1}{R1 + R2 + R3 + R4}\right) = \frac{2}{C}\left(\frac{1}{R2 + R4 + 2000}\right)$$

Let's assume a 1μF capacitor is used for C. If both potentiometers (R2 and R4) are adjusted to their minimum setting, the output frequency will be equal to:

$$F = \frac{2}{0.000001}\left(\frac{1}{0 + 0 + 2000}\right)$$
$$= 2000000\left(\frac{1}{2000}\right) = 2000000 \times 0.0005$$
$$= 1000 \text{ Hz}$$

Turning both potentiometers up to their maximum setting decreases the output frequency to:

$$F = \frac{2}{0.000001}\left(\frac{1}{100000 + 100000 + 2000}\right)$$
$$= 2000000\left(\frac{1}{202000}\right) = 2000000 \times 0.000005$$
$$= 10 \text{ Hz}$$

Table 24-6. The Component Values to Use for the XR-2206 Sawtooth Wave Oscillator.

C	F (minimum)	F (maximum)
1 μF	10 Hz	1000 Hz
0.1 μF	100 Hz	10,000 Hz
0.01 μF	1000 Hz	100,000 Hz
0.001 μF	10,000 Hz	1,000,000 Hz

This circuit has wider output ranges for each value of C than the circuits in Figs. 24-22 through 24-24. The output ranges for this sawtooth wave generator are summarized in Table 24-6.

These basic waveform generator circuits can easily be combined into a multi-waveform function generator. A simple switching arrangement can make the necessary changes in the circuitry to allow the operator to select from the various output waveforms.

The XR-2206 can also be connected as a *Voltage-Controlled Oscillator*, or *VCO*, in which the output voltage is determined by an externally applied control voltage. A number of other applications are also possible, although we do not have the space to discuss them here.

XR-2208 Operational Multiplier

Another type of analog IC is the operational multiplier, which is closely related to the operational amplifiers discussed earlier. The XR-2208 Operational Multiplier is shown in block diagram form in Fig. 24-26. It consists of a four-quadrant multiplier/modulator, a high frequency buffer amplifier, and an op amp in a single package. This device is used to perform mathematical operations such as multiplication, division, and square roots with a minimum of external components.

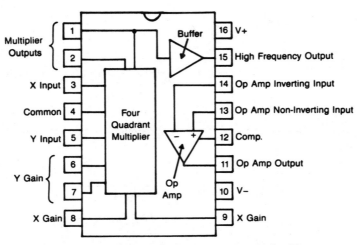

Fig. 24-26. The pin-out diagram for the XR-2208 operational multiplier IC.

SAD 1024 Analog Delay

The SAD 1024 is used to delay analog signals. The SAD portion of its name stands for *Serial Analog Delay*. It uses what is known as the bucket brigade. Imagine ten men passing buckets of water from a well to a fire. Man number 1 passes the bucket to man number 2 and fills a new bucket. Man number 2 passes the first bucket to man number 3, and gets the second bucket from man number 1, and so on. The buckets are passed down the line to the end man (man number 10). A bucket brigade delay system uses a similar approach. Instantaneous signal levels are passed through a string of capacitors.

The SAD 1024 features two independent 512-stage analog delay sections. The clock controlled delay can cover a range from less than 200 μs (microseconds) to about 0.5 second. The analog signal fed through the delay can be anywhere within a wide range from 0 Hz (dc) to more than 200,000 Hz (200 kHz). The SAD-1024 analog delay introduces a very small amount of distortion (less than 1%) to the delayed signal.

This IC can be used for reverberation, chorus, and echo circuits in electronic music and sound effects systems. It can also be employed for variable signal control of equalization filters, or amplifiers, time compression of telephone conservations or other analog signals, or voice scrambling systems. The pin-out diagram for the SAD-1024 analog delay IC is shown in Fig. 24-27.

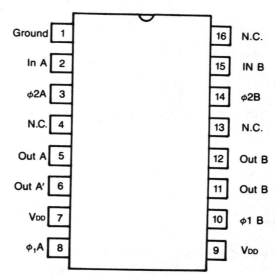

Fig. 24-27. The pin-out diagram for the SAD-1024 analog delay IC.

And Many More

We have barely been able to scratch the surface. Other common analog ICs include FM demodulators, TV modulators, video games, temperature transducers, audio amplifiers, precision current sources, and many others.

Self-Test

1. Which type of ICs are the least complex?

A *SSI*
B *MSI*
C *LSI*
D *CPU*
E *None of the above*

2. What is another name for "linear"?

A *Integration*
B *Digital*
C *Analog*
D *Dip*
E *None of the above*

3. Which of the following is the input of an op amp that will invert a signal 180°?

A *Inverting*
B *Noninverting*
C *Analog*
D *Digital*
E *None of the above*

4. What is the gain of an inverting amplifier circuit when R1 equals 2.2 k and R2 equals 18 k?

A *8.2*
B *−0.12*
C *1.2*
D *−8.2*
E *None of the above*

5. Assuming a gain of 1, what is the output of an operational amplifier when 0.27 volt is applied to the inverting input and −0.33 volts is applied to the noninverting input?

A *0.06 volts*
B *−0.60 volts*
C *−0.06 volts*
D *0.60 volts*
E *None of the above*

6. Which of the following is *not* an op amp IC?

A *741*
B *555*
C *709*
D *748*
E *None of the above*

7. How many op amps are contained in a 324 IC?

A *One*
B *Two*
C *Three*
D *Four*
E *None of the above*

8. What is the name for op amp circuit using the inverting input with a resistor as the input component and a capacitor as the feedback component?

A *Inverting amplifier*
B *Integrator*
C *Differentiator*
D *Difference amplifier*
E *None of the above*

9. What type of circuit is the XR-2208?

A *Op amp*
B *Function generator*
C *Analog delay*
D *Operational multiplier*
E *None of the above*

10. What is the gain of a differentiator when the input frequency is 375 Hz? The resistor is 22 k and the capacitor is $0.01\mu F$.

A *− 0.0005*
B *− 0.5*
C *− 2.2*
D *− 0.035*
E *None of the above*

25

Timers

Another popular and versatile type of integrated circuit is the *timer*. This device is given a chapter of its own, because it is somewhere between analog and digital circuits. It can be used with either.

555 BASICS

The most common timer is the 555, which is currently being manufactured by several different companies. There are a number of other timer ICs, but they all work in basically the same way, so we will confine our discussion in this chapter to the 555.

The 555 integrated circuit is available in a number of package styles, but the 8 pin DIP version is the most frequently used. Figure 25-1 shows the pin-outs for this IC.

We will now examine the 555 integrated circuit pin by pin.

Pin #1 is simply the ground terminal for the circuit. The 555 must be used with a negative ground. No below ground voltages should be applied to this device. That is, the voltage on pin #1 should be the lowest voltage applied to any of the integrated circuit's pins.

Pin #2 is the *trigger* input. Normally this terminal is held at a constant value that is at least one third of the source voltage. If the trigger voltage drops below the one third source point, it will trigger the circuit, that is, the timing cycle will be initiated.

The output of the timer circuit is usually (but not always) taken from pin #3.

Pin #4 is labeled *reset*. As the name suggests, the signal on this terminal is used to return the device to its original rest state at the end of the timing cycle.

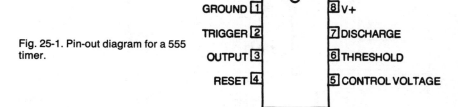

Fig. 25-1. Pin-out diagram for a 555 timer.

Pin #5 is left unused in most circuit applications, but in some special cases, it can be extremely useful. It is a *control voltage* input. The trigger voltage point can be determined by an external voltage applied to this terminal, rather than the normal one third source voltage trigger point. When the voltage control mode is not used, pin #5 should preferably be connected to ground via a 0.01 μF capacitor.

Pin #6 is called the *threshold* input. The voltage on this terminal tells the timer when to end its timing cycle. A *timing resistor* is connected from pin #6 to the positive terminal of the voltage source.

Pin #7 is called the *discharge* pin. This terminal is also used to determine the length of the timing cycle. A *timing capacitor* is connected from pin #7 to ground.

Pin #8 is used to power the circuit. The positive terminal of the voltage source is connected to this pin. In schematic diagrams this pin is usually labeled either "V+", or "VCC". The voltage source for a 555 timer IC should be between 5 and 15 volts, with 15 volts generally preferred for best results.

In passing, the 556 integrated circuit should be mentioned. It consists of two entire 555 timers in a single package. A 556 can be used to replace any two 555's. The pin-out diagram for a 556 integrated circuit is shown in Fig. 25-2.

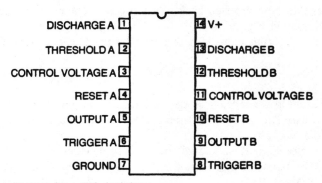

Fig. 25-2. Pin-out diagram for a 556 dual timer.

MONOSTABLE MULTIVIBRATORS

Figure 25-3 shows the most basic circuit that is built around a 555 timer IC. This circuit is called a *triggered monostable multivibrator*. Monostable means the circuit has one stable output state (at about ground level in this case). When this circuit is triggered, the output jumps to an unstable state (near the source voltage level in this circuit). The output holds this unstable level for a specific time period determined by the values of

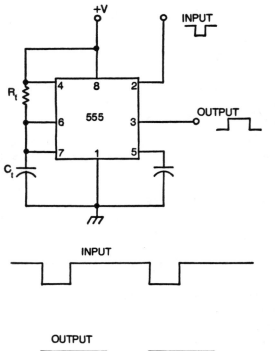

Fig. 25-3. A triggered monostable multivibrator circuit.

INPUT

Fig. 25-4. Input and output signals for a triggered monostable multi-vibrator.

OUTPUT

R_t and C_t. Then the output snaps back to its original, stable state. This process is illustrated graphically in Fig. 25-4.

The time of the output pulse is determined by the following formula:

$$T = 1.1(R_t \, C_t) \qquad \textbf{Equation 25-1}$$

T is the time period in seconds, R_t is the resistance from pin #6 to VCC in megohms (one megohm equals 1,000,000 ohms), and C_t is the capacitance between pin #7 and ground in microfarads (μF).

For reliable operation, R_t should generally be kept within the range of about 0.01 megohm (10,000 ohms) to 10 megohms (10,000,000 ohms). C_t should be kept between 100 pF (0.0001μF) and 1000 μF.

Using equation 25-1 you can see that this circuit can produce timing periods from 0.0000011 second (1.1 microseconds) to 11,000 seconds (about 183 minutes, or just over 3 hours). Clearly, this is quite a wide range device.

This timing pulse will always be the same length regardless of the time of the triggering signal—providing, of course, that the trigger signal is less than the desired output pulse.

A triggered monostable multivibrator is an extremely useful circuit in countless timing and control applications. For example, a photographer might want his strobe light to flash on for 0.2 second when his shutter is opened. He would use his shutter pulse to trigger a monostable multivibrator. In designing this circuit, first arbitrarily select a convenient capacitance value. We'll use a 0.5 μF capacitor. Since T = 1.1(R_t C_t), then R = T/(1.1 $\times$ C_t). Filling in the known variable for our example, we find that R_t = 0.2/(1.1 $\times$ 0.5) = 0.2/0.55, or about 0.36 megohms (360,000 ohms).

Using a 0.5 μF capacitor and a 360,000 ohm resistor, the light can be turned on for precisely 0.2 second each time the unit is triggered.

Of course, the precision of the timing period will be dependent on the tolerance of the timing components. If both the resistor and the capacitor have 20% tolerances, the final value may be off from the calculated value by as much as 40%. Obviously, low tolerance components should be used.

What component values would you use to generate a timing period of 1 second?

ASTABLE MULTIVIBRATORS

Figure 25-5 shows a similar 555 timer circuit. The two most important differences between this, and the previous circuit are that R_t has been split into two separate resistors (R_a and R_b), and there is no input for a trigger signal. This circuit is self-triggering.

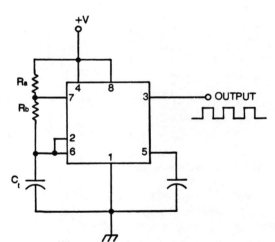

Fig. 25-5. An astable multivibrator circuit.

This circuit's output fluctuates between the same two states as the triggered monostable version, but in this circuit, neither output state is stable. This circuit is called an *astable multivibrator.*

The time the output is at its high voltage state is determined by C_t, R_a, and R_b:

$$T_1 = 0.693 \times (R_a + R_b) \times C_t \qquad \textbf{Equation 25-2}$$

While the time the output is in its low, or grounded, state depends on only C_t and R_b

$$T_2 = 0.693 \; R_b \; C_t \qquad \textbf{Equation 25-3}$$

Obviously the total time of the complete cycle is simply the sum of the high time and the low time, or:

$$T = T_1 + T_2 \qquad \textbf{Equation 25-4}$$

These three equations can be combined and rewritten as:

$$T = 0.693 \times (R_a + 2R_b) \times C_t \qquad \textbf{Equation 25-5}$$

The output *oscillates* between the two states at a fixed rate. The circuit is also called a *rectangular wave oscillator*.

As Fig. 25-6 indicates, the output waveform is rectangular in shape. Notice that for this circuit, T_1 is always at least slightly longer than T_2. The relationship between the high state time and the low state time is called the *duty cycle*. For example, if the signal is in the high state for three quarters of each cycle, the duty cycle is 75%.

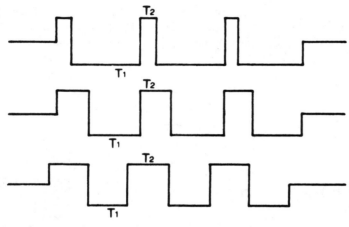

Fig. 25-6. Rectangular waves.

If both times are equal, the duty cycle is 50%, and the signal is called a square wave. A perfect square wave can't be achieved with this particular circuit, but you can come quite close by using a very small value for R_a and a large value for R_b.

Since we have a repeating cycle, we can speak of it in terms of frequency. The frequency of a repeating cycle is the number of complete cycles that take place within a second. The frequency can be found by taking the reciprocal of the time required to complete one cycle. In algebraic terms this is:

$$F = \frac{1}{T} \qquad \textbf{Equation 25-6}$$

Combining equation 25-6 with equation 25-5, we can define frequency as;

$$F = \frac{1.44}{(R_a + 2R_b) C_t} \qquad \textbf{Equation 25-7}$$

F, of course, is the frequency in Hertz for both equations.

Essentially the same range limitations are placed on the resistor and capacitor values as with the triggered monostable multivibrator circuit.

Most circuits built around the 555 integrated circuit timer are variations on one or the other of these two basic circuits. But these basic circuits can find applications in a vast number of electronics systems. A dual 555 in a single 14-pin DIP IC is also available. This is called the 556. It can be used in place of two separate 555 ICs.

THE 2240 PROGRAMMABLE TIMER

A more deluxe and complicated timing device is the 2240 programmable timer. The pin-out diagram for this device is shown in Fig. 25-7. Notice that there are 8 outputs (pins 1 through 8). The output at pin 1 is the basic timing interval. Each of the successive pins multiply the time constant by a factor of two. That is:

PIN 1	T
PIN 2	2T
PIN 3	4T
PIN 4	8T
PIN 5	16T
PIN 6	32T
PIN 7	64T
PIN 8	128T

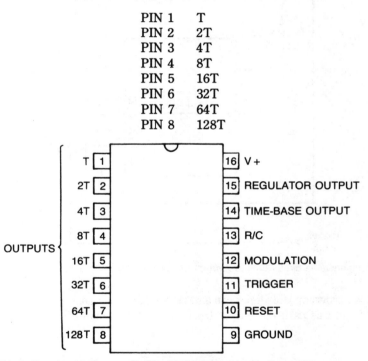

Fig. 25-7. This is the pin-out diagram for the 2240 programmable timer IC.

For example, if the circuit's basic time constant is 0.75 second, the time constant for each output will be as follows:

PIN 1	0.75 second
PIN 2	$2 \times 0.75 = 1.50$ second
PIN 3	$4 \times 0.75 = 3.00$ seconds
PIN 4	$8 \times 0.75 = 6.00$ seconds

PIN 5   16 × 0.75 = 12.00 seconds
PIN 6   32 × 0.75 = 24.00 seconds
PIN 7   64 × 0.75 = 48.00 seconds
PIN 8   128 × 0.75 = 96.00 seconds

A monostable multivibrator circuit using the 2240 programmable timer is shown in Fig. 25-8. The formula for the basic time constant (length of output pulse at pin 1) is simply:

$$T = R_t C_t$$

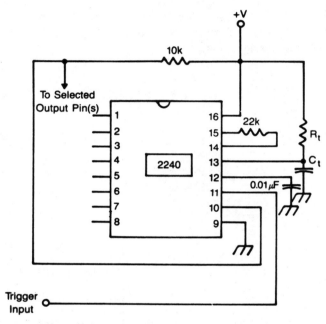

Fig. 25-8. A monostable circuit built around the 2240 programmable timer IC.

The 2240 programmable timer can also be used for an astable multivibrator, as shown in Fig. 25-9. In this case the output frequency will be equal to:

$$F = \frac{1}{2nR_t C_t}$$

where n is the multiple of the appropriate output pin. For example, if R_t = 15k and C_t = 0.22 μF, the output frequency at each of the output pins will work out to the following values:

$$\text{PIN 1} \quad F = \frac{1}{2 \times 1 \times 15000 \times 0.00000022}$$

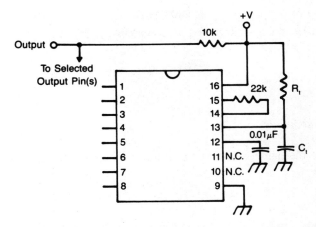

Fig. 25-9. A 2240 programmable timer IC can also be used in this astable multivibrator circuit.

$$\text{PIN 1} \quad F = \frac{1}{2 \times 1 \times 15000 \times 0.00000022}$$

$$= \frac{1}{1 \times 0.0066} = \frac{1}{0.0066} = 151.5$$

$$\text{PIN 2} \quad F = \frac{1}{2 \times 0.0066} = \frac{1}{0.0132} = 75.8 \text{ Hz.}$$

$$\text{PIN 3} \quad F = \frac{1}{4 \times 0.0066} = \frac{1}{0.0264} = 37.9 \text{ Hz}$$

$$\text{PIN 4} \quad F = \frac{1}{8 \times 0.0066} = \frac{1}{0.0528} = 18.9 \text{ Hz}$$

$$\text{PIN 5} \quad F = \frac{1}{16 \times 0.0066} = \frac{1}{0.1056} = 9.5 \text{ Hz}$$

$$\text{PIN 6} \quad F = \frac{1}{32 \times 0.0066} = \frac{1}{0.2112} = 4.7 \text{ Hz}$$

$$\text{PIN 7} \quad F = \frac{1}{64 \times 0.0066} = \frac{1}{0.4224} = 2.7 \text{ Hz}$$

$$\text{PIN 8} \quad F = \frac{1}{128 \times 0.0066} = \frac{1}{0.8448} = 1.2 \text{ Hz}$$

Each successive output pin has a frequency that is exactly half of its immediate predecessor. The apparent inaccuracies in the examples given above are due to rounding off of the values.

Self-Test

1. How many stable states does a monostable multivibrator have?

A *None*
B *One*
C *Two*
D *Three*
E *None of the above*

2. What is the name of the multivibrator circuit with no stable states?

A *Astable*
B *Monostable*
C *Bistable*
D *Unstable*
E *None of the above*

3. How long is the output pulse of a monostable multivibrator when R_t = 47 k and C_t = 0.1 μF?

A *4.7 second*
B *0.0047 second*
C *0.0052 second*
D *0.00052 second*
E *None of the above*

4. What is approximately the longest time that can be achieved with a standard 555 monostable multivibrator circuit?

A *1 second*
B *1 hour*
C *183 seconds*
D *11,000 seconds*
E *None of the above*

5. Assume an astable multivibrator circuit built around a 555 IC uses the following component values: R_a = 33 k, R_b = 27 k, C_t = 0.1 μF. What is the output frequency?

A *534 Hz*
B *437 Hz*
C *166 Hz*
D *240 Hz*
E *None of the above*

6. Which of the following is the correct formula for finding the frequency of a 555 astable multivibrator?

A $F = 1.44 \times (R_a + R_b) \times C_t$

B $F = \dfrac{1.44}{(R_a + 2R_b)C_t}$

C $F = 1.1(R_a + 2R_b)C_t$
D $F = 0.693(R_a + 2R_b)C_t$
E   None of the above

7. What is the waveshape of the output of a 555 astable multivibrator?

A *Rectangle wave*
B *Sine wave*
C *Ramp wave*
D *Square wave*
E *None of the above*

8. If the basic time constant (pin 1) of a timer circuit built around a 2240 is 3.46 seconds, what is the length of the timing pulse at pin 5?

A *17.30 seconds*
B *55.36 seconds*
C *110.72 seconds*
D *27.68 seconds*
E *None of the above*

9. What is the output frequency at pin 6 of a 2240 astable multivibrator circuit in which $R_t = 12$ k and $C_t = 0.033$ μF?

A *78.9 Hz*
B *1262.6 Hz*
C *39.5 Hz*
D *19.7 Hz*
E *None of the above*

10. How many outputs does a 2240 programmable timer IC have?

A *One*
B *Four*
C *Eight*
D *Sixteen*
E *None of the above*

26

Digital Gates

We have discussed analog integrated circuits. The other major category of integrated circuits is called *digital*. All input and output signals are converted into numerical (digital) equivalents and these numbers are treated by the integrated circuits in various ways.

THE BINARY SYSTEM

Ordinarily, we do our counting in the *decimal system,* which has ten digits (0, 1, 2, 3, 4, 5, 6, 7, 8, and 9). If we need to represent a number larger than 9 (the highest digit), additional columns of digits are added. Each column's unit value is equal to the *base* (number of digits in the system) times the unit value of the previous column. In other words, "2873" represents in decimal $2 \times 1000 + 8 \times 100 + 7 \times 10 + 3 \times 1$.

In electronic circuits it is more convenient to use the *binary system,* which has only two digits (0 and 1). If a specific voltage is present we can say we have a 1. If the voltage is absent, we have a 0. There are no intermediate values.

This can also be reversed. A present voltage could be called a 0 and no voltage could represent a 1. This would be called *negative logic*. There is really no difference, except sometimes it is more conceptually convenient to think in terms of negative logic.

Since there are no digits greater than 1 in the binary system, obviously we will need more than one column of digits to represent any number greater than one. The values of these columns are increased in the same way as in the decimal system, except each new column has a unit value of two times its predecessor. The first column is times 1, the second column is times 2, the third column is times 4, the fourth column is times 8, and so forth.

Table 26-1. Counting in Binary, Decimal, Octal, and Hexadecimal.

BINARY	DECIMAL	OCTAL	HEXADECIMAL
00001	1	1	1
00010	2	2	2
00011	3	3	3
00100	4	4	4
00101	5	5	5
00110	6	6	6
00111	7	7	7
01000	8	10	8
01001	9	11	9
01010	10	12	A
01011	11	13	B
01100	12	14	C
01101	13	15	D
01110	14	16	E
01111	15	17	F
10000	16	20	10
10001	17	21	11
10010	18	22	12
10011	19	23	13
10100	20	24	14
10101	21	25	15
10110	22	26	16
10111	23	27	17
11000	24	30	18
11001	25	31	19
11010	26	32	1A
11011	27	33	1B

For example, the binary number 1101 consists of $1 \times 8 + 1 \times 4 + 0 \times 2 + 1 \times 1$, or 13 in decimal. Table 26-1 compares counting in the decimal and binary systems.

OTHER NUMBER SYSTEMS

Notice that Table 26-1 has two additional columns labeled octal and hexadecimal. These additional systems are used as an intermediate level between decimal and binary.

While binary numbers are quite simple for electronic circuits to handle—even when the values are quite large, they are rather unwieldy for humans. For instance, a binary number like 11010001101 is quite difficult to remember.

Unfortunately, human operators have to put numbers into the circuits in the first place, and understand the numbers the circuit comes up with at its output. And electronically converting a decimal input directly into a binary number is rather a complex job.

A compromise is reached if the operator learns to use the octal or the hexadecimal system, which can be quickly converted to binary, because the bases of these systems are squared multiples of the binary base (2).

The octal system is built on the base of eight. Only digits 0-7 are used. Notice in Table 26-1 that each column in an octal number corresponds to three columns in a binary number.

Similarly, the hexadecimal system uses a base of sixteen: the digits 0, 1, 2, 3, 4, 5, 6, 7, 8, 9, A, B, C, D, E, and F. A single column in a hexadecimal number corresponds to four columns in a binary number.

Simple switching circuits can easily convert octal or hexadecimal numbers to binary numbers.

Of course, it is always important to identify which system you are working in. The number "111" represents seven in binary, seventy-three in octal, one hundred and eleven in decimal or two hundred and seventy three in hexadecimal. Obviously confusion can be a major problem if everything isn't carefully marked.

Sometimes numbers are identified by a *subscript* of their base. In the above example we could write 111_2 (binary), 111_8 (octal), 111_{10} (decimal), or 111_{16} (hexadecimal).

In most practical electronics work, you probably won't have to convert from one system to another very often, but you should have an idea of how it is done. And in digital circuits it is essential for you to understand the basics of the binary system.

BINARY ADDITION

The rules for combining or adding binary numbers are really quite simple. For any given column of digits there are only two possibilities. If you are adding two binary digits, there are only four possible combinations. If both digits to be added are 0's, the total is 0.

If one digit is a 0, and the other is a 1, the total is a 1. (Notice that this covers two possible combinations, simply by exchanging the position of the digits. $0 + 1 = 1 + 0$.) If both digits are 1's, then the total is 0, with a 1 carried into the next column.

Here are some examples; $0 + 0 = 0: 1 + 0 = 1: 0 + 1 = 1: 1 + 1 = 10: 1010 + 1100 = 10110$.

Now add $11 + 10: 101 + 1110: 1111 + 10: 1001 + 111$.

In electronic circuits digital *gates* are used to combine binary digits in various ways.

AND GATES

One basic digital gate is called an *AND gate*. There are generally four AND gates in a single integrated circuit package (called a *quad AND gate*).

The schematic symbol for an AND gate is shown in Fig. 26-1. Notice that there are two inputs and a single output. The output will be a 1 if, and only if, both inputs are 1's.

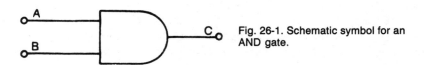

Fig. 26-1. Schematic symbol for an AND gate.

Since there are two inputs, each with two possible signals, there are four possible input conditions. If both inputs are 0's, the output will be 0. If input A is 0 and input B is 1, the output will be 0. If input A is 1 and input B is 0, the output will still be 0. But if both input A and input B are 1's, the output will be a 1. These are the only four possible input combinations.

The relationship between a gate's inputs and outputs are usually shown in a chart called a *truth table*. The truth table for a standard, two-input AND gate is given in Table 26-2A.

While most AND gates have just two inputs, you will sometimes find gates with more inputs. They work in essentially the same way. For instance, Table 26-2B gives the truth table for a four input AND gate. Notice that the output is a 1 if, and only if, all four inputs are 1's. If any input is a 0, the output is a 0.

Now, look at the truth table in Table 26-2C. It is the exact opposite of that for an AND gate. The output is 0, if and only if, both inputs are 1's. If either or both of the inputs goes to 0, the output goes to 1. Not surprisingly, this is a *Not AND* gate, or, more properly, a *NAND gate*.

The schematic symbol for a NAND gate resembles that of an AND gate, except a small circle is added at the output to indicate *state inversion*. See Fig. 26-2.

Two Input AND Gate

IN		OUT	
A	B	C	
0	0	0	**A**
0	1	0	
1	0	0	
1	1	1	

Four Input AND Gate

A	B	C	D	OUT	
0	0	0	0	0	
0	0	0	1	0	
0	0	1	0	0	
0	0	1	1	0	
0	1	0	0	0	
0	1	0	1	0	
0	1	1	0	0	**B**
0	1	1	1	0	
1	0	0	0	0	
1	0	0	1	0	
1	0	1	0	0	
1	0	1	1	0	
1	1	0	0	0	
1	1	0	1	0	
1	1	1	0	0	
1	1	1	1	1	

Table 26-2. Truth Tables: A. AND Gate; B. Four Input AND Gate; C. Two Input NAND Gate.

NAND Gate

IN		OUT	
A	B	C	
0	0	1	**C**
0	1	1	
1	0	1	
1	1	0	

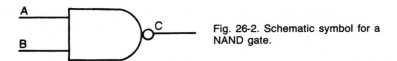

Fig. 26-2. Schematic symbol for a NAND gate.

In digital circuits, this small circle always represents state inversion. If the signal was a 1, it becomes a 0. If it was a 0, it becomes a 1.

OR GATES

Another basic digital gate is the *OR gate*. Its schematic symbol is shown in Fig. 26-3, and its truth table is given in Table 26-3A.

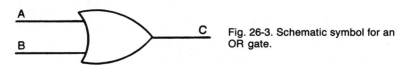

Fig. 26-3. Schematic symbol for an OR gate.

Again, there are two inputs, and a single output that is dependent on the logic states of both inputs. As long as at least one of the inputs is a 1, the output of an OR gate will be a 1. The output will be a 0 only if both inputs are 0's.

A variation of the basic OR gate is the *exclusive OR gate*. As the name implies, the output is a 1 if one or the other input is a 1, but not if the inputs are both 1's, or both 0's. In other words, the output is a 1 if the two inputs are different. The output is a 0 if the two inputs are at the same level. The schematic symbol for an exclusive OR gate is shown in Fig. 26-4, and its truth table is shown in Table 26-3B.

Table 26-3C is the truth table for the inversion of an OR gate. This device (shown schematically in Fig. 26-5) is called a *NOR gate*. The output will be a 1 only if neither input is a 1. If either or both inputs are at a logic 1 level, the output will be a 0.

Table 26-3. Truth Tables: A. OR Gate; B. Exclusive-OR Gate; C. NOR Gate.

Ⓐ OR Gate

IN		OUT
A	B	C
0	0	0
0	1	1
1	0	1
1	1	1

Ⓑ Exclusive OR Gate

IN		OUT
A	B	C
0	0	0
0	1	1
1	0	1
1	1	0

Ⓒ NOR Gate

IN		OUT
A	B	C
0	0	1
0	1	0
1	0	0
1	1	0

Fig. 26-4. Schematic symbol for an
Exclusive-OR (X-OR) gate.

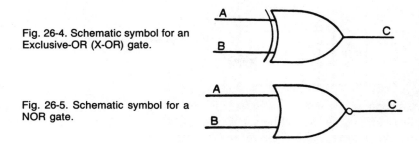

Fig. 26-5. Schematic symbol for a
NOR gate.

BUFFERS AND INVERTERS

Figure 26-6 shows two single input/single output digital devices. Figure 26-6A is an inverter. Obviously, this device inverts or reverses its input. If the input is a 0, the output is a 1, or if the input is a 1, the output is a 0. These are the only possible states.

The device shown in Fig. 26-6B might not seem particularly useful at first, since its output is the same as its input. This device does not affect the logic state in any way.

Any digital gate's output can feed only a limited number of inputs to other digital gates (or other devices). This is called the *fan-out*. The typical fan-out for *TTL* gates (see Chapter 28) is usually about 10. That is, each gate's output can be used as the input for up to 10 other gates.

But, what do you do if you need to drive more than 10 gates from a single output? (This problem often crops up when digital gates need to be *interfaced* with other devices that may have a *fan-in* greater than one.)

Here is where the device in Fig. 26-6B comes in. It is called a *buffer*. A buffer typically has a fan-out of about 30. By passing the desired signal through a buffer, it can be used to drive many more gates (or other devices). Inverters also act as buffers in addition to their state reversal function.

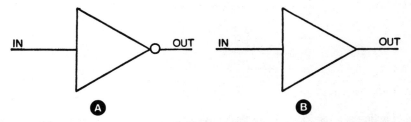

Fig. 26-6. Schematic symbols for single input/single output digital devices: A. inverter; B. buffer.

COMBINING DIGITAL GATES

These seven basic digital gates (AND, NAND, OR, exclusive OR, NOR, inverters and buffers) can be combined to perform almost any logic function.

There are a number of specialized digital IC gates for other logic functions, but these are not always readily available. Fortunately, any function can be built up from combinations of the basic gates.

330

a	b	Output
0	0	0
0	1	0
1	0	1
1	1	0

Table 26-4. A Non-standard Truth Table.

Let's suppose we need to generate the logic pattern shown in the truth table in Table 26-4. The output should be a 1 if, and only if, input A is a 1, and input B is 0. This could be achieved with either of the circuits shown in Fig. 26-7.

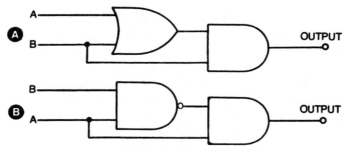

Fig. 26-7. Possible circuits for generating the truth table in Table 26-4.

Of course, more than two inputs may be employed for certain applications. For example, Fig. 26-8A shows a three input AND gate. The output will be a 1 only if all three inputs are 1's. If any of the inputs is a 0, the output will be a 0.

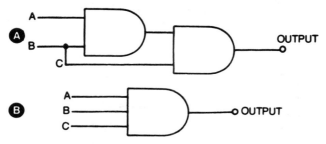

Fig. 26-8. Three input AND gates.

This particular function is sometimes available in a single, ready-made package. If a single package unit is used, it is usually shown schematically as illustrated in Fig. 26-8B.

Table 26-5 shows a complex 4 input logic function. There are a number of ways this can be achieved using the basic gates described in the previous sections. One possible solution is shown in Fig. 26-9. There are 9 AND gates, 4 NOR gates, and 5 exclusive OR gates used in this circuit. Since there are usually four identical two input gates in each integrated circuit package, this circuit would require a minimum of five IC packages, with some of the available gates either left unused, or used in a separate circuit.

Can you come up with a simpler circuit to generate the same truth table? Can you determine the truth table for the logic circuit shown in Fig. 26-10?

a	b	c	d	Output
0	0	0	0	1
0	0	0	1	0
0	0	1	0	0
0	0	1	1	1
0	1	0	0	0
0	1	0	1	0
0	1	1	0	1
0	1	1	1	0
1	0	0	0	0
1	0	0	1	0
1	0	1	0	1
1	0	1	1	0
1	1	0	0	0
1	1	0	1	0
1	1	1	0	0
1	1	1	1	0

Table 26-5. A Complex Four Input Truth Table.

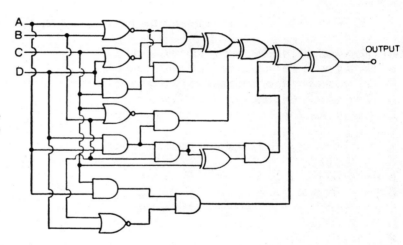

Fig. 26-9. Possible circuit for generating the truth table in Table 26-5.

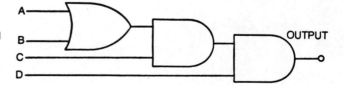

Fig. 26-10. Another complex digital gating circuit.

Self-Test

1. What is 1101 0010 binary in decimal?

A *202*
B *210*
C *110*
D *220*
E *None of the above*

2. What is a 175 decimal expressed in binary?

A *1010 1101*
B *1100 1111*
C *1010 1111*
D *1010 0101*
E *None of the above*

3. Which of the following expresses the functioning of an AND gate?

A *The output is 1 if and only if all inputs are 1's*
B *The output is 1 if and only if all inputs are 0's*
C *The output is 1 if at least one of the inputs is a 1*
D *The output is 0 if and only if all inputs are 1's*
E *None of the above*

4. How can a NAND gate be made from more basic gates?

A *An AND gate with inverters at the inputs*
B *An AND gate with an inverter at the output*
C *An OR gate with an inverter at the output*
D *An AND gate with a buffer at the output*
E *None of the above*

5. Which of the following truth tables represents an OR gate?

	Inputs	Output
A	0 0	1
	0 1	0
	1 0	0
	1 1	1
B	0 0	0
	0 1	1
	1 0	1
	1 1	1
C	0 0	0
	0 1	1

```
        1 0      1
        1 1      0
D   0 0      0
        0 1      0
        1 0      0
        1 1      1
```
E *None of the above*

6. If the input to an inverter is a logic 1, what is the output?

A *0*
B *1*

7. What is the base of the octal numbering system?

A *Two*
B *Eight*
C *Ten*
D *Sixteen*
E *None of the above*

8. Which of the following does *not* produce a 1 at the output when all inputs are 1's?

A *Buffer*
B *AND gate*
C *Exclusive OR gate*
D *OR gate*
E *None of the above*

9. Which of the following input combinations will produce a logic 1 output from a three-input NOR gate?

A *010*
B *111*
C *110*
D *000*
E *None of the above*

10. If both inputs of an OR gate are inverted, the result will be the same as which of the following basic gate types?

A *NAND*
B *AND*
C *NOR*
D *X-OR*
E *None of the above*

27

Other Digital
Integrated Circuits

Gates aren't the only type of digital devices. Some additional digital ICs will be discussed in this chapter.

FLIP-FLOPS

In Chapter 25 we talked about monostable multivibrators (which have one stable state) and astable multivibrators (which have no stable state). As might be expected, there is a third type of multivibrator, which has two stable states. A trigger signal is required to switch the output from one state to the other. This device is called a *bistable multivibrator*, or *flip-flop*. Sometimes it is also called a *latch*, or a *one bit memory*.

Figure 27-1 shows the circuit for a very simple flip-flop. It primarily consists of two inverters connected from input to output.

If S1 is closed, the input to inverter B is pulled down to ground potential; for a positive logic system, of course, this is a logic 0. Inverter B reverses the logic state to a 1. Besides being the output of the circuit, this signal is also fed back to the input of inverter A, which inverts the logic state back to a 0 and feeds this signal back into the input of inverter B. The circuit output will be a logic 1, even if S1 is reopened. The circuit latches itself into a logic 1 output state.

Similarly, if S2 is closed, the input of inverter A is forced to a logic 0. The output of inverter A is a 1, which is inputted to inverter B. This second state inversion would produce a circuit output of 0. And, of course, this signal is fed back to the input of inverter A, to latch the circuit into the 0 state, even after S2 is reopened.

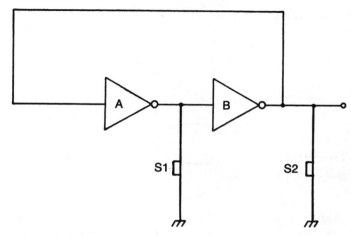

Fig. 27-1. A simple flip-flop circuit.

Either output state (1 or 0) will be held stable until the reverse state is initiated.

Closing both switches simultaneously would force the input and output of both inverters to logic 0. It is impossible to predict what the output of the flip-flop would be when the switches are opened. This is a *disallowed state*, and should be avoided.

Figure 27-2 is an adaptation of the basic circuit that prevents the disallowed state from occurring. For the best results, the switch should be a momentary action, center off SPDT switch.

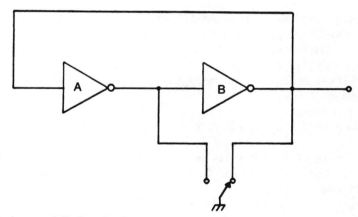

Fig. 27-2. An improved flip-flop circuit.

The disadvantage of both of these circuits is that they can be *triggered* (forced to change states) only by mechanical switches. Relays could provide some form of automatic operation, but they are expensive and bulky. Fortunately, there is a better way of electronically triggering a flip-flop.

Figure 27-3 is an improved version of the flip-flop circuit that can be triggered by the outputs of other logic gates.

336

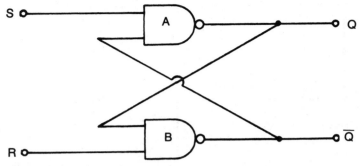

Fig. 27-3. RS flip-flop.

Notice that there are two outputs from this circuit, labeled Q and $\overline{Q}$ (read "Q and not Q"). The line over $\overline{Q}$ indicates that it is the *complement* (the opposite logic state) of Q. If Q is 1, $\overline{Q}$ must be 0, and if Q is 0, then $\overline{Q}$ must be 1.

This circuit also has two logic inputs. One is labeled *S (set)*, and the other is labeled *R (reset)*.

Let's assume that when we first turn on this circuit both S and R are at logic 1, and output Q is 0. Of course output $\overline{Q}$ must be a logic 1.

Gate A's inputs are S (logic 1), and $\overline{Q}$ (logic 1). Gate A is a NAND gate, so its output (Q) remains at logic 0.

Similarly, gate B's inputs are R (logic 1) and Q (logic 0), so its output $(\overline{Q})$ remains a logic 1. The circuit will be latched in this state as long as both inputs (S and R) are kept at logic 1.

If S is changed to a logic 0, the situation changes. The output of Gate A (Q) is changed to a logic 1, and this, in turn, converts the output of gate B $(\overline{Q})$ to a logic zero.

Even if S is now changed back to a logic 1 state, the output states will not change, because of the way they are fed back to the inputs of the gates. The circuit is latched into this new state, and any changes in the signal on the S input will have no effect on the output signal.

Now, if R is changed to a 0, it will change gate B's output $(\overline{Q})$ back to a logic 1, and gate A's output (Q) back to a 0. Further changes in the R input will have no effect on the output.

In other words, a 0 input at S sets the flip-flop (Q = 1, and $\overline{Q}$ = 0), while a 0 input at R resets the flip-flop (Q = 0, and $\overline{Q}$ = 1). A 1 at both inputs will latch the circuit in its present state.

Notice that a 0 at both inputs is once again a disallowed state. The output will be unpredictable under this condition. The truth table for this circuit is shown in Table 27-1. This type of circuit is called a *set-reset flip-flop*, or an *RS flip-flop*.

JK Flip-Flops

Another common type of bistable multivibrator is the *JK Flip-Flop*. The "JK" is used simply to distinguish this type of flip-flop from the RS type. It doesn't stand for anything in particular.

Table 27-1. Truth Table for the RS Flip-flop.

INPUTS		OUTPUTS	
R	S	Q	$\overline{Q}$
0	0	NO CHANGE—PREVIOUS STATE	
0	1	1	0
1	0	0	1
1	1	DISALLOWED STATE	

The inputs and outputs of a JK flip-flop are illustrated in Fig. 27-4. Again, there are two outputs—Q and $\overline{Q}$. $\overline{Q}$ is always the complement of Q. This device also has a total of five inputs—*preset, preclear, J, K,* and *clock.*

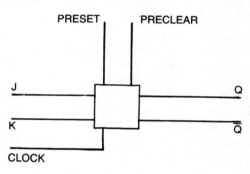

Fig. 27-4. JK flip-flop.

The preset and preclear inputs work rather like the set and clear inputs on an RS flip-flop. A 0 input on the preset terminal immediately forces the Q output to a 1 state (and $\overline{Q}$ to a 0 state). Similarly, a 0 input on the preclear terminal immediately forces the Q output to a 0 state (and $\overline{Q}$ to a 1 state).

If both of these inputs are logic 1's, the output will be determined by the other three inputs.

Putting both preset and preclear in a logic 0 condition simultaneously is a disallowed state. These inputs and outputs are summarized in the truth table given in Table 27-2.

**Table 27-2. Truth Table for the
Preset and Preclear Inputs of a JK Flip-flop.**

INPUTS		OUTPUTS	
PRESET	PRECLEAR	Q	$\overline{Q}$
0	0	DISALLOWED STATE	
0	1	1	0
1	0	0	1
1	1	DETERMINED BY CLOCKED INPUTS	

The J and K inputs are *clocked inputs*. This means they can have no effect on the output until the clock input receives the appropriate signal.

There are two basic types of clocking—*level clocking* and *edge clocking*. In a level clocking system, the clock is triggered by the logic state of the input signal. It may be designed to trigger on either a 1 or a 0 (but not both). The clock will remain activated for as long as the input is held at the appropriate logic level.

Edge clocking, on the other hand, is triggered by the transition from one state to the other. Either the 0 to 1 (positive edge) transition, or the 1 to 0 (negative edge) transition can be used (but not both), depending on the design of the specific circuit. Obviously an edge triggered clock is activated for a much shorter time period than a level triggered device. For most digital work, edge triggering is usually preferred.

Clocked circuits have a number of advantages, especially within large systems. First, by triggering all of the sub-circuits in a large system with the same clock signal, all operations can be forced to stay in step with each other throughout the system, preventing erroneous signals.

Another frequent source of errors in an unclocked circuit is noise on the input lines. The input is most likely to be noisy during the transition from one state to the other, especially when mechanical switches are used. No mechanical switch is perfect, and some bounce will be exhibited every time a switch is used. The slider will rapidly make and break contact many times before finally settling into position. See Fig. 27-5. For most applications this simply doesn't matter, but in digital circuits, it can be quite troublesome. If a flip-flop operated on each and every input pulse, the output would be very noisy and erratic and many undesired operations would take place. Later in this chapter we will discuss a *bounceless switch*.

Fig. 27-5. Mechanical switch bounce as it appears to a digital circuit.

In spite of this problem, if the inputs only change state during the time the circuit is held inactive by the clock, and the circuit is activated only when the inputs have had time to stabilize (only a tiny fraction of a second is needed), very clean, reliable outputs can be achieved. In the truth table shown in Table 27-3, "T" means the clock is triggered, and "N" means the clock is not triggered.

Table 27-3. Truth Table for the Clocked Inputs of a JK Flip-flop.

INPUTS			OUTPUTS	
J	K	CLOCK	Q	Q̄
0	0	N	NO CHANGE	
0	0	T	NO CHANGE	
0	1	N	NO CHANGE	
0	1	T	0	1
1	0	N	NO CHANGE	
1	0	T	1	0
1	1	N	NO CHANGE	
1	1	T	OUTPUT STATES REVERSE (0 BECOMES 1 AND 1 BECOMES 0)	

Notice that if both the J input and the K input are at logic 1, the output will reverse states each time the clock is triggered. For example, let's assume the Q output starts out as a 0 (Q = 1). On the first clocking pulse Q will be a 1 ($\overline{Q}$ = 0). The second clocking pulse will change Q back to a 0 ($\overline{Q}$ = 1). The third clocking pulse will change Q to a logic 1 again ($\overline{Q}$ = 0), and so forth.

The D Type Flip-Flop

The JK flip-flop is quite useful and versatile, and there is no disallowed state for its clocked inputs (preset and preclear aren't used in most applications). But the JK flip-flop's requirement for two inputs in addition to the clock is sometimes inconvenient in certain applications.

The problem can be solved by using a *D Type Flip-Flop* ("D" stands for *data*). Figure 27-6 shows how a D type flip-flop can be made from a JK flip-flop and an inverter. The J and the K inputs will always be at the opposite logic states. If J is a 1, K is a 0, and vice versa. Table 27-4 shows the truth table for a D type flip-flop.

D type flip-flops also have preset and preclear inputs that function in the same way as in a JK flip-flop.

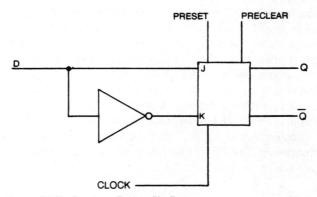

Fig. 27-6. Converting a JK flip-flop to a D-type flip-flop.

Table 27-4. Truth Table for a D-type Flip-flop.

D (INPUT)	OUTPUTS	
	Q	$\overline{Q}$
0 1	0 1	1 0
The D input is functioned only when the clock is triggered.		

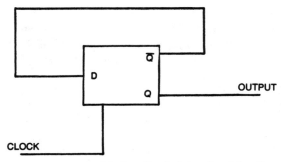

Fig. 27-7. A D-type flip-flop is operated solely from its clock input and functions as a one digit binary counter.

Figure 27-7 shows a flip-flop circuit that is operated solely off of the clocking input. The D input is fed by the Q output. The clock is edge triggered so there is only time for a single state change during each clock pulse.

We'll assume that the Q output starts off at a logic 1 state. This means $\overline{Q}$, and therefore D, must be at logic 0. Nothing happens until the clock is triggered, then the circuit looks at the data on the D input. Since this is a 0, the truth table tells us Q should become a 0, and $\overline{Q}$ should become a 1. This 1 is fed back to the D input, but by this time, the clocking pulse is gone, so the flip-flop waits until the next trigger signal arrives. When the clock is triggered, the 1 on the D input changes Q back to 1, and $\overline{Q}$ back to 0, and we're back where we started.

The input and output signals are illustrated in Fig. 27-8. Notice that both the input and the output are square waves (the output can be any rectangular wave, but the output will always be a square wave). Notice also that it takes two complete input (clock) cycles to produce one complete output cycle. This means the output frequency is exactly one half of the input frequency. For this reason, this type of circuit is often called a *frequency divider*. It divides the clock frequency by two.

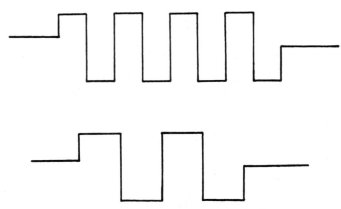

Fig. 27-8. Input and output signals for the circuit shown in Fig. 27-7.

COUNTERS

The circuit in Fig. 27-7 is also a binary counter. It counts the incoming clock pulse in binary form.

But, since the binary system has only two digits, and this counter can only handle a single column of digits, the count starts over after every second input pulse. 0-1-0-1-0-1-0-1 . . .

The counting range can be extended to handle larger numbers by adding extra flip-flop stages, as shown in Fig. 27-9. This is a three stage counter. Let's look at how it works.

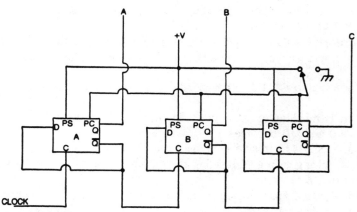

Fig. 27-9. A three stage counter.

The preset and preclear inputs of each of the three flip-flops are tied together. The preset inputs are connected to the voltage source, so they are always held at logic 1. Remember, each integrated circuit package must also have its power supply pins connected to the power supply, even if this is not shown on the schematic diagram. It is always assumed.

The three preclear inputs go to a single SPDT switch, preferably with momentary contacts for one NC connection (shown in the diagram), and one NO connection. In its normal position, the switch sends a logic 1 to the preclear inputs. Remember that according to the truth table (Table 27-4) if both the preset and the preclear inputs are logic 1's, the output is determined by the clock and D. But, if the switch is moved to its other position, a logic 0 is applied to the preclear inputs, forcing each of the Q outputs to a 0 state (and each $\overline{Q}$ to a 1), regardless of their previous states. This switch clears the counter by forcing it to a 000 output state.

We'll start at this initial 000 condition and look at what happens on each clocking pulse. We'll assume the clock inputs trigger on the positive edge (i.e., the 0 to 1 transition).

☐ **Clock pulse #1.** Flip-flop A has a 1 on its D input (since D is tied to $\overline{Q}$). When the first clock pulse is received QA goes to 1, and $\overline{Q}$A goes to 0.

Flip-flop B is clocked by the 0 to 1 transition of flip-flop A's $\overline{Q}$ output. Since $\overline{QA}$ went from 1 to 0, no clocking pulse will be applied to flip-flop B. Its state will remain the same.

Similarly, flip-flop C is clocked by $\overline{QB}$. This signal is not changed during this clock pulse, so flip-flop C's output state is also unchanged. Reading the outputs from right to left (i.e., C - B - A), we have a binary count of 001, or one.

☐ *Clock pulse #2.* $\overline{QA}$ is a 0. This is fed back to DA. When the clock pulse is received, QA goes to 0 and $\overline{QA}$ goes to 1.

Since $\overline{QA}$ has a 0 to 1 transition flip-flop B is also triggered. Originally its outputs were QB = 0 and $\overline{QB}$ = 1, so the 1 on the DB input changes QB to a 1, and $\overline{QB}$ to a 0.

Flip-flop C is not triggered. Reading the outputs from right to left, we now have a binary count of 010, or two.

☐ *Clock pulse #3.* This pulse behaves like clock pulse #1, changing QA to a 1, and $\overline{QA}$ to a 0, and leaving flip-flop's B, and C unchanged. The outputs now read 011, or three.

☐ *Clock pulse #4.* DA is being fed a 0, so QA changes to a 0 and $\overline{QA}$ changes to a 1.

The 0 to 1 transition of $\overline{QA}$ triggers flip-flop B. DB is being fed a 0 from $\overline{QB}$, so QB becomes 0 and $\overline{QB}$ becomes 1.

Since $\overline{QB}$ has a 0 to 1 transition on this pulse, flip-flop C is also triggered. QC changes to a 1 and $\overline{QC}$ changes to a 0. Reading the outputs from right to left we now have a binary count of 100, or four.

☐ *Clock pulse #5.* Once again, only flip-flop A is triggered. Notice that on any odd numbered pulse only the first flip-flop is triggered. QA switches to a logic 1 and $\overline{QA}$ goes to a logic 0. The output is now 101, or five.

☐ *Clock pulse #6.* QA changes from 1 to 0, and $\overline{QA}$ changes from 0 to 1. This triggers flip-flop B, causing QB to go from 0 to 1 and $\overline{QB}$ to change from 1 to 0. Flip-flop C is not triggered, so its output states remain the same. The binary count is now 110, or six.

☐ *Clock pulse #7.* As with the other odd numbered clock pulses, only flip-flop A is triggered. We now have a binary count of 111, or seven.

☐ *Clock pulse #8.* QA changes from 1 to 0, and $\overline{QA}$ changes from 0 to 1, triggering flip-flop B. QB changes from 1 to 0, and $\overline{QB}$ changes from 0 to 1, triggering flip-flop C. QC also changes from 1 to 0, and $\overline{QC}$ goes from 0 to 1.

The binary count after the eighth clock pulse is 000 once more. The counter is reset for another series of count pulses. This pattern will continue repeating as long as there are incoming clock pulses.

Figure 27-10 compares the input (A's clock) and the three output signals (A, B, and C). Notice that A is one half the original clock frequency. B is one half of A's frequency, or one quarter of the original clock frequency. C is one half of frequency B, or the original clock frequency divided by eight.

If, for example, the clock frequency was 6,000 Hz, output A would be 3,000 Hz, output B would be 1,500 Hz, and output C would be 750 Hz. In other words, a counter can be used as a higher level frequency divider.

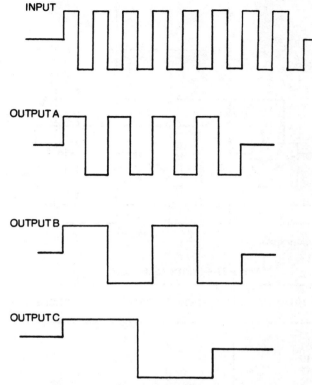

Fig. 27-10. Input and output signals for the circuit shown in Fig. 27-9.

The count (or frequency division) can be extended to any column of binary digits. That is, you can build counter circuits that count to 2, 4, 8, 16, 32, 64, 128, or so forth.

This maximum number of counts is called the *modulo* of the counter. Remember, a binary counter actually counts from 0 to one less than the modulo. For instance, a modulo four counter would count 00 - 01 - 10 - 11 - 00 - 01 - 10 - 11 - 00 - 01 - 10 - 11 . . . Or, in decimal, the count would translate to 0 - 1 - 2 - 3 - 0 - 1 - 2 - 3 - 0 - 1 - 2 - 3 . . .

Suppose you needed a counter with a modulo that was not part of the regular binary series. For example, let's say we need a counter with a modulo of six. Since six in binary is 110, this count cannot be achieved with a whole number of flip-flops. We need some way to set the count back to 000 after 101 (five).

Figure 27-11 shows the solution. The output to the NAND gate is fed back to the preclear input. Remember, a logic 0 level on this input line will force all of the Q inputs back to 0.

The output of the NAND gate will be a logic 1, unless both QA and QC are 1's. Whenever this happens, the gate's output goes to logic 0, allowing the preclear input to set the counter back to 000. As Table 27-5 shows, the binary count 101 (five) is the earliest point in the cycle where A and C are both at logic 1. Notice that the state of

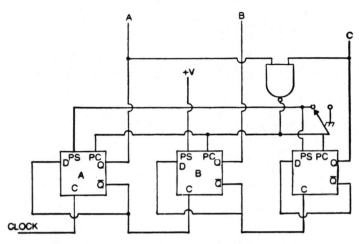

Fig. 27-11. A Modulo-six counter.

Table 27-5. Truth Table for Fig. 27-11.

BINARY OUTPUT	GATE OUTPUT	DECIMAL NUMBER
000	1	0
001	1	1
010	1	2
011	1	3
100	1	4
101	0	5
110*	1	6
11*	0	7
		*These States Will Never Occur

B doesn't matter in this case. Theoretically, the preclear input could also be triggered by binary 111 (seven) but this count will never be reached.

Thus, by using the appropriate number flip-flops and the correct gates, we can produce a counter for any whole number we choose.

Because our ordinary counting system is built on a base of ten, a modulo 10 counter is particularly useful. Such a device would consist of a string of four flip-flops that are reset after a count of 1001 (nine).

This particular counter configuration is available in a number of single integrated circuit packages (with only slight variations from type to type). A modulo 10 binary counter is often called a *BCD* (short for *Binary Coded Decimal*). Table 27-6 shows the sequence of output states for a BCD.

BCDs are especially useful to convert binary data to decimal numbers in display circuits. The four line counter output is fed into another integrated circuit device called a *decoder/driver*, which rearranges the signals into seven outputs. These seven outputs are connected to the segments of an LED seven-segment display which will light up the decimal equivalent to the binary number at the decoder's input.

Table 27-6. Output of a BCD.

Binary Input	Decimal Equivalent
0000	0
0001	1
0010	2
0011	3
0100	4
0101	5
0110	6
0111	7
1000	8
1001	9
1010	Disallowed State
1011	Disallowed State
1100	Disallowed State
1101	Disallowed State
1110	Disallowed State
1111	Disallowed State

The driver section boosts the fan out so it can comfortably handle the relatively high current drawn by the LEDs. Some decoder/drivers are designed for use with common cathode displays. Others are intended for use with common anode displays. See Fig. 27-12 for the standard arrangement of these devices. The decoder and the driver section are usually (but not always) contained within a single integrated circuit package.

SHIFT REGISTERS

Figure 27-13 shows a variation on the basic counter circuit. This circuit does not necessarily count in a sequential manner, because on any specific clock pulse, the initial D input can be made either a 0 or a 1 via an external signal.

We'll assume the outputs all start at 000. On clock pulse #1 we will feed a logic 1 to the input. For all other clock pulses 0's will be inputted. Let's look at what happens on each clock pulse.

☐ **Clock pulse #1.** Notice that the Q outputs are not used (although they could be, if complementary outputs were needed for some specific function).

Remember, there is only time for a single state change to take place during a given clock pulse. This is essential for proper operation, and this is why only edge clocking is used for this type of circuit. In our example we will assume positive edge clocking (0 to 1 transition) is used, although negative edge (1 to 0 transition) can be employed just as well with only minor circuit changes.

QB is a 0, so DC is a 0, leaving QC at a 0 level when the clock pulse is received. Similarly, the initial 0 state of QA is fed to DB, holding QB at 0. But the input DA is being fed a logic 1 from an external source. This changes QA to 1. The binary output from right to left (C - B - A) is 001.

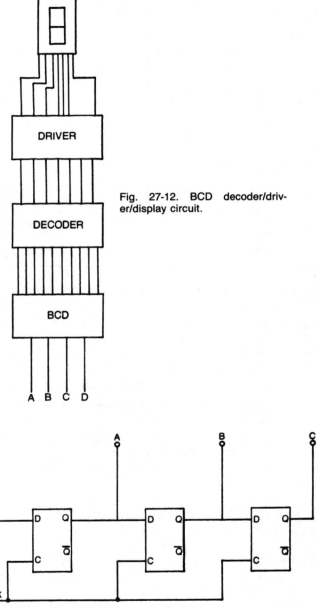

Fig. 27-12. BCD decoder/driver/display circuit.

Fig. 27-13. An SIPO shift register.

☐ **Clock pulse #2.** QB is still 0 so there is still no change in flip-flop C's output state on this clock pulse. QA is a 1, and this is fed into DB, so QB is now changed to a 1. DA (system input) is now receiving only 0's, so QA reverts back to logic 0. The binary output (from right to left) is now 010.

□ **Clock pulse #3.** QB, and thus, DC, is a logic 1, so QC changes to a 1. QA/DB is a logic 0, so flip-flop B's output goes back to 0. DA is receiving only 0's, so QA remains at logic 0. The binary output (from right to left) is now 011.

□ **Clock pulse #4.** QB/DC is at logic 0, so QC also becomes a 0. The first two stages (A and B) also have 0's on their inputs, so their outputs stay at the 0 level.

The output has returned to the 000, initially cleared condition. The outputs will all stay 0's until another logic 1 is inputted to the system.

Notice how the 1 is shifted through the outputs. This circuit is called a *shift register*. This type of circuit is useful when selected digital information needs to be applied to different subcircuits at slightly different times in a specific sequence. It could be called a *digital delay circuit*.

One digit of digital information (a single 0 or 1) is called a *bit*. Bits are usually grouped into *words*, or *bytes*. Bytes are typically four, eight, or sixteen bits long. All bytes within a given system, of course, will be of equal length. Most shift registers are designed to hold a full byte of digital information.

The shift register shown in Fig. 27-13 accepts data in a serial fashion (one bit after another), and the output is parallel (all bits are available simultaneously). It is called a *SIPO* (Serial In—Parallel Out) shift register.

The SIPO is probably the most commonly used type, but other combinations are also possible. Figure 27-14 shows a *SISO* (Serial In—Serial Out) circuit. Figure 27-15 is a *PISO* (Parallel In—Serial Out) device, and Fig. 27-16 is a *PIPO* (Parallel In—Parallel Out) circuit. It is also possible to build a shift register with both serial and parallel inputs and/or outputs.

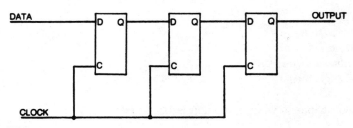

Fig. 27-14. An SISO shift register.

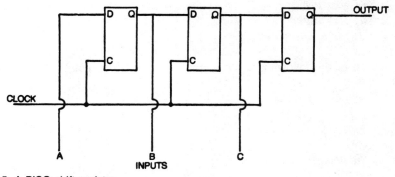

Fig. 27-15. A PISO shift register.

348

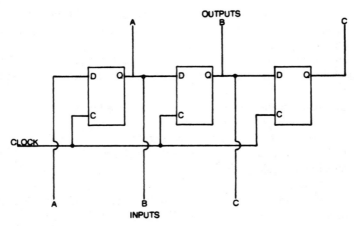

Fig. 27-16. A PIPO shift register.

There are many other digital devices, but most are made up of combinations of the basic devices described in the last two chapters. Many complex digital functions (incorporating several of these basic devices) are available in single chip MSI and LSI integrated circuits.

Self-Test

1. What is another name for a flip-flop circuit?

A *Digital gate*
B *Astable multivibrator*
C *Bistable multivibrator*
D *Monostable multivibrator*
E *None of the above*

2. What happens when a flip-flop is triggered?

A *The output reverses states*
B *The output goes low*
C *The output goes high*
D *The output remains unchanged*

3. Which of the following is *not* a standard type of flip-flop?

A *JK*
B *RS*
C *QS*
D *D*
E *None of the above*

4. What basic type of circuit is used to build up a multistage counter?

A *AND gates*
B *Amplifiers*
C *Flip-flops*
D *Schmitt triggers*
E *None of the above*

5. What is the maximum count of a 5-stage counter in binary?

A *10000*
B *11111*
C *100000*
D *111111*
E *None of the above*

6. What is the maximum count of a 5-stage counter in decimal?

A *32*
B *16*
C *50*
D *31*
E *None of the above*

7. If the Q output of a counter stage is at logic 1, what will be the value at $\overline{Q}$?

A *0*
B *1*
C *Either 0 or 1*

8. A counter stage can divide an input frequency by what factor?

A *0.2*
B *3*
C *4*
D *2*
E *None of the above*

9. Which of the following is a disallowed state for a BCD circuit?

A *0000*
B *1001*
C *0111*
D *1101*
E *None of the above*

10. How many outputs does a six-stage PISO shift register have?

A *One*
B *Two*
C *Five*
D *Six*
E *None of the above*

28

Logic Families

There are a number of basic methods for designing digital circuits. Digital integrated circuits are classified according to their logic families, or basic design approach. The earliest digital integrated circuits were usually either *RTL* or *DTL* devices.

RTL stands for *Resistor-Transistor Logic*. Each digital gate is basically comprised of a resistor and a transistor.

DTL, on the other hand, is *Diode-Transistor Logic*. Each digital gate is made up primarily of a diode and a transistor.

Remember, these are not discrete components, but their electrical equivalents etched into the semiconductor material of an integrated circuit chip.

While RTL and DTL ICs are relatively easy to manufacture, they are rather slow and inefficient in their power consumption. An improved digital family is *TTL*, or, as it is sometimes called, T²L. This stands for *Transistor-Transistor Logic*.

TTL

Thanks to mass production and relative simplicity of design, TTL integrated circuits are quite inexpensive (as low as 25¢ a chip from certain sources). These devices are also capable of high speed operation—20 MHz (20,000,000 Hz) is typical, but some are capable of frequencies as high as 125 MHz (125,000,000 Hz). They have a fair fan-out capability (10 is the average), and a reasonable immunity to noise.

TTL devices are probably the most popular and widely available logic family on today's market. This is due to both the number of manufacturers making TTL ICs, and the large

351

352

Table 28-1. TTL Devices and Their Functions.

7400	QUAD TWO INPUT NAND GATE
7402	QUAD TWO INPUT NOR GATE
7404	HEX INVERTER
7408	QUAD TWO INPUT AND GATE
7410	TRIPLE THREE INPUT NAND GATE
7414	HEX SCHMITT TRIGGERS
7417	HEX BUFFER
7432	QUAD TWO INPUT OR GATE
7442	BCD TO 1 OF 10 DECODER
7447	BCD TO SEVEN SEGMENT DECODER/DRIVER
7473	DUAL JK LEVEL-TRIGGERED FLIP-FLOP
7474	DUAL D EDGE-TRIGGERED FLIP-FLOP
7486	QUAD EXCLUSIVE OR GATE
7490	DECADE COUNTER
7493	BINARY COUNTER (÷ 16)
74 95	FOUR BIT PIPO SHIFT REGISTER
74121	MONOSTABLE MULTIVIBRATOR
74150	1 OF 16 DATA SELECTOR

number of different devices available. Table 28-1 lists some of the more popular TTL devices and their basic functions.

Usually TTL integrated circuits will be numbered in the 74xx format, but there are a few exceptions. Often you will see devices numbered 54xx. These two numbering systems are generally interchangeable. For instance, a 7404 is essentially identical to a 5404. The only significant difference is that the 54xx devices can operate over a wider temperature range. This is usually only needed for military and satellite applications. For most general applications the 74xx/54xx line of devices is ideal.

The power supply for standard TTL integrated circuits must be a tightly regulated 5 volts. The supply voltage should not be allowed to vary more than ±0.5 volts.

When a TTL gate switches output states (goes from a 0 to a 1, or vice versa) it draws a large surge or current. This can cause sharp high voltage spikes to appear on the power supply lines. These spikes could damage the delicate semiconductor chips.

To protect the integrated circuits, *despiking capacitors* should be connected between the +5 volt and ground lines as close to the IC packages as possible. One 0.01 μF to 0.1 μF capacitor is usually sufficient to protect up to four gate packages, if they are closely placed.

For TTL circuits a logic 1 is usually defined as a signal of about + 2.4 volts, while logic 0 is about half a volt above absolute ground. Any signal between these levels is undefined, and may be interpreted by the gate as a 0, a 1, or simply noise.

One important factor to bear in mind is that an unconnected TTL input will usually pull itself up to a logic 1 level. This can cause quite a bit of confusion to a circuit designer who is not prepared for it. Also, under certain conditions, an unused gate can influence the output of another gate within the same IC package. All unused inputs should be tied to one or the other fixed logic state, preferably unused inputs should be connected to ground.

There are a number of subfamilies within the basic TTL group. These trade off various advantages and disadvantages.

Low Power TTL

Regular TTL integrated circuits use a relatively small amount of power individually, but in moderate to large systems, using many IC packages, the power drain can very quickly add up. To alleviate this problem, special *Low Power TTL* ICs are sometimes used.

These low power units are numbered in the same way as regular TTL devices, but with a "L" added in the middle to indicate the low power status. For example, a low power version of the 7400 quad NAND gate is the 74L00.

A low power TTL integrated circuit generally consumes only about a tenth as much power as the standard TTL version.

This lower power consumption does not come without disadvantages, however. Low Power TTL chips can operate only about a tenth as fast as regular TTL devices.

High Speed TTL

Another subfamily of TTL devices is *High Speed TTL*. A High Speed 7400 would be numbers 74H00. This class of devices operates about twice as fast as regular TTL, but consumes about twice as much power.

High Speed TTL integrated circuits are becoming less and less common as time goes on. They are being replaced by Schottky devices, which are generally superior, although somewhat more prone to noise problems.

Schottky TTL

By using *Schottky diodes* in the circuit design, a better speed/power trade-off can be achieved. Schottky diodes are very fast switching diodes, so *Schottky TTL integrated circuits* can be used for very high frequency operation. Generally, a Schottky TTL gate can operate about three and a half times as fast as a regular TTL gate, but the power consumption is only doubled.

A Schottky version of the 7400 would be numbered 74S00.

Low Power Schottky

By combining the design techniques of Low Power TTL and Schottky TTL devices, integrated circuits can be manufactured that operate at the same speed as regular TTL, but at only a fifth of the power consumption. Obviously devices in this group are called *Low Power Schottky TTL*. A Low Power Schottky equivalent to the 7400 would be numbered 74LS00.

The various TTL subfamilies are compared in Table 28-2. Regular TTL is used as the comparison standard.

CMOS

Another major category of digital devices are called *CMOS integrated circuits*. CMOS standards for *complementary metal-oxide silicon*. We have already discussed Metal Oxide Semiconductors in the section on MOSFETs.

CMOS integrated circuits can operate at very high speeds with relatively low power consumption. CMOS devices are also much more tolerant of power supply fluctuations

Table 28-2. Comparing the TTL Sub-families.

Subfamily	Gating Speed	Power Consumption
Standard TTL	1	1
Low Power	0.1	0.1
High-Power	2	2
Schottky	3.5	2
Low Power Schottky	1	0.2

Table 28-3. Comparing a 7400 TTL Quad NAND
Gate with a CD4011 CMOS Quad NAND Gate.

Device	CD4011	7400
DC SUPPLY VOLTAGE RANGE POWER DISSIPATION	− 0.5 to + 15 volts 500 mW	+ 4.5 to 5.5 volts 60 mW

Device	CD4001 +5 volt supply	CD4001 +10 volt supply	7400
PROPAGATION DELAY TIME	50 nS	25 nS	22 nS
OUTPUT VOLTAGE			
LOGIC 0	0	−0.5	0.4
LOGIC 1	4.95	9.95	2.4

than are TTL units. In fact, a CMOS IC can be powered by any voltage between 3 and 12 volts. Surges to 15 volts can be withstood, but should be avoided, if possible.

However, CMOS integrated circuits tend to be somewhat more expensive than their TTL counterparts. Also, they are susceptible to damage from bursts of static electricity. Anything that touches a CMOS chip should be properly grounded, and that includes the human hand. Newer units are less sensitive than older devices, but problems can still arise unless precautions are taken.

There are two common numbering systems for CMOS integrated circuits. One system reflects TTL numbering. A 7400 TTL quad NAND gate could be replaced by a 74C00 CMOS unit. The equivalent in the other numbering systems would be a CD4011. The CD40xx system is more common and the 74Cxx devices are gradually disappearing from the market.

Table 28-3 compares a 7400 TTL quad NAND gate with a CD 4011 CMOS quad NAND gate.

ECL

A somewhat less common but still important logic family is *ECL*, or *Emitter-Coupled Logic*. ECL devices are capable of extremely high switching speeds. The output transistors in an ECL gate are biased so that they are in their active region at all times. The transistors are not driven into saturation, so there is no problem with stored base-charge. This minimizes propagation delay times.

The ECL family of digital devices differs from the more common TTL and CMOS devices in that the logic levels are negative. A logic 1 is represented by a voltage of about -0.9 volt. A logic 0 is approximately -1.75. This can introduce some problems when interfacing ECL devices with circuitry from the other logic families.

The power supply for ECL devices should also be negative (with respect to ground). Acceptable supply voltages range from -8 volts to -3 volts. The preferred nominal supply voltage is -5.2 volts.

The high speed capability of ECL devices comes at a price, of course, Power dissipation is quite high because the transistors within the gate are continuously on. However, at high operating speeds, the differences in current consumption between ECL and other logic families becomes increasingly less significant.

Of course, when the high speed capabilities of ECL devices is taken advantage of, great care must be taken in circuit layout. Interconnecting leads must be kept as short as possible. Shielding may be necessary in some circuits.

The inputs to ECL gates exhibit very high impedances. This means, that to ensure reliable and predictable operation, no inputs should be left floating. All inputs must be fed an unambiguous logic 1 or logic 0.

ECL outputs are open emitters. A small external load resistor is required. The value will typically be between 250 to 500 ohms. The load resistor provides a current path for the output transistor.

Self-Test

1. Which of the following is *not* a standard logic family?

A *VMOS*
B *ECL*
C *TTL*
D *RTL*

2. Which of the following logic families is considered obsolete?

A *TTL*
B *CMOS*
C *DTL*
D *ECL*
E *None of the above*

3. Which of the following logic families is most commonly used by hobbyists?

A *DTL*
B *TTL*
C *ECL*

D *RTL*
E *None of the above*

4. What is a good supply voltage for TTL devices?

A *−5.2 volts*
B *+12 volts*
C *+25 volts*
D *+5 volts*
E *None of the above*

5. What is a good supply voltage for CMOS devices?

A *−5.2 volts*
B *+12 volts*
C *+25 volts*
D *+5 volts*
E *None of the above*

6. Which of the following TTL devices use Schottky diodes in its internal circuitry?

A *74LS04*
B *7400*
C *74L17*
D *74H86*
E *None of the above*

7. Which of the following logic families has the highest switching speed capability?

A *RTL*
B *TTL*
C *CMOS*
D *ECL*

8. Which of the following can damage CMOS devices?

A *Static electricity*
B *A 0-volt input*
C *A supply voltage greater than five volts, but less than ten volts*
D *A 1-volt input*
E *None of the above*

9. Which of the following is not a CMOS device?

A *74C10*
B *CD4001*

C *7404*
D *4049*
E *None of the above*

10. Which of the following logic families can stand the greatest degree of variation in the supply voltage?

A *TTL*
B *CMOS*
C *Schottky TTL*
D *DTL*
E *None of the above*

29

Microprocessors

Digital circuits were initially designed primarily for use within computers. They made the modern personal computer practical. The first computers used tubes and relays and filled buildings. When the transistor was developed, a computer only filled a good-sized room. Using low-level ICs, a computer the size of a desk could be built. Today, we have small, desk-top computers, thanks to specialized LSI integrated circuits. The most sophisticated and complicated ICs developed have been microprocessors.

CPUS AND MICROPROCESSORS

The "brain" of a computer is the *CPU*, or *Central Processing Unit*. This circuit accepts digital data and commands. It interprets what the commands mean, and performs mathematical operations on the data. In modern equipment, the CPU circuitry is almost always contained within a single IC chip, known as a *microprocessor*.

A microprocessor is sometimes called a "computer on a chip," but this isn't entirely accurate. In most cases, a number of additional external ICs are required to form a practical, operating computer.

A CPU contains a number of *registers*, or temporary storage spaces. In its simplest form, one number for a given equation is put into one register. When the next number is entered, it is mathematically combined with the number already in the register. The register now contains the result, which is fed out to the display. This type of register is called the *accumulator register*. The circuitry for performing these mathematical functions is called the *arithmetic logic unit*, or *ALU*. Most calculators use CPUs that are not much more than simple ALUs.

The CPUs used in computers, however, are much more complex and have a number of additional registers. These registers are used for various purposes, such as storing an instruction, or keeping track of its place in the program. The 8080 (a popular CPU IC), for example, contains no fewer than 13 registers.

In addition to the registers, there must also be some way to get data in and out of the CPU to be manipulated. This is done via the *data bus*. The data bus may consist of four, eight, or sixteen lines for digital data. Eight is probably the most popular number.

Each line carries a single *bit* (a 0 or a 1) of information. All of the lines together simultaneously carry a *word* of digital information. For a given system, the words are always of the same length. A eight-bit, or a sixteen-bit word is called a *byte*. A four-bit word is called a *nibble*. The size of the words used, determines the amount of data they can communicate. A nibble, for example, has only 16 possible combinations. An eight-bit byte can have over 65 thousand different values. Obviously, the greater the word length, the more powerful and versatile the computer.

Usually the same data bus is used for both inputting and outputting data. This means there must also be an additional single line input for telling the CPU whether data should be coming in or going out at any given instant.

There is also a *memory address bus* that is used to find a specific location in the memory. This typically consists of sixteen lines, so over 65 thousand separate memory locations that can be individually addressed. Newer microprocessors use sixteen bit commands and thirty-two bit address lines, for very large memory access.

COMMANDS AND DATA

A CPU or microprocessor "understands" a number of commands. These commands must be in binary form. For example, 01001101 or 11101001. Notice that the commands are in the same form as the data. The CPU uses position (what came just before the current number) and timing to distinguish between commands and data. A command tells the CPU what to do with the data.

For commands, each binary number has a specific operational meaning for the CPU. A series of commands to perform a specific task is called a *program*.

If you program the microprocessor directly using the binary number commands, this is called *machine-language programming*.

Translation programs are available to allow the user to program the microprocessor with a more convenient (more English-like) set of commands.

The next step up from machine-language programming is *assembly-language programming*. Each binary number command is replaced with an easy to remember *mnemonic*. For instance, "ADD A,B" for adding values A and B instead of 10011100. Each assembly-language command corresponds to exactly one machine-language command.

For higher level languages, such as *BASIC*, or *Pascal*, each user-entered command can be translated into a sequence of several machine-language commands in sequence. A special program (a *compiler* or an *interpreter*) is used to convert the English-like commands into the binary form understandable by the CPU. Higher level languages will be discussed in Chapter 40.

COMPONENTS OF A COMPUTER

A CPU or microprocessor is not a full computer in itself. Figure 29-1 shows the basic structure of a typical computer. Two or more of these stages can be included on a single IC, or separate ICs are used for each stage. Some stages are made up of multiple ICs. This especially tends to be true for the memory stage. For our purpose here, we will consider the various stages of the computer to be separate and distinct.

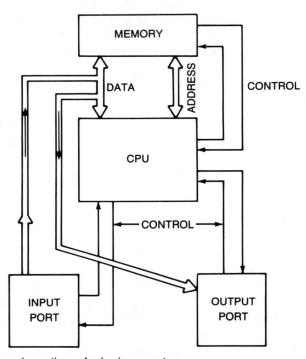

Fig. 29-1. The four main sections of a basic computer.

There are four primary sections of a computer;

    Processor (CPU)
    Memory
    Input Port
    Output Port

The processor does the actual computing. A microprocessor is just a miniaturized processor. The memory stores the program commands and data used in executing the program.

The input and output ports allow the microprocessor to communicate with the outside world (anything that is not an integral part of the computer itself). The input port permits data from some external device (such as a keyboard or a modem) to be fed into the computer. Similarly, the output port lets the computer feed its results out to some external device (such as a printer or monitor screen).

Communication between the various internal sections of the computer are accomplished via buses. These buses are digital signal lines that can carry coded binary data either serially (one bit at a time) or in parallel (several bits at once). The binary data on a bus can represent numerical values, alphanumeric characters, or machine-language commands. The only difference is in how the CPU is instructed to interpret the binary information.

The *data bus* connects the microprocessor to everything else. The data bus goes from the CPU to the memory, the input port, and the output port. Data flows from the input port to the CPU, from the CPU to the output port, or in either direction between the memory and the CPU. The CPU determines which piece of data goes where.

The *address bus* goes from the CPU to the memory. The data on this bus determines which portion of the memory the microprocessor wants to use. The concept is simple enough to understand if you think that the CPU needs to know the address to find a friend's home.

The third bus is used for system control and synchronization. The signals on this bus keep the various sections of the computer functioning simultaneously. For example, the input port uses this bus to let the microprocessor know when there is some incoming data available from an external input device. Without proper synchronization, the data exchanged between the various sections of the computer would become garbled and meaningless.

MEMORIES

Even with several registers, the CPU cannot store very much data, and it can only handle a single program instruction one step at a time. For this reason, some sort of external memory storage device is required for the computer to be useful.

There are two basic types of memory. They are called *ROMs* and *RAMs*. A ROM is a *read-only memory*. Data is permanently stored in this type of device. The data is either placed in the memory by the manufacturer, or it can be entered by the user if a PROM (*programmable read-only memory*) is used. In either case, once the data is entered, it cannot be altered.

There is also a device called an *EPROM*, or *erasable programmable read-only memory*, which can be cleared by exposing it to a strong ultra-violet light. It can then be reprogrammed like an ordinary PROM.

No data in any kind of ROM can be altered in any way while it is in use in the computer. If power is removed, the data in the ROMs will not be lost.

The other type of memory is called RAM, or *random access memory*. A better name would be *read-write memory*, since data can be written into a read out of this type of memory by the CPU. If power is interrupted, the data in a RAM will be lost.

With either type of memory, the data can be randomly accessed. This means the CPU can call out any number on the memory address bus, and look at (or, in the case of a RAM, write) the data at the specified memory location.

ROMs are typically used to store *language translation programs*, while RAMs are used to store individual programs and data. Computer languages will be discussed in the next section.

The total number of possible memory locations (both ROM and RAM) is limited by the number of possible address codes the CPU generate. In complex programs, or where amounts of data must be stored, there quite often will not be enough memory space available.

Besides, memory circuits are rather expensive, so the maximum possible number is not included in most small computer systems. Memory space is usually identified as so many K. Ordinarily K stands for 1,000, but in digital circuits it is actually 1,024 (2^{10}), because 1,000 is not a convenient binary number. Small computer systems generally have 4K or 16K of memory. Usually the maximum possible is 64K.

Another problem is that a frequently used program must be re-entered each time it is used. This is time consuming and inconvenient at best.

For all of these reasons, some sort of external memory storage system is highly desirable, if not absolutely essential for certain programs.

The earliest systems used punch cards, or punched paper tape. See Fig. 29-9. The presence or absence of a given punched hole identifies a specific bit's logic state. This approach is often adequate, or, at least, better than nothing, but it is non-erasable, bulky and inconvenient to store, and the holes can snag and tear.

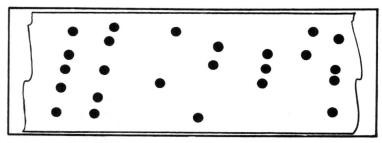

Fig. 29-2. Computer punch tape.

External memory can also be stored on magnetic tape by digitally encoding a tone. Large, commercial systems generally use reel to reel tape, on large reels, while smaller systems often use ordinary cassettes, which are less expensive, easier to use, and more convenient to store.

The biggest disadvantage of using magnetic tape for storing data (especially cassettes) is that it is relatively slow, and not randomly accessible. The computer must search through the entire tape until it locates the data it needs. Much of the computer's time is wasted searching for data.

Floppy discs are the answer to this problem. These are flexible discs that are coated with magnetic particles. Data can be recorded by these particles, in much the same way as ordinary magnetic tape. The surface of the disc is divided into *sectors* that are randomly accessible. Floppy discs offer large amounts of storage space, and allow for extremely fast data transfer.

Thanks to improved technology, improved reliability, and steadily decreasing costs, hard disc drives are becoming more common in many small computer systems. A hard disk stores much more data than a floppy disk. The storage capacity of a hard disk is measured in megabytes (millions of bytes).

Self-Test

1. Which section of a computer is the "brain"?

A *CPU*
B *Memory*
C *Input Port*
D *Bus*
E *None of the above*

2. What is the difference between a data byte and a command byte?

A *Data bytes always begin with 1*
B *Command bytes have more bits*
C *Command bytes have fewer bytes*
D *There is no difference*

3. How many bits are there in a standard byte?

A *One*
B *Two*
C *Four*
D *Eight*
E *Ten*

4. What is a "program"?

A *A series of commands to perform a specific task*
B *A language for communicating with a computer*
C *A single instruction*
D *A data file*
E *None of the above*

5. Which language(s) can a CPU understand directly?

A *BASIC*
B *Machine-language*
C *Assembly-language*
D *Pascal*
E *English*

6. Which of the following is *not* part of a basic computer?

A *Microprocessor*
B *Output port*
C *Printer*
D *Input port*
E *Memory*

7. What is a bus?

A *A digital signal line*
B *An input device*
C *An output device*
D *A type of memory storage*
E *None of the above*

8. Which type of memory is *not* user programmable?

A *RAM*
B *ROM*
C *EPROM*
D *None of the above*

9. What happens to data stored in RAM when power is interrupted?

A *Nothing*
B *It is inverted*
C *It is stored in the CPU's registers*
D *It is moved to ROM*
E *It is lost*

10. Which of the following is *not* a type of long term data storage?

A *Floppy disk*
B *Magnetic tape*
C *Punch cards*
D *Modem*
E *None of the above*

30

Sensors and Transducers

Sensors and *transducers* are devices that allow electronic circuits to communicate with the outside world. Sensors sense some external condition and produce an electrical signal, or alter an existing electrical signal in response. Transducers transduce, or change, one form of energy into another. If a sensor produces an electrical signal in response to some external condition it is actually a transducer as well as a sensor.

In Chapter 22 we dealt with light sensitive devices, or photosensors. These devices convert light energy into electrical energy (or a related electrical parameter). The LEDs discussed back in Chapter 19 are also transducers, but they work in the opposite manner. They normally convert electrical energy into light energy. Under certain special circumstances, LEDs can be used as light sensors.

In this chapter we will examine a number of sensors and transducers that detect or produce energy other than light.

HAND CAPACITANCE AND RESISTANCE

An interesting way to control an electronic circuit is by using the operator's body itself as a component. Like all matter, your hand exhibits a certain amount of resistance. The value is relatively high, but it is quite variable. For example, lie detectors work on the principle that skin resistance changes with emotional states.

Typical application is the *touch switch*. This is simply two conductive plates separated by a small distance. See Fig. 30-1.

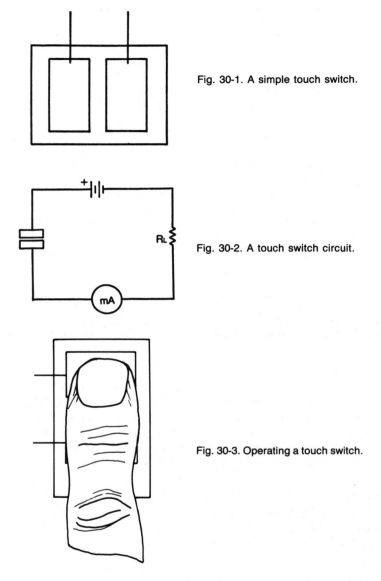

Fig. 30-1. A simple touch switch.

Fig. 30-2. A touch switch circuit.

Fig. 30-3. Operating a touch switch.

If a touch plate is included in a circuit like the one illustrated in Fig. 30-2, ordinarily no current will flow. But if the operator touches both plates simultaneously (see Fig. 30-3) current can flow through his finger to complete the circuit.

For safety's sake this should only be done with low power dc circuits!

Because the resistance of the human hand is relatively high, it can be used as the dielectric of a capacitor. The circuit board shown in Fig. 30-4 is used for this purpose. The dark areas are strips of copper foil mounted on an insulating board (this is called a *printed circuit board*). This board can be connected into a circuit like an ordinary capacitor. Ordinarily the dielectric would be air and the insulating board material. But, if

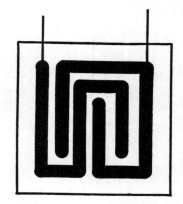

Fig. 30-4. A hand capacitance board.

the operator laid his hand across the various copper traces, his hand becomes the dielectric (since the hand's resistance is much lower than that of the board, the board can now be ignored). By changing the pressure of his hand on the board, the operator can vary the amount of contact area between the dielectric (hand) and the plates (copper traces). Of course, this affects the effective capacitance.

Another interesting body capacitance effect takes place without any physical contact at all. If a hand is brought near an *antenna* (discussed in a later chapter), the antenna will act like a *proximity* detector that senses the distance to the hand.

An intriguing application of this effect is an instrument called the *Therimin*. This is a musical instrument that can be played without touching it. It generally has two antennas—one controlling pitch, and the other controlling volume. By moving one's hands around the two antennas, different tones can be produced.

THE PIEZOELECTRIC EFFECT

Back in the chapter on crystals, we discussed how the piezoelectric effect can cause a crystal slab to oscillate. Remember that electrical stress along the X axis causes the mechanical stress to be produced along the Y axis. Similarly, a mechanical stress along the Y axis will produce an electric signal along the X axis. This means a crystal can also be used as a *pressure sensor*.

Perhaps the most common application for this effect is the *ceramic cartridge* used in record players. Records are plastic discs with a spiral groove cut into their surface. This groove undulates in a specific pattern that corresponds to the music recorded. A *needle*, or *stylus* is connected to the crystal element in the crystal. This needle rides in the grooves cut on the record. As the needle is forced back and forth by the fluctuations of the groove different mechanical stresses are put upon the crystal. This is converted to an electrical signal that can be amplified and treated by the rest of the circuitry.

THERMISTORS

Another electronic sensor device is the *thermistor*. This is a resistor whose value changes in response to temperature. Some obvious applications for thermistors include electronic thermometers and thermostats for precise temperature control. They can also be used in fire alarms. The schematic symbol for a thermistor is shown in Fig. 30-5.

Fig. 30-5. Schematic symbol for a thermistor.

MICROPHONES

Microphones are transducers that convert audio energy (sound waves) into electrical energy. There are a number of basic methods for accomplishing this. Most microphone types are represented by the schematic symbol shown in Fig. 30-6. The word "microphone" is often shortened to *"mike."*

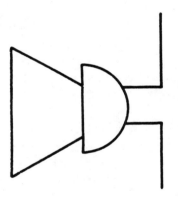

Fig. 30-6. Schematic symbol for a microphone.

Carbon Microphones

Perhaps the simplest type of microphone is the carbon microphone. The basic construction of this device is illustrated in Fig. 30-7. Basically a carbon microphone consists of a small container filled with carbon granules. This container has a carbon disc at either end. One of these discs is rigidly held in a fixed position, while the other is movable. The movable disc is connected to the *diaphragm*. Sound is caused by fluctuations in air pressure. This pressure moves the diaphragm (and thus the movable carbon disc) back and forth. This puts greater or lesser pressure on the carbon particles within the container. The changes in the density of these particles changes their effective resistance. If this assembly is in series with a dc voltage source (such as a battery, as shown in Fig. 30-8) the voltage drop will vary along with the changes in the resistance of the carbon particles, which is, in turn, caused by the changing sound pressure. Thus, the voltage output varies in step with the sound waves striking the diaphragm. We have an electrical equivalent of the acoustic energy.

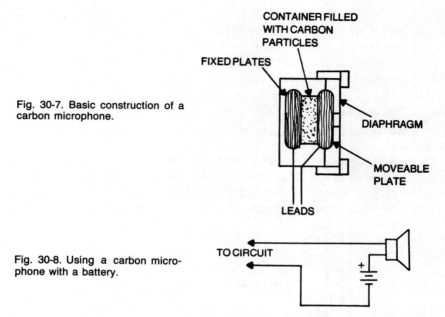

Fig. 30-7. Basic construction of a carbon microphone.

Fig. 30-8. Using a carbon microphone with a battery.

The primary advantages of carbon microphones are that they are relatively low in cost, and they provide the highest level output of all commonly available microphone types.

However, there are also a number of significant disadvantages. Carbon microphones require an external voltage source. They have a rather narrow frequency response, and their noise and distortion levels are higher than any other microphone type.

Carbon microphones are typically used in telephone handsets.

Crystal Microphones

Another type of microphone employs the piezoelectric effect. Figure 30-9 shows the basic construction of a *crystal microphone*. The sound pressure put on the diaphragm produces a mechanical stress on the crystal element. This, of course, generates a voltage in step with the mechanical stress.

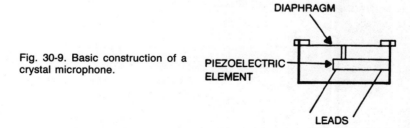

Fig. 30-9. Basic construction of a crystal microphone.

While resonant crystals are generally made of quartz, crystal microphones usually use elements made of Rochelle salt.

The crystal microphone offers a number of advantages. It requires no external voltage, has a fairly high output level and a fair frequency response. However, it is rather fragile. Also, the Rochelle salt element can absorb moisture, ruining it. These two problems can be dealt with by replacing the Rochelle salt with a somewhat more rugged ceramic element. In this case we have a *ceramic microphone*.

Crystal and ceramic microphones are good for general communications applications, but the frequency response is not adequate for high fidelity use.

Dynamic Microphones

Dynamic microphones are probably the most popular type of general purpose microphones.

The basic structure of a dynamic microphone is shown in Fig. 30-10. The diaphragm is connected to a small coil which is suspended, so both can move freely in response to sound pressure. The coil moves within the magnetic field of a permanent magnet. Of course, this induces a voltage in the coil that varies in step with its movement.

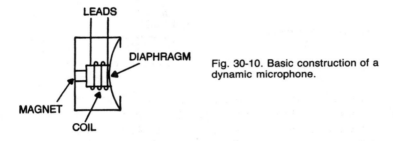

Fig. 30-10. Basic construction of a dynamic microphone.

The output voltage of a dynamic microphone is fairly low, but its frequency response is quite good.

Ribbon Microphones

Another type of microphone that is similar to the dynamic type is the *ribbon microphone*. Figure 30-11 shows the basic construction of this device. In this type of microphone a corrugated-aluminum ribbon is moved through the magnetic field of the permanent magnet. A small voltage will be induced in the ribbon via this process.

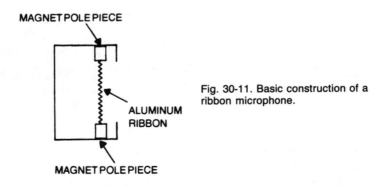

Fig. 30-11. Basic construction of a ribbon microphone.

The output voltage is extremely low, and usually has to be fed through a step-up transformer to reach a usable level. This transformer is often contained within the case of the microphone itself. Even with the transformer the output from a ribbon microphone is very low, but the frequency response is excellent. Ribbon microphones are also quite rugged.

Condenser Microphones

Figure 30-12 shows the basic construction of a condenser microphone. Two small plates are separated by a small amount. One is rigid, and the other is flexible (acting as the diaphragm). The movement of the diaphragm varies the distance between the two plates, which changes the capacitance between them. A small circuit within the microphone's case converts this varying capacitance to a varying voltage.

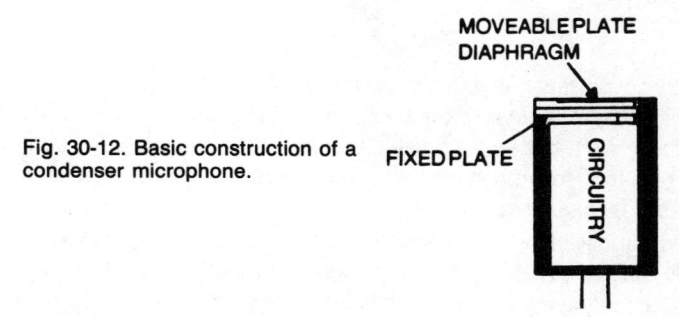

Fig. 30-12. Basic construction of a condenser microphone.

Condenser microphones offer very low distortion and an excellent frequency response. However, they are rather expensive, and require their own dc voltage source.

SPEAKERS

A *speaker* is the opposite of a microphone. It converts electrical energy back to sound pressure in the air. In fact, a small speaker can be used as a low quality dynamic microphone. The schematic symbol for a speaker is shown in Fig. 30-13. Often the term "speaker" is shortened to "SPKR" in print. The basic construction of a speaker is illustrated in Fig. 30-14.

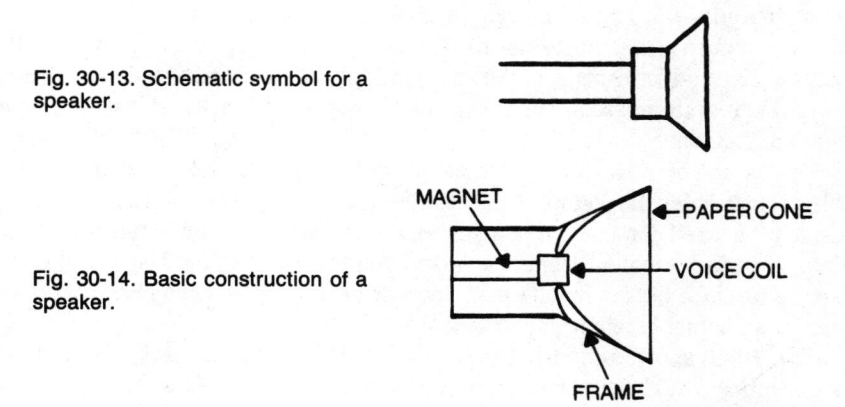

Fig. 30-13. Schematic symbol for a speaker.

Fig. 30-14. Basic construction of a speaker.

The electrical signal is applied to a coil of wire called the *voice coil*. Since the voice coil is suspended within the magnetic field of a permanent magnet, it will move back and forth in step with the applied signal. The voice coil is connected to the center of a paper cone. The outer rim of this cone is firmly attached to a sturdy frame. The cone is forced to move in and out along with the voice coil. See Fig. 30-15. This movement of the cone changes the air pressure in step with the original electrical signal. These pressure fluctuations are perceived by the ear as sound.

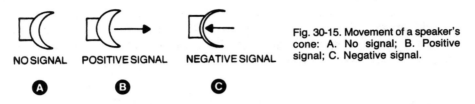

NO SIGNAL   POSITIVE SIGNAL   NEGATIVE SIGNAL

Fig. 30-15. Movement of a speaker's cone: A. No signal; B. Positive signal; C. Negative signal.

A   **B**   **C**

No speaker is perfect. All exhibit some signal loss and distortion. Also the frequency response will always be less than ideal. A small speaker cone will reproduce high frequencies fairly well, but won't be able to handle low frequencies. Conversely, a large cone area is required to reproduce low frequencies, but it won't be able to vibrate fast enough for high frequencies.

This is why most high fidelity speaker systems will contain more than one speaker within a single cabinet. A small *tweeter* reproduces the high frequencies, and a large *woofer* handles the low frequencies. A *crossover network* is used to separate the signals. The crossover network is basically a high pass filter that blocks low frequencies from the tweeter. A large amplitude low frequency signal could damage the delicate voice coil of a tweeter. High frequencies applied to a woofer will essentially be ignored by the speaker itself.

Many speaker systems include a third speaker between the woofer and the tweeter. Not surprisingly, this third speaker is called a *mid-range speaker*.

Some speakers are designed as *full-range* units and are capable of reproducing frequencies from about 100 Hz to 15,000 Hz, but these full range units are not as accurate as separate woofer/tweeter combinations.

While a speaker's frequency range is definable, the frequency response is not. The frequency response of a speaker can be greatly affected by the shape and size of the cabinet, the material the cabinet is made of, the positioning of the speaker within the cabinet, the size and furnishings of the room the speaker is used in, and even the position of the speaker system within the room. Obviously all these variables provide infinite possible combinations.

Speakers are often tested for frequency response in echoless chambers so room variables do not enter the picture. While the frequency responses obtained in this manner can usually be used for comparison purposes, it must be remembered that the test conditions are far removed from real world conditions. It is entirely possible for one speaker to produce better results than another in an echoless chamber, yet not sound as good in an actual listening environment.

Figure 30-16 shows the ideal frequency response curve for a theoretically perfect speaker. Notice that all frequencies are reproduced equally. This is called *flat response*.

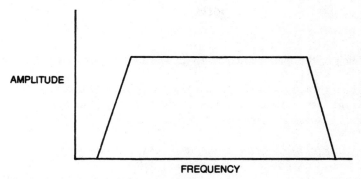

Fig. 30-16. Ideal speaker frequency response curve.

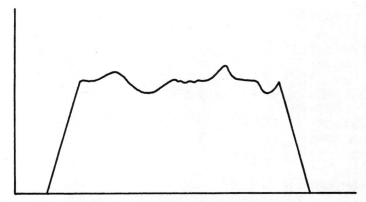

Fig. 30-17. Typical frequency response curve for a practical speaker.

A typical frequency response curve from a practical speaker is illustrated in Fig. 30-17. Notice that certain frequencies are reproduced at a higher level than others. Also remember, this curve would look quite different if the same speaker was placed in a different environment, but the basic shape will probably be more or less the same.

Self-Test

1. Which of the following terms describes a sensor which reacts to light?

A *Photosensitive*
B *Photoemissive*
C *Transducer*
D *Piezoelectric*
E *None of the above*

2. What type of power should be used for a touch switch circuit?

A *Any dc*
B *AC*

C *Low-power dc*
D *Doesn't matter*

3. Which of the following correctly describes the piezoelectric effect?

A *A mechanical stress along the X axis produces an electrical signal along the Y axis*
B *A mechanical stress along the Y axis produces an electrical signal across the X axis*
C *A mechanical stress along either axis produces an electrical signal across the opposite axis*
D *Pressure affects the dielectric value, and thus, the capacitance*

4. Which of the following microphone types requires its own dc voltage supply?

A *Condenser*
B *Dynamic*
C *Ribbon*
D *Crystal*
E *None of the above*

5. What bodily characteristic is used in lie detectors?

A *Skin resistance*
B *Hand capacitance*
C *Heart rate*
D *Body inductance*
E *None of the above*

6. What is a typical application of hand capacitance?

A *Lie detector*
B *Optoisolation*
C *Pressure measurement*
D *Touch switches*
E *None of the above*

7. Which of the following would be used for a pressure sensor?

A *Photodiode*
B *Optoisolator*
C *Crystal*
D *Varactor*
E *None of the above*

8. What electrical characteristic varies with temperature in a thermistor?

A *Capacitance*
B *Resistance*

C *Gain*
D *Inductance*
E *None of the above*

9. Which of the following is not a common type of microphone?

A *Carbon*
B *Ribbon*
C *Condenser*
D *Dynamic*
E *None of the above*

10. What type of speaker is best at reproducing low frequencies?

A *Dynamic*
B *Tweeter*
C *Woofer*
D *Crossover*
E *None of the above*

31

Experiments 3

Table 31-1 lists the equipment and parts needed for this final collection of experiments.

EXPERIMENT #1 INVERTING AMPLIFIER

Breadboard the circuit shown in Fig. 31-1 around a 741 operational amplifier in an 8 pin DIP package. If you use an op amp in another package configuration (say, a 14 pin DIP) the pin numbers in the diagram will not be correct. If you use the specification sheet for your op amp to determine the correct pin-out, the circuit will work.

Notice that three voltage sources are required. Two of these, +9 volts (B2) and −9 volts (B3), provide power for the operational amplifier, while the three volt battery (B1) generates the input signal. R1 is a 10,000 ohm (brown-black-orange) resistor, and R3 is a 10,000 ohm potentiometer. For the first part of this experiment use a 22,000 ohm (red-red-orange) resistor.

Connect your voltmeter between point A (+) and ground (−) and carefully adjust R3 for a reading of exactly 0.5 volt. Now, reverse the polarity of the voltmeter and connect it between point B (−) and ground (+). The output voltage should be about 2.2 times the input voltage and the polarity is reversed. Your exact measured value may vary from this somewhat due to resistor tolerances. Enter the output voltage in the appropriate space in Table 31-2. Repeat this procedure for input voltages of 1, 1.5, and 2 volts. Enter each output voltage into Table 31-2.

Now, replace R2 with a 33,000 ohm (orange-orange-orange) resistor and repeat the experiment.

Table 31-1. Equipment and Parts Needed for the Experiments.

VOM	(voltmeter, AC voltmeter)
Breadboarding system	(oscillator, power supplies- - - +9V*, −9V*, +3V*, and well regulated +5V. The voltages marked "*" may be provided by batteries)
1	270 ohm resistor (yellow-violet-brown)
2	1000 ohm resistor (brown-black-red)
2	4700 ohm resistor (yellow-violet-red)
4	10,000 ohm resistor (brown-black-orange)
1	22,000 ohm resistor (red-red-orange)
1	33,000 ohm resistor (orange-orange-orange)
1	10,000 ohm resistor (brown-black-yellow)
1	1,000,000 ohm resistor (brown-black-green)
1	0.01 µF disc capacitor
1	0.1 µF disc capacitor
1	1 µF 10 volt electrolytic capacitor
1	10 µF 10 volt electrolytic capacitor
1	1N914 diode
1	741 operational amplifier IC (8 pin DIP)
1	555 timer IC (8 pin DIP)
1	7400 TTL quad NAND gate IC
1	7402 TTL quad NOR gate IC
1	7474 TTL dual D type flip-flop IC
2	SPDT switches
1	SPDT momentary contact switch
1	DPDT switch
1	8 ohm miniature speaker
2	LEDs
1	10,000 ohm potentiometer

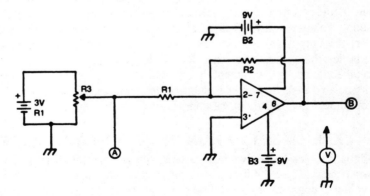

Fig. 31-1. Inverting amplifier circuit.

Now, substitute a 100,000 ohm (brown-black-yellow) resistor for R2 and go through the experiment again.

Finally repeat the entire procedure using a 10,000 ohm (brown-black-orange) resistor for R2.

Table 31-2. Worksheet for Experiment #1.

R2 VALUE	INPUT VOLTAGE	OUTPUT VOLTAGE	GAIN
22,000 ohms	0.5 volt	———	———
22,000 ohms	1.0 volt	———	———
22,000 ohms	1.5 volt	———	———
22,000 ohms	2.0 volt	———	———
33,000 ohms	0.5 volt	———	———
33,000 ohms	1.0 volt	———	———
33,000 ohms	1.5 volt	———	———
33,000 ohms	2.0 volt	———	———
100,000 ohms	0.5 volt	———	———
100,000 ohms	1.0 volt	———	———
100,000 ohms	1.5 volt	———	———
100,000 ohms	2.0 volt	———	———
10,000 ohms	0.5 volt	———	———
10,000 ohms	1.0 volt	———	———
10,000 ohms	1.5 volt	———	———
10,000 ohms	2.0 volt	———	———

Now, go back and calculate the gain for each input-output combination (G = output/input). For the three smaller resistors the gain should be fairly constant for each input value. But when a 100,000 ohm resistor is used for R2, the gain is more than the operational amplifier can handle with these input levels. The theoretical gain in this case is 10 (G = R2/R1), so an input voltage of 1 volt should produce an output of 10 volts. 1.5 volts in should result in 15 volts out, and 2 volts in should give an output of 20 volts. But the supply voltage is only ± 9 volts, so the amplifier saturates on these input levels. Remember, the output voltage can never be greater than the source voltage.

Also, notice the special case when R1 and R2 have the same value (G = 1). The output voltage should be approximately equal to the input voltage, except, of course, the polarity is reversed. Any error is due to component tolerances. Table 31-3 lists the results you should get, assuming all components are exactly on value.

EXPERIMENT #2 NON-INVERTING AMPLIFIER

Breadboard the circuit shown in Fig. 31-2. Notice that it is quite similar to the one used in experiment #1, except in this case we are working with a non-inverting amplifier.

The voltage supply values should be the same as in the last experiment. Similarly, R1 and R3 retain the same values. You will also use the same sequence of resistors for R2.

For each step connect your voltmeter between point A (+) and ground (−) and adjust R3 to the desired input voltage. Then move the positive lead of the voltmeter over to point B and measure the output voltage. Enter this value into Table 31-4.

Table 31-3. Nominal Results for Experiment #1.

R2 VALUE	INPUT VOLTAGE	OUTPUT VOLTAGE	GAIN
22,000 ohms	0.5 volt	− 1.1 volt	− 2.2
22,000 ohms	1.0 volt	− 2.2 volt	− 2.2
22,000 ohms	1.5 volt	− 3.3 volt	− 2.2
22,000 ohms	2.0 volt	− 4.4 volt	− 2.2
33,000 ohms	0.5 volt	− 1.65 volt	− 3.3
33,000 ohms	1.0 volt	− 3.3 volt	− 3.3
33,000 ohms	1.5 volt	− 4.95 volt	− 3.3
33,000 ohms	2.0 volt	− 6.6 volt	− 3.3
100,000 ohms	0.5 volt	− 5 volt	− 10
100,000 ohms	1.0 volt	− 9 volt	(− 9)
100,000 ohms	1.5 volt	− 9 volt	(− 6)
100,000 ohms	2.0 volt	− 9 volt	(− 4.5)
10,000 ohms	0.5 volt	− 0.5 volt	− 1
10,000 ohms	1.0 volt	− 1.0 volt	− 1
10,000 ohms	1.5 volt	− 1.5 volt	− 1
10,000 ohms	2.0 volt	− 2.0 volt	− 1

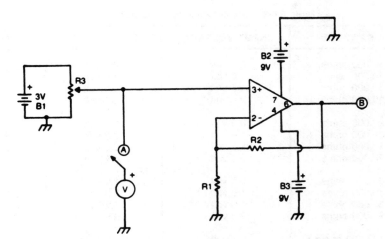

Fig. 31-2. Non-inverting amplifier circuit.

Repeat this procedure for each R2 value and input voltage listed in the table. The nominal (perfect tolerance) results are given in Table 31-5.

Basically your results should be about the same as in the last experiment, except for this circuit the output is the same polarity as the input.

EXPERIMENT #3 DIFFERENCE AMPLIFIER

Now, build the circuit shown in Fig. 31-3. Notice that both the − and the + inputs are used in this circuit. This is a difference amplifier. All three voltage sources should

Table 31-4. Worksheet for Experiment #2.

R2 VALUE	INPUT VOLTAGE	OUTPUT VOLTAGE	GAIN
10,000 ohms	0.5 volt	_____	_____
10,000 ohms	1.0 volt	_____	_____
10,000 ohms	1.5 volt	_____	_____
10,000 ohms	2.0 volt	_____	_____
22,000 ohms	0.5 volt	_____	_____
22,000 ohms	1.0 volt	_____	_____
22,000 ohms	1.5 volt	_____	_____
22,000 ohms	2.0 volt	_____	_____
33,000 ohms	0.5 volt	_____	_____
33,000 ohms	1.0 volt	_____	_____
33,000 ohms	1.5 volt	_____	_____
33,000 ohms	2.0 volt	_____	_____
100,000 ohms	0.5 volt	_____	_____
100,000 ohms	1.0 volt	_____	_____
100,000 ohms	1.5 volt	_____	_____
100,000 ohms	2.0 volt	_____	_____

Table 31-5. Nominal Results for Experiment #2.

R2 VALUE	INPUT VOLTAGE	OUTPUT VOLTAGE	GAIN
10,000 ohms	0.5 volt	0.5 volt	1
10,000 ohms	1.0 volt	1.0 volt	1
10,000 ohms	1.5 volt	1.5 volt	1
10,000 ohms	2.0 volt	2.0 volt	1
22,000 ohms	0.5 volt	1.1 volt	2.2
22,000 ohms	1.0 volt	2.2 volt	2.2
22,000 ohms	1.5 volt	3.3 volt	2.2
22,000 ohms	2.0 volt	4.4 volt	2.2
33,000 ohms	0.5 volt	1.65 volt	3.3
33,000 ohms	1.0 volt	3.3 volt	3.3
33,000 ohms	1.5 volt	4.95 volt	3.3
33,000 ohms	2.0 volt	6.6 volt	3.3
100,000 ohms	0.5 volt	5.0 volt	10
100,000 ohms	1.0 volt	9.0 volt	(9)
100,000 ohms	1.5 volt	9.0 volt	(6)
100,000 ohms	2.0 volt	9.0 volt	(4.5)

have the same values as in the previous experiments. All resistors are 10,000 ohms (brown-black-orange).

Connect your voltmeter between point A (+) and ground (−) and measure the inverting input voltage. Nominally this value is three volts (the value of B1), but your battery might not be exactly on voltage. Enter this voltage in the appropriate spaces in Table 31-6.

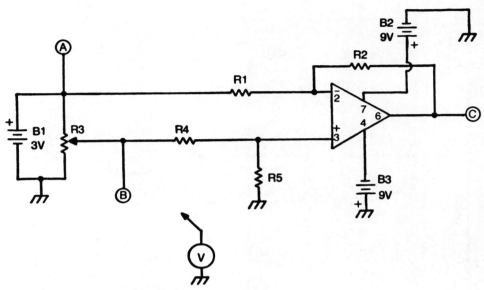

Fig. 31-3. Difference amplifier circuit.

Table 31-6. Worksheet for Experiment #3.

INVERTING INPUT	NON-INVERTING INPUT	OUTPUT
——— ——— ——— ———	1.0 volt 1.5 volt 2.0 volt 2.5 volt	——— ——— ——— ———

Now move the positive voltmeter lead to point B and adjust R3 for a non-inverting voltage of 1 volt.

Reverse the voltmeter leads and connect the meter between point C (–) and ground (+). Since the inverting input voltage is larger than the non-inverting input voltage, the inverting input should predominate. Measure the output voltage and enter your result in Table 31-6.

Repeat this procedure for each of the input values listed in the table. You do not have to measure point A each time—it should be a constant value.

When you are finished, each output voltage should equal the inverting input voltage minus the non-inverting input voltage. Minor variations are due to unequal component tolerances.

EXPERIMENT #4 INTEGRATOR

Breadboard the circuit shown in Fig. 31-4. This is an integrator. B1 and B2 are the 9 volt power supplies for the operational amplifier. R1 and R3 should be 10,000 ohm

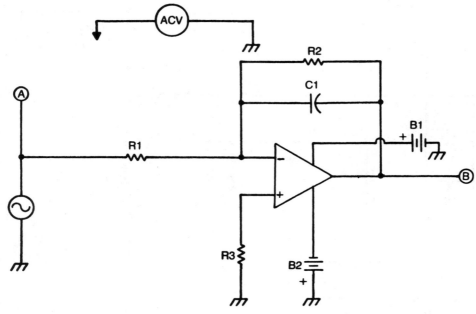

Fig. 31-4. Integrator circuit.

Table 31-7. Worksheet for Experiment #4.

FREQUENCY	INPUT VOLTAGE	OUTPUT VOLTAGE
A (lowest) B C D E (highest)	——— ——— ——— ——— ———	——— ——— ——— ——— ———

(brown-black-orange) resistors, and C1 is a 0.1 μF capacitor. R2 is 100,000 ohms (brown-black-yellow).

Measure the input voltage at point A, then measure the output voltage at point B. Enter both values into Table 31-7. Do this for a number of input frequencies starting with the lowest frequency at the top of the table and increasing the frequency as you move down the column. The gain should decrease as the frequency increases.

If you have an oscilloscope it would also be interesting to compare the input waveshape with the output signal. There should be a dramatic difference.

EXPERIMENT #5 DIFFERENTIATOR

Repeat the last experiment with the differentiator circuit shown in Fig. 31-5. R1 should be a 270 ohm (red-violet-brown) resistor. R2 and R3 should be 10,000 ohm (brown-black-orange) units. C1 is a 0.1 μF capacitor. The power supplies are ±9 volts.

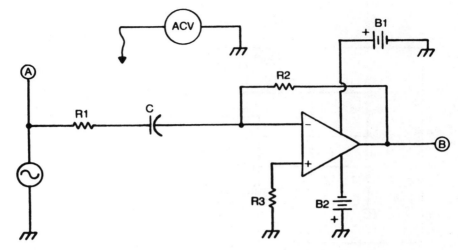

Fig. 31-5. Differentiator circuit.

Table 31-8. Worksheet for Experiment #5.

FREQUENCY	INPUT VOLTAGE	OUTPUT VOLTAGE
A (lowest) B C D E (highest)	——— ——— ——— ——— ———	——— ——— ——— ——— ———

Your results in this experiment should be the opposite of what happened in experiment #4. The gain should increase as the frequency increases. Record your results on Table 31-8. Again, it is interesting to compare the input and output signals with an oscilloscope.

EXPERIMENT #6 A MONOSTABLE MULTIVIBRATOR

Figure 31-6 shows a basic monostable multivibrator circuit built around a 555 timer IC. The pin numbers are given for an 8 pin DIP.

The power supply voltage can be anything between five and fifteen volts. A 9 volt battery is fine. R1 and R2 are 4,700 ohm (yellow-violet-red) resistors, and R3 is 1 megohm, or 1,000,000 ohms (brown-black-green). D1 is 1N914 diode.

For the first part of this experiment, C1 will be a 1 μF electrolytic capacitor. Be sure to observe the correct polarity. C2 is a 0.01 μF disc capacitor.

S1 is obviously an SPDT switch. Preferably it should be a momentary contact type. The diagram shows its normal position.

Push the switch briefly and release it while watching the meter. The pointer should jump up from approximately zero to some value somewhat below the power supply voltage. After a definite period of time the pointer should move back down to near zero. This time should be approximately one second. The nominal value is actually 1.1 second.

384

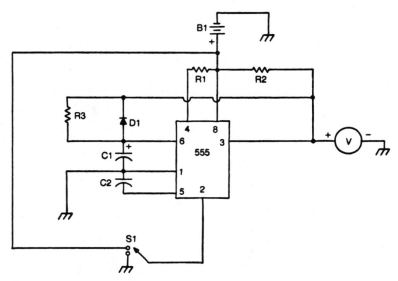

Fig. 31-6. Monostable multivibrator circuit.

Now, replace the 1 μF capacitor (C1) with a 10 μF electrolytic capacitor (observe polarity) and repeat the experiment. How long does the pointer stay at its high voltage position now?

EXPERIMENT #7 AN ASTABLE MULTIVIBRATOR

Breadboard the astable multivibrator circuit shown in Fig. 31-7. Again the circuit is built around an 8 pin DIP 555 timer IC. B1 is 9 volts, R1 is a 10,000 ohm potentiometer and R2 is a 1,000 ohm (brown-black-red) resistor. C1 is a 0.01 μF capacitor. The speaker is a miniature 8 ohm unit.

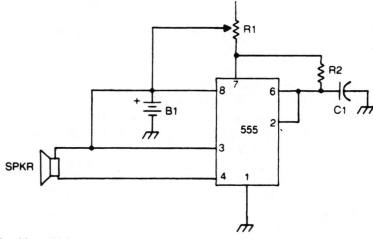

Fig. 31-7. Astable multivibrator circuit.

You'll notice that as shown, as power is applied to this circuit, a tone will be emitted by the speaker.

If you connected an oscilloscope across the speaker's leads you would see that the signal is a rectangular wave.

Move the knob on the potentiometer and listen to the sound coming from the speaker. The pitch (or frequency) should change as you vary the resistance of R1. R1 is the timing resistor, so changing its resistance changes the time constant of the circuit.

EXPERIMENT #8 THE NAND GATE

Connect pin 14 of a 7400 TTL quad 2 input NAND gate integrated circuit to a well regulated 5 volt power supply and connect pin 7 to ground.

Now, set up the rest of the circuit shown in Fig. 31-8. The pin labeled "A" in the diagram is pin 1. Pin "B" is pin 2, and pin "C" is pin 3. R1 is a 1,000 ohm (brown-black-red) resistor, and S1 and S2 are SPDT switches.

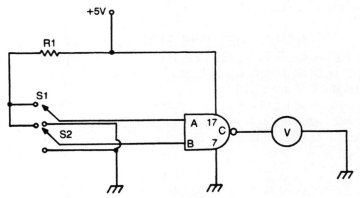

Fig. 31-8. A NAND gate demonstration circuit.

When a switch is making contact with ground, it is providing a logic 0 input to the gate. When a switch is making contact to +5 volts through R1, it is a logic one. The meter's pointer should be near 0 or between 2 and 3 volts, corresponding to logic 0 and logic 1, respectively. Set the switches for each of the combinations shown in Table 31-9.

You should get a logic 0 output only when both switches are in their logic 1 positions. For all other input combinations, the output should be a logic 1. This is a NAND gate.

Table 31-9. Worksheet for Experiment #8.

SWITCH 1	SWITCH 2	OUTPUT
0	0	———
1	0	———
1	1	———
0	1	———

Table 31-10. Pin Numbers for each Gate in a 7400.

INPUTS		OUTPUT
A	**B**	**C**
1	2	3 GATE #1
4	5	6 GATE #2
9	10	8 GATE #3
12	13	11 GATE #4
Pin 14 is connected to the positive voltage source and pin 7 goes to ground for all gates.		

Repeat the experiment for each of the gates in the package, using the pins listed in Table 31-10. You should get the same results from all of the gates, unless one of them happens to be defective.

EXPERIMENT #9 THE NOR GATE

Repeat the procedure of the last experiment with a 7402 TTL quad 2-input NOR gate IC. See Fig. 31-9. The pin connections for each gate are listed in Table 31-11. Record your results in Table 31-12.

In this case you should have a logic 1 output only when both switches are in their logic 0 positions. All other input combinations should result in a logic 0 output. This is a NOR gate.

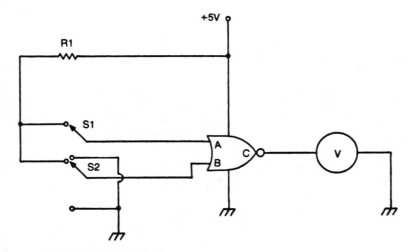

Fig. 31-9. A NOR gate demonstration circuit.

EXPERIMENT #10 COMBINING GATES

Breadboard the circuit shown in Fig. 31-10. IC1 is a 7400 quad NAND gate, and IC2 is a 7402 quad NOR gate.

Table 31-11. Pin Numbers for each Gate in a 7402.

INPUTS		OUTPUT	
A	B	C	
2	3	1	GATE #1
5	6	4	GATE #2
8	9	10	GATE #3
11	12	13	GATE #4
Pin 14 is connected to the positive voltage source and pin 7 goes to ground for all gates.			

Table 31-12. Worksheet for Experiment #9.

SWITCH 1	SWITCH 2	OUTPUT
0	0	———
1	0	———
1	1	———
0	1	———

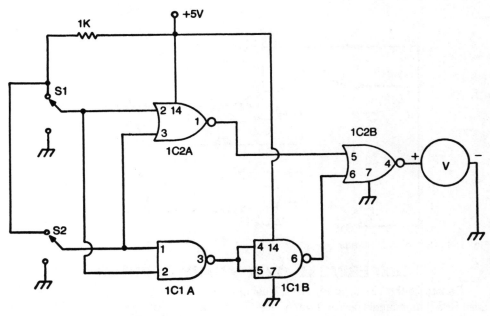

Fig. 31-10. A complex gating circuit.

Before actually performing the experiment try to determine what the output state will be for each of the input combinations in Table 31-13.

Try each of the switch combinations and see if you correctly predicted the output state. Does this circuit resemble one of the basic gate types?

Table 31-13. Worksheet for Experiment #10.

SWITCH 1	SWITCH 2	PREDICTED OUTPUT	ACTUAL OUTPUT
0	0		
0	1		
1	0		
1	1		

EXPERIMENT #11 A BISTABLE MULTIVIBRATOR

Breadboard the circuit shown in Fig. 31-11 using two of the NOR gates in a 7402 IC. S1 is a DPDT switch. Connect the positive voltmeter lead to point A. It should be either close to zero (logic 0) or between 2 and 3 volts (logic 1). Changing the position of S1 should reverse the output state. Move S1 back and forth several times so you can see how it affects the output. Position the switch so there is a logic 1 output at point A. Now move the positive voltmeter lead to point B. You should read a logic 0 here. Now reverse the position of S1. Remember, this caused output A to change from 1 to 0. Output B should change from 0 to 1. A and B are always opposites. Either output state can be held indefinitely.

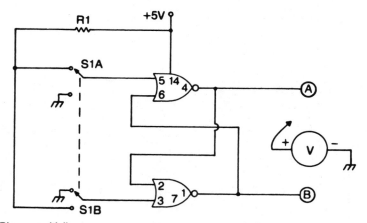

Fig. 31-11. Binary multivibrator.

EXPERIMENT #12 A BINARY COUNTER

Construct the circuit in Fig. 31-12 using a 7474 TTL Dual D type flip-flop IC. S1 is an SPST momentary action switch. The figure shows its normal position. R1 and R2 are 1,000 ohm (brown-black-red) resistors.

When you first turn on the circuit one or both of the LEDs may be lit. Press S1 several times until both LEDs are dark.

We will call a dark LED a 0, and a lit LED a 1. LED A is the ones column. LED B is the twos column. We are starting out with a displayed count of 00 (both LEDs are dark).

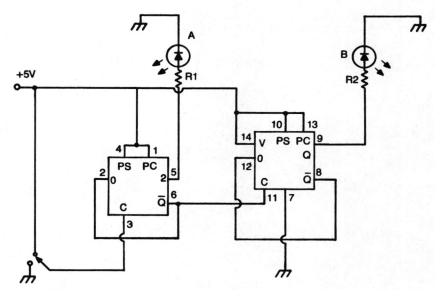

Fig. 31-12. Binary counter.

If you press the switch once LED A will light up, and LED B will stay dark. We now have a binary count of 01, or one.

The second time we press the switch, LED A should go out and LED B should come on. Our binary count has changed to 10, or two.

The third time we press the switch both LEDs should be lit for a binary count of 11, or three.

On the fourth switch pressing both LEDs will go out again, and the count will start over.

What is the modulo for this counter? If you are unsure, re-read the section on Counters in Chapter 27.

32

Power Supplies

Virtually all electronic circuits require some source of voltage. This means *power supply* circuits are extremely important. Fortunately, they are fairly easy to understand.

If a circuit requires an ac voltage, the power supply can simply be a transformer connected to the ac house current. If a circuit requires dc and has low power requirements, batteries can be used. Most practical circuits, however, are dc operated, and require power levels that would make battery operation uneconomical. These devices need a power supply circuit.

Actually, power supply circuits are somewhat misnamed. They do not truly supply power. They are more properly "power converters." Generally they convert ac voltages into dc voltages.

There are a number of methods of accomplishing this function, using diodes. When diodes are used for this type of application, they are generally called *rectifiers*. Rectifiers can be either tube diodes or semiconductor diodes. The process itself is called *rectification*.

HALF-WAVE RECTIFIERS

A single diode power supply circuit is called a *half-wave rectifier*. This is shown, in its most basic form, along with its input and output signals, in Fig. 32-1.

The terminals of the ac voltage source reverse polarity twice each cycle. For one half of each cycle terminal A is positive with respect to terminal B (ground). For the

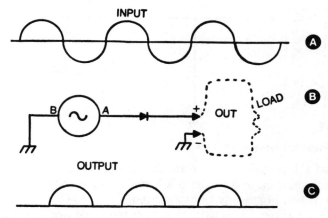

Fig. 32-1. A half-wave rectifier circuit.

other half of each cycle, terminal A is negative with respect to terminal B. When terminal A is positive, the diode is forward biased, the current can flow through it to the output.

However, when terminal A is negative, the diode is reverse biased, so no current can flow. Only the positive half cycles appear at the output.

Of course, this is not true dc power at all. Half of the time there is no output voltage. The rest of the time the voltage is either in the process of rising from zero to the maximum level, or from the maximum level back to zero. It never holds any constant value.

A closer approximation of true dc can be achieved by placing a large capacitor across the diode's output, as shown in Fig. 32-2.

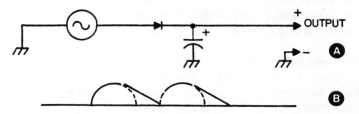

Fig. 32-2. A half-wave rectifier circuit with a filter capacitor.

When the output voltage rises from zero to its peak value, the capacitor is charged. When the voltage drops off, the capacitor starts to discharge through the load. If the capacitor is large enough, it will not be fully discharged before the next positive half cycle starts.

In other words, the capacitor will be charged, partially discharged, charged, partially discharged, and so forth. The final output signal will resemble Fig. 32-2B.

The larger the capacitance, the slower the discharge rate, and therefore, the shallower the discharging angle in the output waveform. Electrolytic capacitors with values of several hundred to a few thousand microfarads are typically used. Figure 32-3 shows an improved, practical half-wave rectifier circuit.

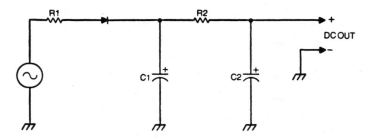

Fig. 32-3. A practical half-wave rectifier circuit.

R2, C1 and C2 comprise a low pass filter that smooths the output signal more efficiently than a single capacitor. There will still be some voltage fluctuations (called *ripple*), but they won't be nearly as pronounced.

R1 is a *surge resistor* that is used to protect the diode from any sudden increase in the current drawn through the circuit. The surge resistor typically has a fairly small value, so normally the voltage drop across it is also rather small. But an increase in the current drawn through the resistor will cause its voltage drop to rise too, since according to Ohm's law, voltage equals current times resistance (E = IR). This brings the voltage applied to the diode down to a level it can comfortably dissipate.

Sometimes R1 is also fused for additional protection. A surge resistor is not only for protection against circuit defects. Often it is also needed for normal operating conditions.

Assume no source voltage at all is being applied to the circuit. Any residual charge on the capacitors will be soon discharged through R2 and the load circuit. The capacitors are completely discharged.

Now, when power is first applied, the capacitors will draw a large amount of current until they are almost completely charged. There isn't sufficient time for them to be completely charged during a single cycle, so it takes a few cycles for ordinary operation to begin.

Of course, this extra current drain will increase the voltage drop across the other components. Again, the surge resistor is used to protect the diode.

Often, for more efficient protection, a thermistor (a temperature sensitive resistor) is used for R1. When the circuit is first turned on, the components, including the thermistor, are cool. The thermistor has a high resistance when it is cold. This means that when power is first applied, there is a relatively large voltage drop across the thermistor, leaving only a relatively small voltage to pass through the diode.

As current passes through the circuit, the components start to dissipate heat. The increased temperature causes the resistance of the thermistor (and thus, its voltage drop) to fall to a low value, and from then on it acts like any ordinary surge resistor.

FULL-WAVE RECTIFIERS

In the half-wave rectifier half of each input cycle is completely unused. Of course this means power is wasted. Figure 32-4 illustrates a more efficient power supply circuit called a *full-wave rectifier*. Notice that a full-wave rectifier must be used with a center-tapped transformer.

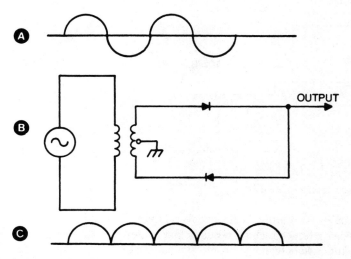

Fig. 32-4. A full-wave rectifier circuit.

Remember, that if the center tap of a transformer secondary is grounded, the lower half of the secondary winding will carry a signal that is equal to, but 180° out of phase with the upper half's signal. This means that when D1 is passing a positive half cycle, D2 is blocking a negative half cycle. And, when D1 is blocking a negative half cycle, D2 is passing a positive half cycle. One of the diodes is conducting and one is non-conducting at all times. This means the output will resemble Fig. 32-4C.

Notice that besides wasting less input power, the output of a full-wave rectifier is easier to filter, because there is less time for the filter capacitor to discharge before it is charged again. See Fig. 32-5. Notice that both the positive and the negative output lines need their own filter and they are isolated from the ac ground.

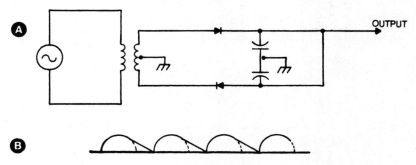

Fig. 32-5. A full-wave rectifier circuit with filter capacitors.

BRIDGE RECTIFIERS

A *bridge rectifier* circuit (see Fig. 32-6) combines the advantages of both full-wave rectifiers and half-wave rectifiers. Like the full-wave rectifier, the bridge rectifier uses the entire input cycle, and is easy to filter.

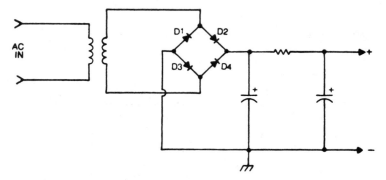

Fig. 32-6. A bridge rectifier circuit.

Like the half-wave rectifier, the bridge rectifier does not require a center-tapped transformer. While a bridge rectifier requires four diodes, it is still usually more economical for semiconductor circuits than most center-tapped transformers. The circuit also requires less space, as a rule, and produces less heat. Bridge rectifiers using tube diodes are not practical.

Also, like the half-wave rectifier, one of the bridge rectifier's output lines may be at ground potential. At any point of the input cycle, two of the diodes in the bridge are conducting and two are reverse biased. For the positive half wave the circuit effectively looks like Fig. 32-7. Figure 32-8 shows the equivalent circuit for the negative half cycle.

Bridge rectifiers may consist of four separate diodes, or they may be encapsulated in a single package, as shown in Fig. 32-9. This is usually done simply to conserve space. Electrically, such a single unit bridge is the exact equivalent to four discrete diodes.

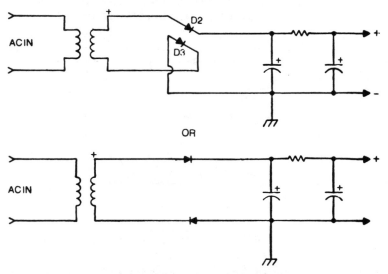

Fig. 32-7. Equivalent circuits for a bridge rectifier circuit during a positive half cycle.

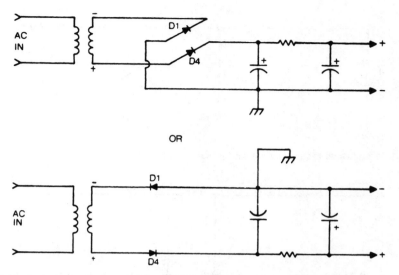

OR

Fig. 32-8. Equivalent circuits for a bridge rectifier circuit during a negative half cycle.

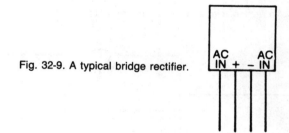

Fig. 32-9. A typical bridge rectifier.

VOLTAGE REGULATION

One problem with all of the power supply circuits discussed so far is that the output voltage is dependent, to a large extent, on the amount of current drawn by the load circuit. If the current drawn by the load circuit increases for any reason, the voltage drop across the components within the power supply circuit will also rise, resulting in a lower output voltage. Of course, just the opposite happens if the current drawn by the load decreases—the output voltage will increase.

A partial solution is shown in the half-wave rectifier circuit in Fig. 32-10. R2 is replaced with a coil called a *choke*. This coil will oppose any change in the voltage passing through it. It also acts as a better filter than an ordinary resistor, this means lower ripple in the output signal. Also, since the dc resistance of a coil is extremely low, there will be very little wasted voltage drop across the choke.

Using a zener diode across the output, as shown in Fig. 32-11 is another basic method of voltage regulation. This is discussed in detail in Chapter 18.

The best voltage regulation can be achieved with the voltage regulator circuit illustrated in Fig. 32-12. This is actually a simplified basic circuit. There are a number of variations possible.

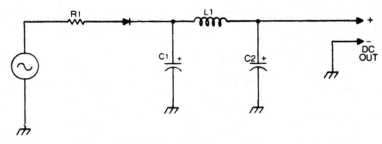

Fig. 32-10. A half-wave rectifier circuit with a filter choke.

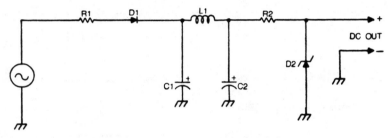

Fig. 32-11. A half-wave rectifier circuit with a zener diode regulation.

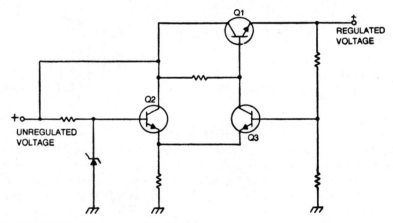

Fig. 32-12. A simplified voltage regulator circuit.

Q1 is what is known as a pass transistor. Q2 and Q3 comprise a difference amplifier that is set up to detect any error in the output voltage. The two resistors between the output and the ground form a voltage divider (explained in the next section of this chapter).

If the output voltage increases even slightly because of an increase in the current drawn by the load circuit, the voltage on the base of Q3 is increased. Notice that the voltage on the base of Q2 is held constant by the zener diode. Ordinarily these two base voltages should be equal. But we now have a condition where Q3's base is at a higher voltage than Q2's base. This is the difference detected by the difference amplifier.

The increase in base voltage on Q3, of course, increased the emitter and collector currents. This means the voltage drop across the shared emitter resistor must also increase (E = IR). Since the emitter of Q2 is now at a somewhat higher voltage, the difference between the base and emitter voltages has decreased. This has the same effect as reducing Q2's base voltage. This means Q2's output must also decrease. The current through Q1 will be forced to compensate for the difference. In practice, the voltage drop across the emitter resistor will remain virtually constant, because the changes take place so quickly. In effect Q1 increases its resistance, which has the effect of dropping the output voltage back to the desired level. If the output voltage drops for any reason, the opposite reactions will take place.

The difference amplifier is extremely sensitive, and even changes of less than one tenth of one percent can be quickly corrected by some circuits.

Obviously the voltage regulator also serves as a superior ripple filter, since ripple is a fluctuation in the output voltage, and is thus corrected by the circuit like any other output error.

Because voltage regulator circuits are so frequently needed, a number of IC versions are available for frequently used output voltages, such as five, twelve, or fifteen volts. Voltage regulators are available for either positive or negative ground operation. These are usually not interchangeable, though the two types can be used together for two polarity supplies, as shown in Fig. 32-13.

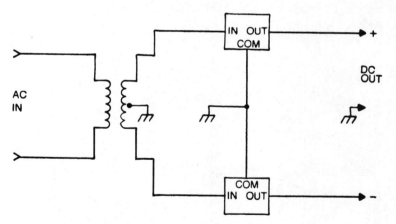

Fig. 32-13. A regulated two polarity power supply circuit.

Voltage regulator integrated circuits have three output leads, input (the unregulated voltage), output (the regulated voltage), and common (the reference point for both the input and the output). The device usually resembles a somewhat over-sized transistor. Figure 32-14 shows two typical package styles.

The input voltage can vary over a large range without affecting the output voltage. One five volt regulator will accept input levels up to thirty-five volts.

Voltage regulators are limited in the amount of current that may be safely drawn from them. If greater current is required, several voltage regulators can be used in parallel.

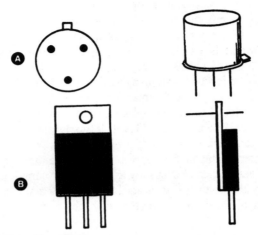

Fig. 32-14. Typical voltage regulator IC packages.

VOLTAGE DIVIDERS

If you connect a string of series resistors across the voltage supply, varying voltages may be tapped off between each resistor pair. This is called a *voltage divider*, because it divides the output voltage into a series of smaller voltages. Figure 32-15 shows a typical voltage divider.

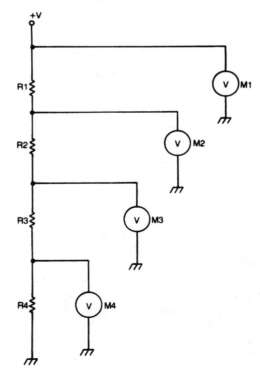

Fig. 32-15. A voltage divider circuit.

Let's suppose all four resistors are 1000 ohm units, and 12 volts dc is being applied between +V and ground. Since we have four 1000 ohm resistors in series, the total resistance is 4000 ohms. The current through the string is determined by Ohm's law. I = E/R = 12/4000 = 0.003 ampere, or 3 milliamperes.

Similarly, we can calculate the voltage drop across each of the resistors. E = IR = 0.003 ampere × 1000 ohms = 3 volts.

M1, of course, will read the full 12 volts. R1 drops 3 volts, so M2 will read 9 volts. R2 drops 3 more volts, so M3 reads 6 volts. R3 drops an additional 3 volts, bringing M4's reading down to 3 volts. The remaining 3 volts is dropped across R4, so the ground point will be at zero volts.

The resistors in a voltage divider do not necessarily have to be of equal values. For instance, R1 might be 1000 ohms, R2 might be 270 ohms, R3 4700 ohms and R4 680 ohms. The total series resistance would be 6,650 ohms. This means the current flowing through the voltage divider with a 12 volt input would be equal to 12/6,650 or 0.0018 ampere (1.8 milliampere).

Under these conditions M1 would read 12 volts. R1 drops 1.8 volt, so M2 will read 10.2 volts. The voltage drop across R2 is about 0.5 volts, leaving 9.7 volts to be read on M3. R3 drops 8.5 volts so M4 must read 1.2 volts. This remaining voltage is dropped by R4.

The voltage divider is an inexpensive circuit, but it is difficult to regulate because the load circuits are in parallel with the resistors, affecting the voltage drops. Voltage divider resistors are usually given the smallest values possible to minimize this effect.

A voltage divider circuit serves another important function. It provides a discharge path for the heavy filter capacitors when the equipment is turned off. If these capacitors are not allowed to discharge to ground, a technician working on the circuit could receive a dangerous shock. Resistors used for this purpose are called *bleeder resistors*.

VOLTAGE DOUBLERS

Sometimes we need a dc voltage that is more than the input ac voltage. Figure 32-16 shows a basic *voltage doubler* circuit.

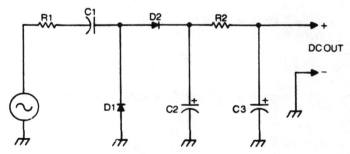

Fig. 32-16. A voltage doubler circuit.

During the negative half cycle, D1 is forward biased and D2 is reverse biased, charging C1. When the polarity is reversed during the positive half cycle the input voltage passes through the now forward biased D2, just as in an ordinary half-wave regulator. D1 is reverse-biased, and effectively out of the circuit. In addition to the regular input voltage,

D2 also passes the voltage from C1 which is discharging through the diode. The charge on C1 is close to the input voltage, so the signal passing through D2 is essentially twice the input signal.

Besides being fed to the output, this voltage also charges C2. On the next negative half-cycle, when D2 is again cut-off, C2 partially discharges through the output, so the output voltage remains more or less constant. C3 is simply for filtering the output ripple.

The exact output is dependent on the values of C1 and C2. The larger they are, the closer the output will equal twice the input's peak voltage. These two capacitors should also be of equal capacitance value for the best results. If they are unequal, some ac may leak through, increasing ripple.

Self-Test

1. What kind of power is supplied by batteries?

A *ac*
B *dc*
C *Pulsating dc*
D *Static electricity*
E *None of the above*

2. What component can be used as a half-wave rectifier?

A *Capacitor*
B *FET*
C *Inductor*
D *Diode*
E *None of the above*

3. What is the biggest disadvantage of a half-wave rectifier circuit?

A *Expensive*
B *Wastes half of each input ac cycle*
C *Complicated circuitry*
D *Unreliable*
E *None of the above*

4. What is the purpose of the output capacitor in a power supply circuit?

A *Filter out ac ripple*
B *Act as a voltage divider*
C *Increase the output current capability*
D *Protect the diode from sudden current surges*
E *None of the above*

5. What is the minimum number of diodes needed for a full-wave rectifier?

A *One*
B *Two*
C *Three*
D *Four*
E *More than four*

6. How many diodes are in a bridge rectifier?

A *One*
B *Two*
C *Three*
D *Four*
E *More than four*

7. What is the advantage of using a bridge rectifier for full wave rectification?

A *Cheaper*
B *Simpler circuitry*
C *No center tap is required on the transformer*
D *A lower input voltage may be used*
E *None of the above*

8. How many leads does a typical voltage regulator IC have?

A *One*
B *Two*
C *Three*
D *Four*
E *More than four*

9. Refer to Fig. 32-15. Assume each of the resistors has a value of 33,000 ohms (33 k), and V+ equals 18 volts. What will the voltage reading on meter M2 be?

A *18 volts*
B *14 volts*
C *13.5 volts*
D *4.5 volts*
E *None of the above*

10. What is the purpose of a zener diode in a power supply circuit?

A *Rectification*
B *Voltage regulation*
C *Filtering*
D *Current amplification*
E *None of the above*

33

Amplifiers

The amplifier is probably the most widely-used circuit type in all of electronics. Practically every complex electronics system or large circuit includes at least one amplifier stage.

In a sense, every transistor used in every circuit is an amplifier. Even when a transistor is used for switching, its operation uses its amplification properties.

DEFINING AMPLIFIERS

In simple terms, an amplifier is a circuit that performs amplification, or increases the amplitude of a signal. In an ideal amplifier, the signal at the output should be absolutely identical to the signal at the input, except for the change in amplitude. Any other changes in the signal are due to distortion effects within the amplifier circuitry. All practical circuits distort to some extent, but in many cases the amount of distortion is negligible. In other cases, it can be quite severe.

Normally, an amplifier increases the signal amplitude, but there are occasional exceptions. Some amplifiers have negative gain. That is, the output signal is at a lower amplitude than the input signal. Strictly speaking, this is an attenuator rather than an amplifier, but since the same sort of circuitry is involved an active attenuator circuit is commonly called an amplifier.

Some amplifiers have unity gain, there is no change in the amplitude. The output signal (ideally) is identical to the input signal. This type of amplifier is called a *buffer*, or a *buffer amplifier*. It is used primarily for impedance matching and stage isolation purposes.

Most amplifier circuits are divided into classes that define their basic operating characteristics. Some amplifier classes have much stronger distortion effects than others. Each amplifier class is suitable for a different set of applications. In this chapter we will take a look at the major amplifier classes.

The most popular amplifier classes are;

Class A
Class B
Class AB
Class C

Additional classes run from D through H. These higher classes are used for special purposes, and some special considerations are involved when designing a circuit in one of these classes. In some cases patents are involved. For this reason, we will concentrate on classes A through C. These are the amplifier classes you are most likely to encounter in general electronics work, whether professionally or on the hobbyist level.

For convenience in our discussion, we will concentrate solely on transistor amplifiers. The internal circuits in ICs work in pretty much the same way. Tube circuits are also basically similar, even though they are rarely encountered in modern electronics.

BIASING

The basic differences between the various amplifier classes lie in how the transistor is *biased*. Biasing is the balancing of circuit polarities for correct operation of the transistor. For a NPN transistor to conduct, the collector must be more positive than the base, while the emitter must be negative with respect to the base. For a PNP transistor, these polarities are simply reversed.

An input signal (usually applied to the base—see Chapter 20) will cause the biasing to fluctuate from its fixed levels. An input signal could throw off the fixed bias enough to cut off the transistor. The fixed bias point in the circuit must be selected with this in mind.

CLASS A AMPLIFIERS

In Class A amplifiers, the transistor is biased so that it conducts during the entire input cycle. This gives us an amplifier with very high linearity (low distortion), but low efficiency. The power supply to an amplifier circuit only supplies a finite amount of power. Any power consumed by the amplifier circuit itself, cannot be used in the actual amplification of the signal. In a Class A amplifier, most of the supplied power is consumed by the circuit. As much as 75 to 80 percent of the supplied power is wasted in this manner. This limits how much amplification can be performed by the circuit. A typical Class A amplifier circuit is shown in Fig. 33-1.

Generally, low power Class A amplifiers are biased at the center of the transistor's load line. That is, if we graph the operation of the transistor, the circuit bias point will be selected right in the middle of the linear portion of the graph.

Ideally, larger Class A power amplifiers are also biased at this mid-point, but as the power increases, there are some additional factors to be considered. It is an inescapable

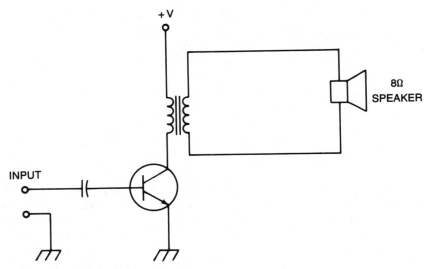

Fig. 33-1. A Class A amplifier circuit.

fact that for any transistor to deliver power, it must dissipate power. Regardless of the size of the load (output circuit), the product of the collector current and the collector-to-emitter voltage at any point of the load line must be less than the maximum power dissipation rating of the transistor. If this maximum power dissipation rating is exceeded, the transistor could self-destruct from excess heat. In low-power circuits, this isn't much of a problem. In a larger power amplifier of the Class A type, the load line should be selected so that it always falls below the maximum permissible power dissipation curve of the transistor used in the circuit.

Class A power amplifiers are widely used to feed power to a loudspeaker in small radios. Most loudspeakers have a relatively low impedance (typically between 4 and 16 ohms), some sort of impedance matching network is usually employed to present a reasonable load to the collector circuit. In most cases, a transformer is used for this task (as shown in Fig. 33-1).

Theoretically, when a transformer is used between the loudspeaker and the transistor in a Class A amplifier, the maximum efficiency of the overall circuit is 50 percent. In other words, the maximum power that can be delivered to the load is equal to half the power supplied to the circuit. The other half of the supplied power is dissipated by the transistor, even if there is no input signal. For example, if you bias the transistor so that it draws a maximum of 10 watts from the power supply, only 5 watts can be applied to the output signal fed to the load. The other 5 watts is dissipated as waste heat.

An efficiency of 50 percent certainly sounds very low. But the situation is usually worse than theory suggests. In a practical Class A amplifier circuit, the actual efficiency is 20 to 40 percent lower because of losses and limits of the transformer, and the amplifier circuitry itself.

In some cases, matters can be worse yet. If the transformer is eliminated, and the load is placed directly in the collector circuit, the load has to dissipate dc power too. This limits the maximum theoretical efficiency to a mere 25 percent, and that 25 percent can be reduced by the same 20 to 40 percent mentioned earlier. The net result is an amplifier circuit with an efficiency of about 20 percent—with luck.

In short, the Class A amplifier circuit has a very high price for its excellent linearity. It wastes an incredible amount of power. Fortunately, there are more efficient ways to amplify ac signals.

Since the Class A amplifier offers the lowest possible distortion, it is used in some audiophile stereo amplifiers, where price (and wasted power) is no object. For the most part, however, Class A amplifiers are normally used only in very low-power applications.

CLASS B AMPLIFIERS

A Class B amplifier conducts for just half of each input cycle, and rests the remainder of the time. This is illustrated in Fig. 33-2. Since the amplifier is active for only half of each cycle, the circuit has much greater efficiency than with a Class A amplifier.

To allow the Class B amplifier to conduct during just one half of each cycle of the input signal, the transistor is biased at the point where the *quiescent* (or idling) *current* is equal to zero. If there is no input signal, the transistor does not conduct—it just sits there, theoretically consuming no power.

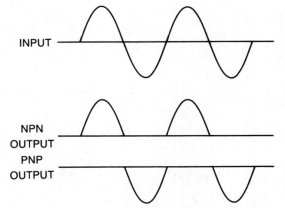

Fig. 33-2. A Class B amplifier conducts for just half of each input cycle.

If you assume that an NPN transistor is being used in a Class B amplifier circuit, the only time current will flow through the transistor is when the positive portion of the cycle is applied to the base. During the negative half-cycle, the transistor is cut off, and no current flows. A PNP transistor works exactly the same way, of course, except that all of the polarities are reversed. A PNP transistor in a Class B amplifier circuit conducts only during negative half-cycles and is cut off during positive half-cycles. A simple Class B amplifier circuit is shown in Fig. 33-3.

This approach to amplification is very efficient compared to a Class A amplifier. Unfortunately, chopping off half of each cycle of the input signal is an extreme form of distortion. This is clearly undesirable in virtually any audio application, especially if any degree of high fidelity is desired. In order for a Class B amplifier to reproduce signals with reasonable accuracy, two parallel transistors must be used in the circuit. One of these parallel transistors is an NPN type to amplify the positive half-cycles, while the other is a PNP device to amplify the negative half-cycles. The two amplified half-cycles

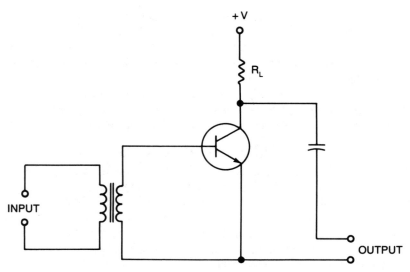

Fig. 33-3. Class B amplifier circuit.

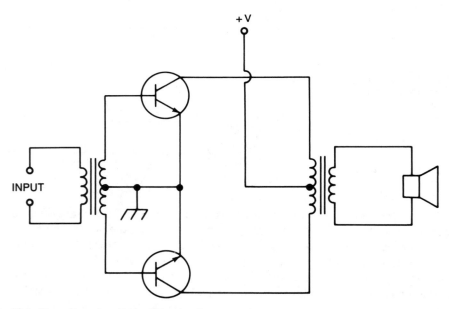

Fig. 33-4. Class B push-pull amplifier circuit.

are then recombined across the load to produce an amplified version of the original total input waveform. This kind of dual Class B amplifier circuit is known as a *push-pull amplifier*. A typical push-pull circuit is illustrated in Fig. 33-4.

This circuit is far more efficient than the Class A amplifier discussed earlier. It can deliver up to 78.5 percent of the supplied power to the load. Because of the push-pull arrangement, the power that can be delivered to the load over the complete cycle is double the *maximum instantaneous power* each individual transistor can pass during the cycle.

Since the two halves of each output cycle are being amplified by two separate transistors, any (even minute) differences between the electrical characteristics of the transistors (or their surrounding components) will result in some degree of nonlinearity (distortion) at the point where the two halves of the cycle are recombined. This is called *crossover distortion*. A signal with greatly exaggerated crossover distortion is shown in Fig. 33-5.

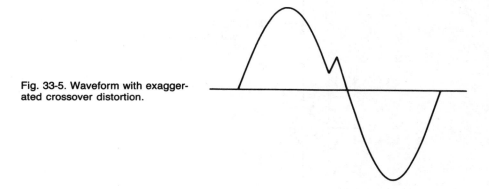

Fig. 33-5. Waveform with exaggerated crossover distortion.

Crossover distortion in Class B audio amplifiers can be reduced by adding a source of bias to the circuit, as shown in Fig. 33-6. This modification actually turns the circuit into a Class AB amplifier, since the collector current now flows for more than half of each cycle. Class AB amplifiers will be discussed in the next section of this chapter.

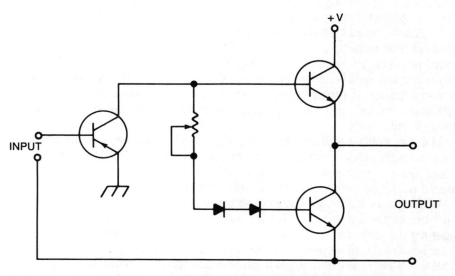

Fig. 33-6. Crossover distortion in Class B push-pull amplifiers can be reduced by adding bias.

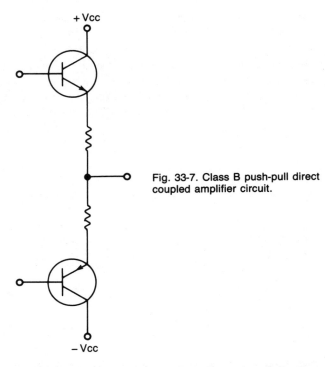

Fig. 33-7. Class B push-pull direct coupled amplifier circuit.

+Vcc

−Vcc

Because of its inherent high efficiency, the Class B, push-pull direct coupled amplifier (as illustrated in Fig. 33-7) is probably the most popular audio amplifier configuration in use today. Early designs (before the mid-1970s) tended to rely more on capacitor coupling (as shown in Fig. 33-8). The chief advantage of capacitor coupling is that such circuits require only a single ended power supply. In addition, since capacitors block dc voltages, speaker damage in the event of problems is unlikely using this approach.

Designers and audiophiles soon learned that capacitor coupling is scarcely ideal. The capacitors tend to cause poor low-frequency response, and instability. Direct coupling is not susceptible to these particular problems, so it is now more popular. However, a direct coupled amplifier does require a dual polarity power supply, and fairly complex circuitry to protect the speakers against the possibility that one of the output transistors might short out and dump the full supply voltage (dc) across the speaker's voice coil, burning it out.

In dealing with Class B amplifiers, you might come across the term *output symmetry*. This term refers to how equally the positive and negative half-cycles are reproduced. An amplifier with poor output symmetry will have *asymmetrical* (unequal) *clipping*, and reduced output power for a given distortion rating.

While a Class B amplifier is highly efficient (up to nearly 80 percent efficiency) when it delivers its maximum power, it isn't quite as efficient at lower amplification levels. Under normal music listening conditions, an audio amplifier is generally driven to deliver full (or nearly full) output power for only a very small fraction of the time it is operating. Signal levels vary quite a bit in most music. Under practical music listening conditions, the actual efficiency of a typical Class B amplifier may be as low as 20 percent.

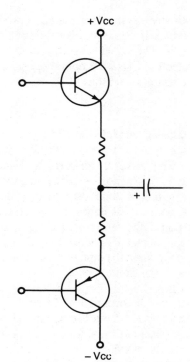

+ Vcc

− Vcc

Fig. 33-8. Class B push-pull capacitor coupled amplifier circuit.

CLASS AB AMPLIFIERS

The Class AB amplifier is a neat compromise between the high linearity (low distortion) of a Class A amplifier, and the efficiency of a Class B amplifier. Practical push-pull circuits are almost always Class AB amplifiers.

As you should recall from the earlier sections in this chapter, in Class A operation, the transistor is biased at the center of the load line, while in Class B operation, the transistor is biased at the point where no collector current flows when there is no input signal. In Class AB operation the transistor is biased to operate somewhere between Class A and Class B.

In the AB mode, some small amount of collector current flows when the transistor is idling (no input signal). This decreases the efficiency of the circuit somewhat. A Class AB amplifier can be useful where good and reliable (low distortion) reproduction of the input signal is required.

The chief advantage of Class AB amplifiers is that this mode of operation significantly reduces crossover distortion. This type of distortion occurs in a Class B push-pull amplifier because no (or very little) *conduction* takes place through a transistor unless a specific minimum voltage is exceeded. Virtually no current will flow through the base-emitter junction of a germanium transistor if less than 0.2 or 0.3 volts is applied. For silicon transistors, the minimum signal voltage is between 0.5 and 0.7 volts.

As the ac input signal passing through a Class B push-pull amplifier goes from positive to negative (or vice versa), it obviously passes through zero. There is a brief delay between the time one transistor is cut off and the other is turned on. This slight delay (along with any electrical mismatch between the two transistors) is what causes cross-over distortion. If a transistor is cut off at a rapid rate, large transient voltages can develop in the circuit. This could cause the transistor to break down. Such problems are pretty much avoided in Class AB operation. One transistor is not cut off until after the other transistor has already started to conduct.

CLASS C AMPLIFIERS

In a Class C amplifier, the transistor is biased so that it conducts for less than one half of each cycle. Only a very small portion of each input cycle is passed through the amplifier. This severely distorts the signal at the output, but in some applications (especially rf or Radio Frequency amplifiers) linearity isn't nearly as important as low power consumption (high efficiency). The output of a Class C amplifier is in the form of pulses, and the efficiency can be as high as 85 percent.

Using an NPN transistor, a Class C amplifier is created by biasing the base so that with no input signal, it is negative with respect to the emitter. This keeps the transistor cut off except when the input signal exceeds the bias voltage, which happens only during the peaks of the input cycle.

The Class C mode of operation is absolutely worthless for audio applications, because of its extreme distortion. But Class C amplifiers are widely used in if (Intermediate Frequency) and rf (Radio Frequency) circuits. A resonant LC circuit (see Chapter 10) is normally placed at the output of a Class C amplifier in a receiver. Each time a signal is applied to the amplifier's input, a full cycle is generated across the LC tank, provided that the circuit is tuned for the frequency (or a multiple of the frequency) of the signal applied to the amplifier's input.

COMPARING THE CLASSES

Each of the amplifier classes can be compared with regard to efficiency and distortion (linearity).

CLASS	EFFICIENCY	LINEARITY
A	poor	very good
B	very good	fair
AB	good	good
C	excellent	poor

The best trade-off will depend on the specific desired application.

Self-Test

1. What does an amplifier do to an input signal?

A *alters the waveshape*
B *changes the frequency*
C *alters the level*
D *changes the linearity*

2. Which of the following is *not* a standard amplifier Class?

A *Class A*
B *Class B*
C *Class BC*
D *Class AB*
E *Class C*

3. What kind of amplifier circuit is subject to crossover distortion?

A *Class A*
B *Push-pull*
C *Class C*
D *RF*

4. What type of gain does a buffer amplifier have?

A *Unity*
B *Zero*
C *Positive*
D *Negative*
E *None of the above*

5. Which of the following best describes a Class A amplifier?

A *High efficiency and low distortion*
B *High efficiency and high distortion*
C *Low efficiency and high distortion*
D *Low efficiency and low distortion*
E *None of the above*

6. Which of the following best describes a Class C amplifier?

A *High efficiency and low distortion*
B *High efficiency and high distortion*
C *Low efficiency and high distortion*
D *Low efficiency and low distortion*
E *None of the above*

7. How is a Class B amplifier biased?

A *In the middle of the load line*
B *So it is cut off for less than half of each cycle*
C *So it is cut off for half of each cycle*
D *So it is cut off for more than half of each cycle*

8. Which class of amplifier offers the highest efficiency?

A *Class A*
B *Class B*
C *Class AB*
D *Class C*

9. Which class of amplifier offers the lowest distortion?

A *Class A*
B *Class B*
C *Class AB*
D *Class C*

10. What type of transistors should be used in a push-pull circuit?

A *Both NPN*
B *Both PNP*
C *One NPN and one PNP*
D *Doesn't matter as long as both transistors are the same type.*

34

Oscillators

An *oscillator* is a circuit which produces an ac signal at some specific frequency (which may or may not be variable).

Often a distinction is made between oscillators and *waveform generators*. Under these definitions, the ac signal produced by an oscillator is a sine wave (see Fig. 34-1) which is a very pure signal that (theoretically) contains only the nominal frequency. A generator, on the other hand, can produce other, more complex waveforms which contain the nominal (or *fundamental*) frequency and one or more of this frequency's *harmonics* (or whole number multiples) in some specific proportion. Common complex waveforms will be discussed later in this chapter.

For our purposes here, we can consider the terms "oscillator" and "generator" to be essentially synonymous. We will define an oscillator as any electronic circuit that produces a repeating ac signal.

SINE WAVE OSCILLATORS

Producing a truly pure, harmonic free sine wave is actually quite difficult, but practical circuits can generate reasonable approximations that are adequate for most applications.

Most sine wave oscillators are built around parallel resonant LC circuits. The fundamental operations of an LC oscillator are illustrated in Fig. 34-2.

When the switch is closed, as in Fig. 34-2A, the voltage through coil L1 will rapidly increase from zero to the source voltage. Of course this means the current through the coil must also be increasing. A change of current through a coil will generate a magnetic

413

414

Fig. 34-1. A sine wave.

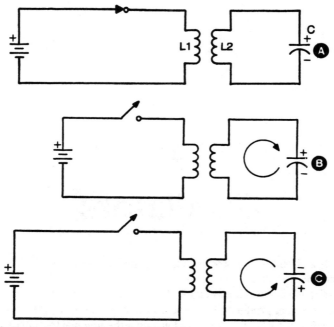

Fig. 34-2. Oscillation in a parallel resonant LC circuit.

field. This magnetic field will induce a similar voltage in L2, which will be stored in capacitor C.

When the current through L1 reaches a stable point and stops increasing, the magnetic field collapses. This means no further voltage is induced into L2 and C can discharge through the coil in the opposite direction from the original voltage.

This discharge voltage will cause L2 to induce a voltage into L1 which in turn induces the voltage back into L2, charging C in the opposite direction. This is shown in Fig. 34-2B.

Once the induced voltage in the coils collapses, this whole discharge-charge process repeats with the polarities again reversed, as in Fig. 34-2C.

Theoretically, this cycling back and forth between the capacitor and the coil will continue indefinitely. In real world components, however, the coil and capacitor will have some dc resistance that will decrease the amplitude on each oscillation. The signal will look like Fig. 34-3. This is called a *damped* sine wave.

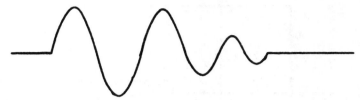

Fig. 34-3. A damped sine wave.

In addition to dc resistance, any energy that is tapped out of the circuit to be used by another circuit will subtract from the available energy within the LC circuit. Eventually a point will be reached when the signal level is too weak to feed back and sustain oscillation.

For this reason practical oscillator circuits always include some kind of amplifier stage. The output of the oscillator is continuously fed back to the input of the amplifier, and so it will be maintained at a usable level. Positive feedback is used.

It might at first seem that the amplifier would keep increasing the output amplitude infinitely, but all amplifiers have natural limitations that will prevent any further increases beyond some specific output level. This level is usually linked with the source voltage.

In an oscillator this *saturation point* will be reached within a few cycles of power on, and the amplitude of the output signal will remain essentially constant. This is just a natural characteristic of amplifiers.

Another natural characteristic of amplifier circuits is often taken advantage of to start the oscillation in the first place. All amplifiers have some internal noise which will produce a tiny output signal even if the input is perfectly grounded. This noise is fed back through the amplifier until it is increased to a level that can start oscillations within the LC circuit, or *tank*.

The frequency of the oscillations will be determined by the resonant frequency of the specific coil-capacitor combination used. The formula is:

$$F = \frac{1}{6.28 \sqrt{LC}} \qquad \textbf{Equation 34-1}$$

where F is the frequency in Hertz, L is the inductance in Henries, and C is the capacitance in Farads. 6.28 is two times pi (π).

As you should remember from Chapter 8, this is the frequency at which the reactance of the capacitor equals the reactance of the coil. As frequency is increased, inductive reactance increases, while capacitive reactance decreases. Only at resonance (equal reactances) can oscillations be sustained.

The most common types of sine wave oscillators are the *Hartley oscillator*, the *Colpitts oscillator*, and the *crystal oscillator*.

The Hartley Oscillator

A typical Hartley oscillator circuit is shown in Fig. 34-4. You should notice that the coil L has a center tap. In effect, it acts like two separate coils in close proximity. A current through coil AB will induce a signal into coil BC. For obvious reasons, this type of circuit is also known as a *split-inductance oscillator*.

416

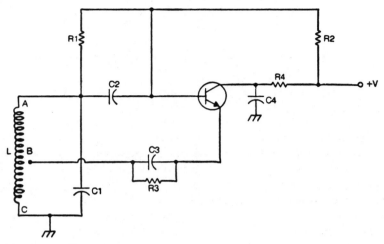

Fig. 34-4. A Hartley oscillator.

When power is first applied to this circuit, R2 places a small negative voltage on the base of the transistor, allowing it to conduct. Internal noise will build up within the transistor amplification stages, as discussed above. When this noise signal reaches a usable level, current from the collector will pass through R4, R2, and R1, finally reaching coil section AB. This rising current will induce a voltage into coil section BC. This voltage is stored by capacitor C1.

C2 is selected so that it has an extremely low impedance at the oscillating frequency, so the base of the transistor is effectively connected directly to C1. (The base voltage provided by R2 is quite low, so it can be ignored once oscillations begin. It is only needed to initiate oscillation.)

As C1 charges, it will increase the bias on the transistor. This in turn will increase the current through coil section AB, and the induced voltage through coil section BC. At the same time, the charge on both C1 and C2 is increased.

Eventually the voltage from C1 will equal the R1/C2 voltage, but with the opposite polarity. In other words, these voltages cancel out. At this point the transistor is saturated. Its output will stop rising, so coil section AB's magnetic field collapses. C1 starts to discharge through coil section BC, allowing C2 to discharge through R1, cutting off the transistor until the next cycle begins.

It takes some finite time for C1 to discharge through coil section BC, so as the current through the coil increases, it builds up a magnetic field. Once the capacitor is discharged, the coil will tend to oppose the change in current flow, so it continues to conduct, charging the capacitor in the opposite direction, and the entire process is repeated.

In some applications the low impedance of the transistor may load the tank circuit excessively, increasing power loss, and possible decreasing stability. This problem can be readily dealt with simply by using a high impedance active device in place of the transistor. Figure 34-5 shows a Hartley oscillator built around an FET instead of an ordinary bipolar transistor. Except for the increased impedance, operation is the same.

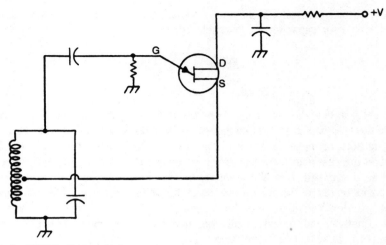

Fig. 34-5. An FET Hartley oscillator.

The Colpitts Oscillator

The *Colpitts oscillator* is very similar to the Hartley oscillator, except it is a *split-capacitance* instead of a split-inductance device. A typical Colpitts oscillator is shown in Fig. 34-6.

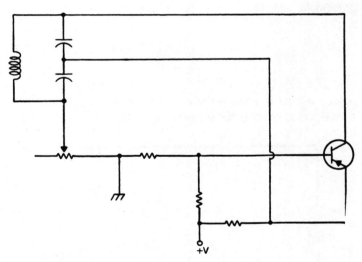

Fig. 34-6. A Colpitts oscillator.

With the two capacitors in series, they will work like a single capacitor as far as the LC resonant circuit is concerned, but the center tap (the connection between the two capacitors) provides a feed-back loop to the transistor's emitter.

If the two capacitors are of equal value, the total effective capacitance within the LC network (determining the frequency of oscillation) will be equal to one half the value of either capacitor separately. If they are unequal, the total may be calculated using the standard formula for capacitors in parallel, i.e.,:

$$\frac{1}{C_T} = \frac{1}{C_1} + \frac{1}{C_2}$$

Equation 34-2

In actual practice the two capacitors are usually of unequal values, because the strength of the feed-back signal is dependent on the ratio between these two capacitances. By changing both of these capacitors in an inverse fashion the feed-back level can be varied, while the resonant frequency remains constant. That is, when C1 is increased, C2 would be decreased by a like amount, or vice versa.

This points up one of the chief problems of the Colpitts oscillator. When the frequency is changed, we don't want the feed-back signal to vary or the level of the output signal will not be constant, and, in fact, oscillations may not be sustained. This means we have to change both capacitances simultaneously.

True, we could make the coil adjustable instead of the capacitors, but in most applications this would be just as impractical. Generally speaking, it is preferable to use an adjustable capacitor rather than an adjustable coil.

A common solution to this problem with the Colpitts oscillator is illustrated in Fig. 34-7. This method keeps the C1, C2 ratio constant, while the resonant frequency is readily adjustable. The total effective capacitance in the tank circuit is determined by the following formula:

$$C_T = C_3 + \frac{1}{\left(\dfrac{1}{C_1} + \dfrac{1}{C_2}\right)}$$

Equation 34-3

The Colpitts oscillator is a very popular circuit, because it offers very good frequency stability at a reasonable cost. It also appears in a number of variations.

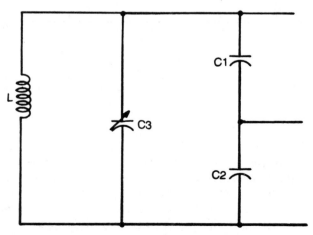

Fig. 34-7. A variable frequency tank circuit for a Colpitts oscillator.

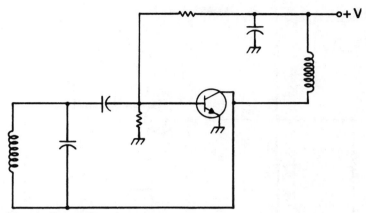

Fig. 4-8. An Ultra-Audion oscillator.

One common variation of the Colpitts oscillator is the *Ultra-Audion oscillator*, which is well suited for vhf (Very High Frequency) applications. A typical Ultra-Audion oscillator circuit is shown in Fig. 34-8.

Internal capacitances within the transistor itself provide the feed-back paths. These internal capacitances give the effect of small capacitors from the base to the emitter, and from the emitter to the collector. In the vhf range these internal capacitances have quite a low reactance. For operation at lower frequencies larger external capacitors could be added, as shown in Fig. 34-9. The Ultra-Audion oscillator is frequently used as the local oscillator in television tuners.

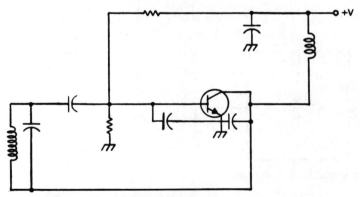

Fig. 34-9. A low-frequency Ultra-Audion oscillator.

Another variation of the Colpitts oscillator is illustrated in Fig. 34-10. This is the *Clapp oscillator*, which is tuned by a series resonant LC circuit, rather than the more common parallel resonant tank. Feedback for the Class oscillator is provided by the voltage divider formed by the two smaller capacitors in the emitter circuit. Figure 34-11 shows a practical Colpitts oscillator that can be used as the variable frequency ac source in the experiments. The parts list is given in Table 34-1.

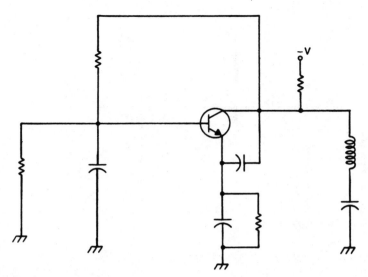

Fig. 34-10. A Clapp oscillator.

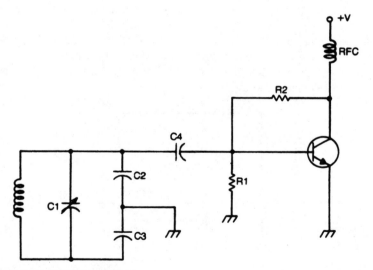

Fig. 34-11. A practical Colpitts oscillator.

**Table 34-1. Parts List for the Practical
Colpitts Oscillator Circuit of Fig. 34-11.**

C1	365 pF variable capacitor
C2, C3, C4	0.1 μF capacitor (experiment with other values)
RFC	radio frequency choke
L1	0.1 μH coil
Q1	almost any npn transistor (2N3904, 2N2222, etc. . .)
R1, R2	experiment should be at least 1,000 ohms

Crystal Oscillators

While the frequency stability of LC oscillators like the Hartley and the Colpitts can be quite good, it often isn't precise enough for certain critical applications, such as in broadcast transmitters (see Chapter 37), or the local color oscillator in a television receiver.

Oscillators built around quartz crystals can be extremely accurate and stable, especially if a constant temperature is held. Most broadcast stations (both television and radio) keep their carrier frequency oscillators in special crystal ovens to maintain a precisely constant temperature.

The fundamental operation of a crystal was described in Chapter 11. Remember that the thickness of the crystal slab determines the resonant frequency. Obviously, there is a practical limit to just how thin a crystal can be cut. Since the thinner the crystal, the higher the resonant frequency, this puts a theoretical upper limit on the frequency of a crystal oscillator.

Higher than normal frequencies can be obtained from crystal oscillators by passing the signal through special circuits called *frequency doublers*. For example, if a 5 MHz (5,000,000 Hz) oscillator signal is fed through a frequency doubler, the output will be 10 MHz (10,000,000 Hz). If this signal is passed through a second frequency doubler, the signal will be raised to 20 MHz (20,000,000 Hz).

For crystal oscillator circuits, an ac signal is applied between the plates, and thus, through the crystal slab itself. This causes the crystal to vibrate, due to the piezoelectric effect. Depending on the thickness of the crystal slab, it will be mechanically resonant (most willing to vibrate) at some specific frequency.

If the applied ac voltage is equal to the crystal's resonant frequency (or some exact harmonic of it) the amplitude of the vibrations will be quite large, and oscillations will be sustained.

Figure 34-12 shows the schematic diagram for an oscillator using a crystal as a parallel resonant circuit. Figure 34-13 shows an example of the series resonant mode. The parallel resonant form is somewhat more commonly used.

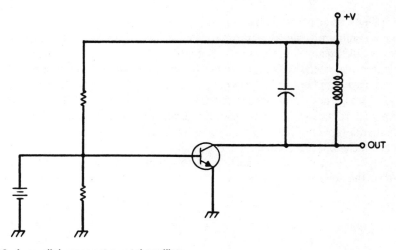

Fig. 34-12. A parallel resonant crystal oscillator.

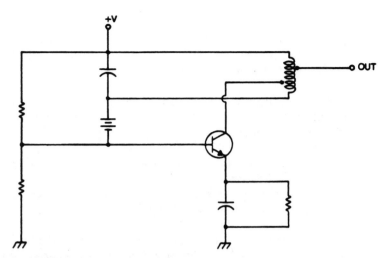

Fig. 34-13. A series resonant crystal oscillator.

There are a number of other types of sine wave oscillators, but these are the basic types you are most likely to encounter.

TRIANGLE WAVE OSCILLATORS

Figure 34-14 shows another common waveshape, the *triangle wave*. This is also sometimes called a *delta wave*.

Fig. 34-14. A triangle wave.

The triangle wave is somewhat similar to a sine wave in that the instantaneous level rises from a minimum value to a maximum value, then immediately reverses direction, without ever holding any constant value. But where the signal level in a sine wave varies according to a sinusoidal curve, the level in a triangle wave varies linearly.

The primary result of this is the presence of the odd harmonics. Besides the nominal fundamental frequency, the signal contains weaker frequencies at three times the fundamental (third harmonic), five times the fundamental (fifth harmonic), seven times the fundamental (seventh harmonic), and so forth. No even harmonics (those divisible by two) are included in a triangle wave signal.

The amplitude of the various harmonics relate to the amplitude of the fundamental according to the following formula:

$$A_h = \frac{Af}{h^2}$$

Equation 34-4

where A_h is the amplitude of the harmonic in question, h is the number of the harmonic, and A_f is the amplitude of the fundamental frequency.

If we arbitrarily assign A_f a value of 1, the amplitude of the third harmonic would be $1/3^2$, or $1/9$, the amplitude of the fifth harmonic would be $1/5^2$, or $1/25$, the amplitude of the seventh harmonic would be $1/7^2$ or $1/49$, and so forth.

A triangle waveform (or any other complex waveform) can be created by combining sine waves of appropriate frequencies and amplitudes—one for the fundamental, and one for each of the harmonics. This is called *additive synthesis*.

Similarly, if we filter a triangle wave (or any complex waveform) so there is no output except for the fundamental frequency, or one specific harmonic frequency, we would be left with a sine wave. This is called *subtractive synthesis*.

If we wanted to build up a 200 Hz triangle wave using additive synthesis starting with an 18 volt peak to peak fundamental at 200 Hz, we'd need to add a 2 volt peak to peak 600 Hz sine wave (third harmonic), a 0.72 volt peak to peak 1000 Hz sine wave (fifth harmonic), a 0.37 volt peak to peak 1400 Hz sine wave (seventh harmonic), a 0.22 volt peak to peak 2200 Hz sine wave (eleventh harmonic). For most purposes any additional harmonics beyond the eleventh would be at too low a level to be of significance.

As a second example, a triangle wave with a 1,450 Hz, 30 volt peak to peak fundamental would have a 4,350 Hz, 3.33 volt peak to peak third harmonic, a 7,250 Hz, 1.20 volt peak to peak fifth harmonic, a 10,150 Hz, 0.61 volt peak to peak seventh harmonic, a 13,050 Hz, 4.07 volt peak to peak ninth harmonic, and a 15,950 Hz, 0.25 volt peak to peak eleventh harmonic.

Notice that the fundamental frequency is always at a much larger voltage than all of the harmonic frequencies taken together. For many applications the harmonics can simply be ignored, and the triangle wave used in place of a sine wave. This is often done because it is generally easier and more economical to produce a stable triangle wave than a comparable sine wave.

Figure 34-15 shows a typical triangle wave oscillator circuit built around two 741 operational amplifier integrated circuits. IC1 is a square wave oscillator (discussed in Chapter 24). An integrator converts a square wave signal into a triangle wave. The fundamental frequency of the output signal is determined by C1 and R3. Of course, a single 747 dual op amp could be used in place of the two separate 741's.

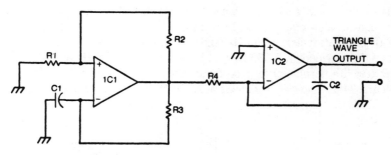

Fig. 34-15. A triangle wave oscillator.

RECTANGLE WAVE OSCILLATORS

The ac voltage at the output of a sine wave oscillator or a triangle wave oscillator varies continuously between its minimum and maximum values. A rectangle wave, however, consists of only two voltage levels—a maximum level, and a minimum level. Ideally, the voltage switches instantly from one level to the other. A real world circuit takes some finite time to change states. This is measured as a specification called the *slew rate*.

The relationship between the high voltage time and the low voltage time is called the *duty cycle*. The duty cycle is important because it determines the harmonic content of the signal. A rectangle wave with a duty cycle of 1:X will contain all harmonics, except those which are multiples of X. The amplitude of each harmonic relates to the amplitude of the fundamental by the formula:

$$A_h = \frac{A_f}{h} \qquad\qquad \textbf{Equation 34-5}$$

For example, let's consider a rectangle wave with a 1:3 duty cycle (see Fig. 34-16), and a 10 volt peak to peak 1000 Hz fundamental. The second harmonic would be equal to $2 \times F = 2 \times 1,000$, or 2,000 Hz. The amplitude equals A/h = 10/2, or 5 volts peak to peak. The third harmonic is absent. The fourth harmonic is at $4 \times 1,000$, or 4,000 Hz with an amplitude of 10/4, or 2.5 volts peak to peak. The fifth harmonic is at $5 \times 1,000$, or 5,000 Hz, and its amplitude equals 10/5, or 2 volts peak to peak. The sixth harmonic is absent. The seventh harmonic is $7 \times 1,000$, or 7,000 Hz. Its amplitude is 10/7, or 1.43 volts peak to peak. The eighth harmonic is at $8 \times 1,000$, or 8,000 Hz, and its amplitude equals 10/8 or 1.25 volts peak to peak. The ninth harmonic is missing. And so forth. Every third harmonic is absent from the signal.

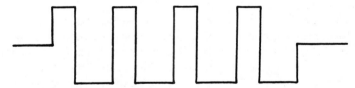

Fig. 34-16. A 1:3 duty cycle rectangle wave.

Similarly, let's take a rectangle wave with the same fundamental, but a duty cycle of 1:5. The harmonics would be 2,000 Hz (second harmonic), 3,000 Hz (third harmonic), 4,000 Hz (fourth harmonic), 6,000 Hz (sixth harmonic), 7,000 Hz (seventh harmonic), 8,000 Hz (eighth harmonic), 9,000 Hz (ninth harmonic), 11,000 Hz (eleventh harmonic), 12,000 Hz (twelfth harmonic), and so forth. Every fifth harmonic is missing from the signal. Figure 34-17 shows a rectangle wave with a 1:5 duty cycle.

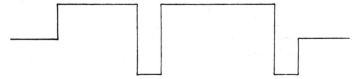

Fig. 34-17. A 1:5 duty cycle rectangle wave.

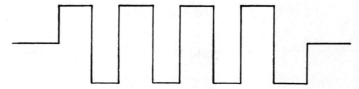

Fig. 34-18. A 1:2 duty cycle rectangle wave (square wave).

Table 34-2. Comparing a Sine Wave with a Triangle Wave.

HARMONIC NUMBER	FREQUENCY	AMPLITUDE IN A SQUARE WAVE	AMPLITUDE IN A TRIANGLE WAVE
FIRST (FUNDAMENTAL)	500 Hz	50 volts	50 volts
THIRD	1.500 Hz	16.67 volts	5.56 volts
FIFTH	2.500 Hz	10.00 volts	2.00 volts
SEVENTH	3.500 Hz	7 14 volts	1.02 volts
NINTH	4.500 Hz	5.56 volts	0.62 volts
ELEVENTH	5.500 Hz	4.54 volts	0.42 volts
THIRTEENTH	6.500 Hz	3.85 volts	0.30 volts
FIFTEENTH	7.500 Hz	3.33 volts	0.22 volts
SEVENTEENTH	8.500 Hz	2.94 volts	0.17 volts
NINETEENTH	9.500 Hz	2.63 volts	0.14 volts
TWENTY-FIRST	10.500 Hz	2.38 volts	0.11 volts
TWENTY-THIRD	11.500 Hz	2.17 volts	0.09 volts
TWENTY-FIFTH	12.500 Hz	2.00 volts	0.08 volts

A special case is a rectangle wave with a 1:2 duty cycle (see Fig. 34-18). This is called a *square wave*. Sometimes other rectangle waves are called square waves, but this is incorrect. A square wave has only the odd harmonics, like a triangle wave, but the proportions are different. Table 34-2 compares the harmonic content of a triangle wave and a square wave with 500 Hz, 50 volt peak to peak fundamentals. You can readily see that the square wave has considerably more harmonic content than the triangle wave.

Rectangle wave oscillators are also called multivibrators. We discussed multivibrators in earlier chapters. Remember that there are three basic types of multivibrators—the monostable, the bistable, and the astable. In this context we are only concerned with astable multivibrators.

Figure 34-19 repeats the astable multivibrator built around a 555 timer integrated circuit discussed in Chapter 25. Figure 34-20 shows another rectangle wave oscillator, using discrete transistors. When power is first applied to this circuit, one of the two transistors will start conducting a little faster than its mate, because no two transistors are ever exactly identical. It could be either transistor, and it doesn't really matter which one it is. For our discussion, we will assume Q1 conducts first.

As Q1 draws more current, the voltage dropped across R1 will increase (since E = IR), pulling the collector of Q1 negative, and charging C1 so that the end connected to the collector of Q1 is negative and the end connected to the base of Q2 is positive. This keeps the second transistor cut off for the time being.

At some point C1 will be completely charged, and the current flow to the base of Q2 will stop. Q2 will now be able to start conducting, with R4 and C2 mirroring the

426

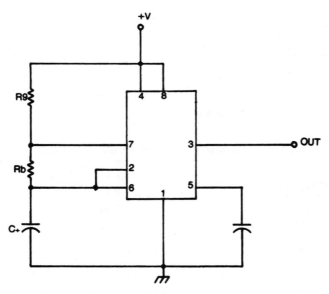

Fig. 34-19. A 555 timer IC rectangle wave oscillator.

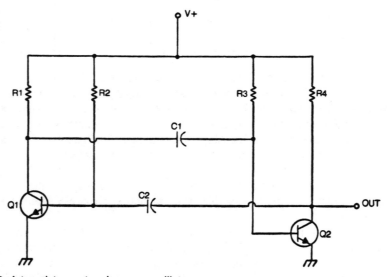

Fig. 34-20. A transistor rectangle wave oscillator.

earlier action of R1 and C1. As C2 is being charged, it will cut off Q1, allowing C1 to discharge.

This entire process continues back and forth indefinitely. The way the transistors are wired, they are either cut off completely, or saturated (operating at their maximum level) with a very small transition time between the two states (i.e., the slew rate). For most applications we can consider the switching time to be effectively zero, so the output is always either high, or low—switching back and forth at some specific frequency. Of course, when one transistor is cut off the other is always saturated.

The output could be taken from the collector of either transistor. In fact, in some cases, both outputs are used. Obviously these two outputs are mirror images, or complements of each other—when one is high, the other is low, and vice versa.

Four components determine the frequency of this oscillator. These are R1, C1, R4 and C2. R1 and C1 determine the one time for Q1 (time = R × C), and R4 and C2 control Q2.

If these two lines are equal (i.e., R1 = R4, and C1 = C2) the output will be a square wave.

The rectangle wave oscillator shown in Fig. 34-21 works in essentially the same way, except most of the circuitry is contained within the digital NOR gates. Notice that there is only one capacitor. It is used for both halves of the cycles. This means the on and off times are automatically equal. The output of this circuit is always a square wave.

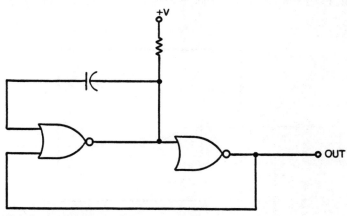

Fig. 34-21. A digital square wave oscillator.

Another simple square wave oscillator circuit is illustrated in Fig. 34-22. This circuit uses four sections of a hex inverter integrated circuit, such as a TTL 7404, and a single external capacitor—just two parts!

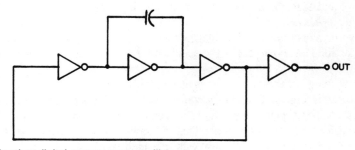

Fig. 34-22. Another digital square wave oscillator.

The capacitor can have a value anywhere between about 300 pF (giving an output frequency of about 1 MHz—1,000,000 Hz) to about 300 μF (giving an output frequency of about 1 Hz).

There is one important precaution to be taken with this circuit. The capacitor must *not* be an electrolytic type. In this circuit current must be able to pass through the capacitor in both directions, and, of course, this can't be done with an electrolytic capacitor. If the polarity across an electrolytic capacitor changes damage will result both to the capacitor itself, and (possibly) to other components in the circuit (in this case, the integrated circuit).

Semiconductors (such as transistors and integrated circuits) are particularly well suited for use in multivibrator circuits because they can switch from cut-off to saturation and back very rapidly, giving excellent slew rates. Tubes can also be used, but not as efficiently.

An operational amplifier, or a timer IC is a perfect point for designing a rectangle wave oscillator. Figure 34-23 shows a square wave oscillator built around a 741 operational amplifier.

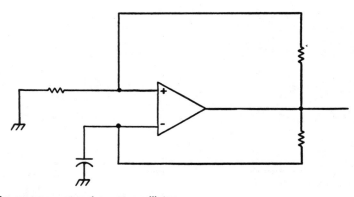

Fig. 34-23. An op amp rectangle wave oscillator.

SAWTOOTH WAVE OSCILLATORS

Figure 34-24 shows an *ascending sawtooth*, or *ramp wave*. The signal starts as a minimum level, builds linearly up to a maximum, then drops back down to its original minimum level and starts over. An ascending sawtooth wave is also sometimes called a *positive sawtooth wave*.

This waveform contains all of the harmonics (second, third, fourth, fifth, sixth, seventh, and so forth). The amplitudes of the successive harmonics decrease in an exponential manner.

The mirror image of this waveform is also sometimes used. A *descending sawtooth wave* is illustrated in Fig. 34-25. This waveform is also called a *negative sawtooth wave*.

Generally speaking, positive sawtooth waves are more commonly used. If the type is unspecified (i.e., when a circuit is called just a "sawtooth wave oscillator")it is usually safe to assume a positive sawtooth wave is intended.

Figure 34-26 is a simple sawtooth wave oscillator built around a unijunction transistor. This circuit is extremely versatile, because all of the components can be virtually any

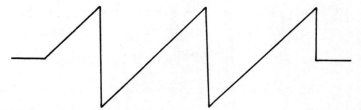

Fig. 34-24. An ascending sawtooth wave.

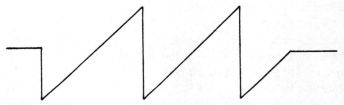

Fig. 34-25. A descending sawtooth wave.

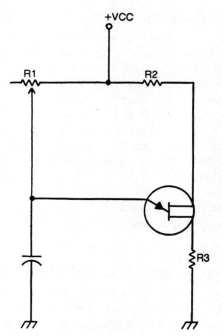

Fig. 34-26. A sawtooth unijunction oscillator.

value. Typically, however, R2 and R3 are between 15 ohms and 1,500 ohms. These resistors do not have to be of identical values.

The frequency of the oscillator is determined by R1 and C1. C1 charges through R1 at a specific fixed rate (i.e., the time constant). At some point enough voltage is built up at the emitter for the transistor to fire. When this happens, the capacitor quickly

430

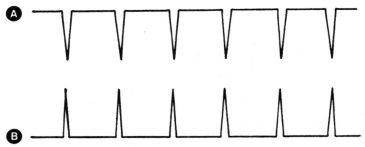

Fig. 34-27. Additional signals for Fig. 34-26.

discharges, and the whole cycle starts over again. This process produces a positive sawtooth wave signal on the emitter of the transistor.

It is also possible to take a signal off of either base. B1 produces a string of negative spikes, as shown in Fig. 34-27A. B2's signal is shown in Fig. 34-27B. This is the mirror image of B1's output—a string of positive spikes. All three signals are produced at the same frequency, of course.

FUNCTION GENERATORS

A *function generator* is actually nothing more than an oscillator that can produce more than one type of waveform. The sawtooth oscillator in Fig. 34-26 is actually a function generator of sorts. Most function generators produce triangle waves and rectangle waves, and perhaps one or two other waveforms, such as a sine wave or a sawtooth wave.

Figure 34-28 shows a simple function generator circuit built around operational amplifiers. IC1 is a square wave oscillator. IC2 is an integrator that converts the square wave to a triangle wave. IC3 is a filter that converts the triangle wave to a sine wave. S1 selects which of the three waveforms will appear at the output. IC4 is simply an amplifier. This circuit could be built using a single 324 quad operational amplifier integrated circuit. Another simple function generator circuit is given in Fig. 34-29. This circuit uses discrete transistors.

Function generators are extremely useful devices for electronic testing because of the wide range of possible output signals that can be experimentally fed through other electronic circuits.

VOLTAGE CONTROLLED OSCILLATORS

All of the oscillator circuits described so far either have a fixed frequency output, or an output whose frequency may be altered by manually adjusting some variable component (a potentiometer, or a variable capacitor). For many applications, however, it is preferable, or even necessary to have some way to automatically adjust the frequency via purely electronic means.

A *voltage controlled oscillator* (abbreviated *VCO*) is an oscillator whose output frequency will vary in step with an input voltage. This input voltage is called the *control voltage*, or *C.V.*

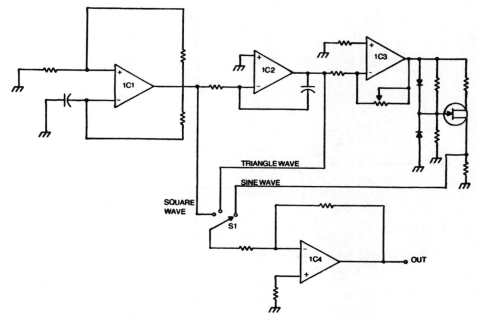

Fig. 34-28. An op amp function generator.

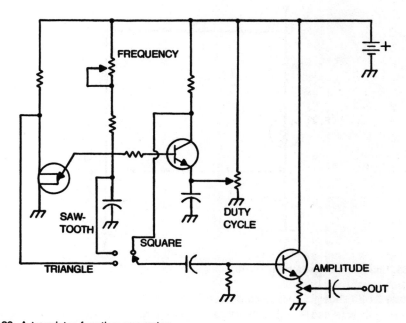

Fig. 34-29. A transistor function generator.

432

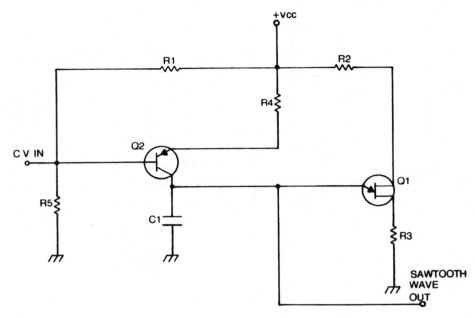

Fig. 34-30. A VCO.

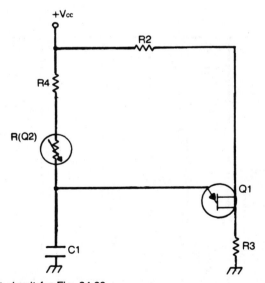

Fig. 34-31. Equivalent circuit for Fig. 34-30.

Figure 34-30 shows a voltage controlled version of the sawtooth wave oscillator from Fig. 34-26. Q2 acts like a voltage variable resistor—the equivalent circuit is shown in Fig. 34-31.

R4 and R(Q2) take the place of R1 in the original circuit. Since the output frequency is determined by R1 (R4 and R(Q2)), varying R(Q2) will vary the output frequency. Since

Q2's effective resistance is determined by the external voltage applied to its base, we have a voltage controlled oscillator.

Self-Test

1. What is the simplest, purest type of waveform?

A *Square wave*
B *Sine wave*
C *Triangle wave*
D *Sawtooth wave*
E *None of the above*

2. Which of the following is *not* a common type of sine wave oscillator?

A *Hartley*
B *Crystal*
C *Multivibrator*
D *Colpitts*
E *None of the above*

3. Which of the following is an oscillator with a split inductance?

A *Hartley*
B *Crystal*
C *Multivibrator*
D *Colpitts*
E *None of the above*

4. If the two capacitors in the LC tank of a Colpitts oscillator have values of 0.01 μF and 0.0047 μF, what is the total effective capacitance?

A *0.0147 μF*
B *0.0053 μF*
C *0.0032 μF*
D *0.000047 μF*
E *None of the above*

5. What is another name for a triangle wave?

A *Delta wave*
B *Pulse wave*
C *Ramp wave*
D *Angular wave*
E *None of the above*

6. In a rectangle wave with a duty cycle of 1:4, which of the following is a correct summary of the lower harmonic content?

A *Fundamental, third, fifth, seventh, ninth*
B *Fundamental, second, fourth, fifth, seventh, eighth, tenth*
C *Fundamental, second, third, fifth, sixth, eighth, ninth, tenth*
D *Fundamental, second, third, fifth, sixth, seventh, ninth, tenth*
E *None of the above*

7. What is the duty cycle of a square wave?

A *1:2*
B *1:5*
C *1:4*
D *1:1.5*
E *None of the above*

8. Is the fifth harmonic stronger in a triangle wave, or a square wave?

A *Triangle*
B *Square*
C *They are equal*
D *It varies*
E *None of the above*

9. What is the output frequency range of a square wave oscillator made up of four inverter stages and a capacitor (Fig. 34-22)?

A *1 Hz to 100 kHz*
B *10 Hz to 1 MHz*
C *100 Hz to 100 kHz*
D *1 Hz to 1 MHz*
E *None of the above*

10. What is the name for a waveform that begins at a low level, smoothly builds up to a maximum level, then quickly snaps back down to the minimum and starts over?

A *Descending sawtooth*
B *Delta*
C *Spike*
D *Ascending sawtooth*
E *None of the above*

35

Filters

A *filter* is a circuit that removes selected frequencies from a signal. There are four basic types of filters—*low pass, high-pass, band-pass,* and *band-reject.*

LOW-PASS FILTERS

Figure 35-1 shows the basic frequency response graph of a typical low-pass filter. Notice that low frequencies are able to pass through to the output, but higher frequencies are increasingly blocked. The steeper the cut-off slope, the better the filter. The point marked "X" is the *cut-off frequency* of the filter.

The simplest possible low-pass filter is shown in Fig. 35-2. This consists of just a resistor and a capacitor. The cut-off frequency is determined by the formula:

$$F = \frac{159,000}{RC}$$ **Equation 35-1**

where F is the cut-off frequency, in Hertz, R is the resistance in ohms and C is the capacitance in microfarads.

Suppose we have a filter consisting of a 10,000 ohm (10 k) resistor and a 0.1 μF capacitor. The cut-off frequency would be equal to 159,000/(10,000 × 0.1) = 159,000/1,000 = 159 Hz.

Similarly, if R is 330 ohms and C equals 0.5 μF, then F equals 159,000/(330 × 0.5) = 159,000/165, or approximately 964 Hz.

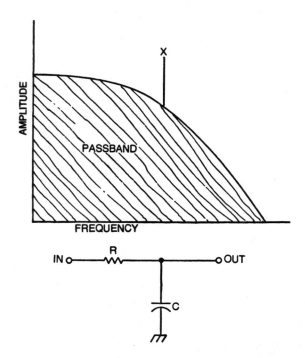

Fig. 35-1. Frequency response graph for a typical low-pass filter.

Fig. 35-2. Passive low-pass filter circuit.

A 0.0022 μF capacitor and a 470 ohm resistor would result in a cut-off frequency of 159,000/(470 × 0.0022) = 159,000/1.034, or about 153,772 Hz.

Notice that reducing the value of the capacitance and/or the resistance increases the cut-off frequency. If you have to design a filter with a specific cut-off frequency, the formula can be rewritten as:

$$R = \frac{159,000}{FC} \qquad \textbf{Equation 35-2}$$

As an example, suppose we need a low-pass filter with a cut-off frequency of 1,000 Hertz. First we arbitrarily select a convenient value for C. We'll use a 0.1 μF capacitor. Now, we can calculate that the required resistance equals 159,000/(1000 × 0.1) = 159,000/100 = 1,590 ohms. Unless the application is extremely critical, we can select the closest standard resistance value. In this case we would probably use a 1,500 ohm resistor.

The same cut-off frequency could be just as easily achieved with a different combination of components. For instance, if we started with a 0.033 μF capacitor, the resistor would have to be equal to 159,000/(1000 × 0.033) = 159,000/33 = about 4,818 ohms. A standard 4,700 ohm resistor could be used.

Notice that, for a given frequency, as one component increases in value, the other component must decrease in value.

Of course it is possible to start with an arbitrarily selected resistance value and solve for the capacitance using the formula in this form:

$$C = \frac{159,000}{RF}$$

Equation 35-3

but there are more readily available standard resistor values than standard capacitor values, so it is generally more convenient to select a convenient capacitor and solve for resistance.

The simple filter shown in Fig. 35-2 has two major disadvantages. Its cut-off slope is extremely gradual—quite a bit of the output signal could consist of frequencies above the cut-off frequency and, since only passive components are used, the entire signal will be attenuated to some degree. Even the desired frequencies will be partially decreased. This is called *insertion loss*.

Both of these problems can be taken care of by using an active filter instead of a passive one. An active filter includes an amplifier stage to compensate for any insertion loss. The disadvantages of an active filter are increased cost and circuit complexity, and the need for a power supply.

Figure 35-3 shows a basic low-pass filter built around an operational amplifier. This circuit is also known as an integrator, and has been discussed earlier in this book. As you'll recall, the formula for determining the gain in an operational amplifier is:

$$G = \frac{R2}{R1}$$

Equation 35-4

But in this circuit the feedback component is a capacitor whose reactance decreases as the frequency increases, according to the formula:

$$X_c = \frac{1}{2\pi FC}$$

Equation 35-5

So we now have to combine the two equations:

$$G = \frac{X_c}{R1} = \frac{1/(2\pi FC)}{R1}$$

Equation 35-6

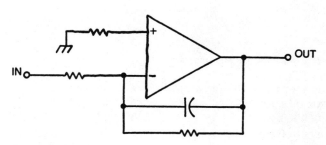

Fig. 35-3. Active low-pass filter circuit.

438

Let's assume R1 is 4,700 ohms, and C is 0.022 μF (0.0000000022 Farad). The basic formula becomes (1/(2 × 3.14 × F × 0.0000000022))/4700, or (1/(0.00000014 × F))/4700 = (7234316/F)/4700.

At 10 Hz, the gain is equal to (7234316/10)/4700 = 72341.6/4700, or about 154. But if the frequency is increased to 1,000 Hz, the gain becomes equal to (7234316(1000)/4700 = 7234.316/4700, or about 1.5. Increasing the frequency to 5000 Hz brings us into the region of negative gain—that is, the output level is less than the input level. The gain equals (7234316/5000)/4700 = 1447/4700 = just over 0.3.

If the frequency is increased to 50,000 Hz, the gain goes down to (7234316/50000)/4700 = 145/4700 = a gain of 0.03. The output level at this frequency is just 3% of its original input level.

While it is a definite step up over the simple passive filter previously described, the cut-off slope of this circuit still isn't particularly steep.

Figure 35-4 shows an improved active filter with a much steeper cut-off slope. The cut-off frequency for the circuit is determined by the following formula:

$$F = \frac{1}{2 \pi (R2\ R3\ C1\ C2)\ 1/2}$$
Equation 35-7

and then gain is determined by:

$$GAIN = \frac{R3}{R1}$$
Equation 35-8

This, of course, is the gain for frequencies below the cut-off frequency—that is, within the *passband*. For unity gain all three resistors should be equal.

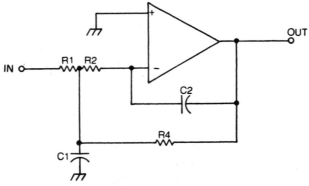

Fig. 35-4. Improved active low-pass filter circuit.

Let's assume we have such a circuit with three 1,000 ohm resistors, C1 equals 0.5 μF, and C2 equals 0.02 μF. The cut-off frequency equals 1/(6.28 × (1000 × 1000 × 0.0000005 × 0.00000002)$^{1/2}$) = 1/(6.28 × (0.00000001)$^{1/2}$) = 1/(6.28 × 0.00001) = 1/0.000628. The cut-off frequency is approximately 1,592 Hertz. The passband gain of course is 1, or unity.

HIGH-PASS FILTERS

A high-pass filter is the mirror image of a low-pass filter. Figure 35-5 shows the frequency response chart for a typical high-pass filter.

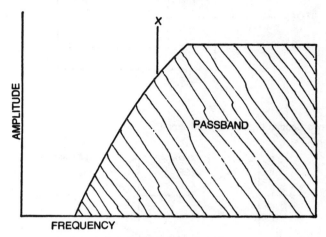

Fig. 35-5. Frequency response graph for a typical high-pass filter.

For a simple passive high-pass filter, the components simply change places, as shown in Fig. 35-6. The equations are the same as the basic low-pass filter.

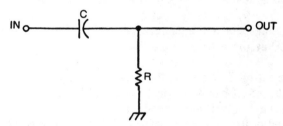

Fig. 35-6. Passive high-pass filter circuit.

Similarly, a simple op amp high-pass filter (differentiator) is the same as the low-pass version, except the resistor and the capacitor swap positions. See Fig. 35-7.

Figure 35-8 shows a steep slope active high-pass filter built around an operational amplifier.

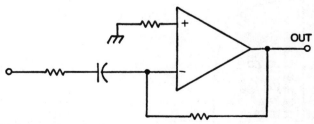

Fig. 35-7. Active high-pass filter circuit.

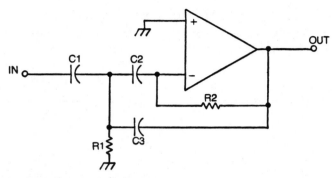

Fig. 35-8. Improved active high-pass filter circuit.

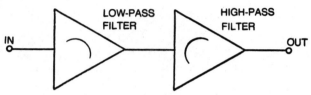

Fig. 35-9. Low-pass filter and high-pass filter in series, forming a band-pass filter.

BAND-PASS FILTERS

If a low-pass filter and a high-pass filter are placed in series, as shown in Fig. 35-9, the result will be a new, unique type of filter.

The low-pass filter blocks the high frequencies, and the high-pass filter blocks the low frequencies. Only those frequencies that are passed by both filters will appear in the output signal.

Obviously the cut-off frequency of the low-pass must be lower than that of the high-pass filter. If their cut-off bands overlap, there will be no output signal at all.

Figure 35-10 shows a typical frequency response graph for this type of filter. Since only a specific band of frequencies is passed, this is called a *Band-pass filter*. Rather than using a separate low-pass filter and high-pass filter, a single band-pass filter circuit

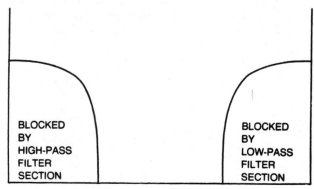

Fig. 35-10. Frequency response graph for a typical low Q band-pass filter.

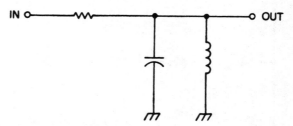

Fig. 35-11. Passive band-pass filter circuit.

can be used. Figure 35-11 shows a basic passive band-pass filter circuit. The center frequency of the pass-band will be determined by the resonant frequency of the capacitor/coil combination.

The bandwidth is found by ignoring the coil, and considering the resistor/capacitor combination as a low-pass filter. If, for example, the low-pass circuit has a cut-off frequency of 10,000 Hz, the bandwidth of the band-pass filter will also be 10,000 Hz, centered around the resonant frequency. Assuming the resonant frequency is 20,000 Hz, the filter will pass frequencies from 15,000 Hz, to 25,000 Hz.

If you know the upper frequency (F_u), and the lower frequency (F_l) required, you can find the required resonant frequency (F_r) using the following formula:

$$F_r = \sqrt{F_u F_l} \qquad \textbf{Equation 35-9}$$

For example, if F_u is 17,000 Hz and F_l is 9,000 Hz, F_r must equal $\sqrt{17,000 \times 9,000}$ = $\sqrt{15,300,000}$ = approximately 12,370 Hz.

The resonant frequency, of course, is solved in the same way as any LC circuit. That is:

$$L = \frac{1}{4 \pi^2 F^2 C} \qquad \textbf{Equation 35-10}$$

The bandwidth can be found with the following formula:

$$BW = \frac{159,000}{RC} \qquad \textbf{Equation 35-11}$$

The bandwidth and resonant frequency of a band-pass filter are interrelated by the *quality factor*, or simply *Q*. Q is the indication of the sharpness of the bandwidth. The graph in Fig. 35-10 is for a fairly low Q filter. Fig. 35-12 shows the frequency response for a high Q filter. Assuming the resonant frequency remains constant, increasing the Q decreases the bandwidth.

Q can be found using either of the two following formulas:

$$Q = \frac{Fr}{BW} \qquad \textbf{Equation 35-12}$$

442

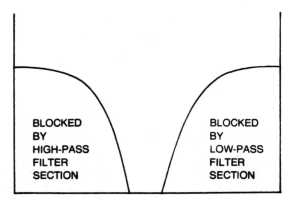

Fig. 35-12. Frequency response graph for a typical high Q band-pass filter.

$$Q = \frac{\sqrt{F_u F_l}}{F_u - F_l}$$ **Equation 35-13**

Figure 35-13 shows an active band-pass filter. Besides the usual advantages of active filters, this circuit requires no coils, as the passive version does.

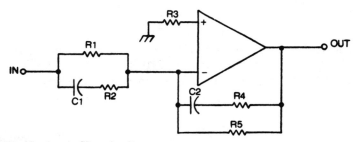

Fig. 35-13. Active band-pass filter circuit.

The lower cut-off frequency is found by the following formula:

$$F_l = \frac{1}{2 \pi C_1 (R_1 + R_2)}$$ **Equation 35-14**

while the formula for finding the upper cut-off frequency is:

$$F_u = \frac{1}{2 \pi C_1 R_2}$$ **Equation 35-15**

The Q of this filter can be found by using the formula given in Equation 35-13.

The resonant frequency can be found by either Equation 35-9, or by the following formula:

$$F_r = Q (F_u - F_l)$$ **Equation 35-16**

BAND-REJECT FILTERS

If a low-pass filter and a high-pass filter are connected in parallel, as shown in Fig. 35-14, any frequency passed by either filter will be present at the output. Only those frequencies blocked by both filters will be blocked by the combination. In this case the cut-off bands must overlap. That is, the cut-off frequency of the low-pass filter must be lower than that of the high-pass filter. As the frequency response graph in Fig. 35-15 indicates, this is the exact opposite of a band-pass filter. Only a specific band of frequencies is blocked, so this is called a *band-reject filter*. It is also sometimes called a *notch filter*.

Figure 35-16 shows a simple passive band-reject filter circuit. Notice that it is simply a rearrangement of the components of a band-pass filter. All of the equations remain the same.

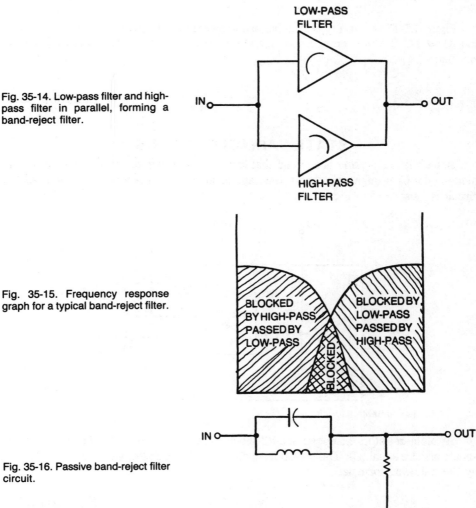

Fig. 35-14. Low-pass filter and high-pass filter in parallel, forming a band-reject filter.

Fig. 35-15. Frequency response graph for a typical band-reject filter.

Fig. 35-16. Passive band-reject filter circuit.

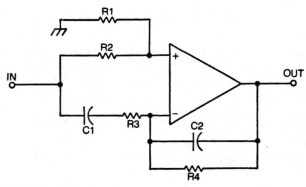

Fig. 35-17. Active band-reject filter circuit.

Figure 35-17 shows a typical active band-reject filter circuit. If R3 = R1, C1 = C2, and R1 = 1/2 R2, the center frequency of the rejected band is found by the following formula:

$$F = \frac{1}{2\pi C1\ R3}$$

Equation 35-17

STATE VARIABLE FILTERS

A *state variable filter* is a circuit that can operate in two or more of the basic filter modes (low-pass, high-pass, band-pass, and band-reject). A typical state variable filter circuit is illustrated in Fig. 35-18.

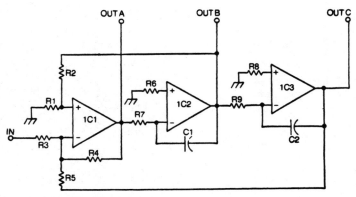

Fig. 35-18. State variable filter circuit.

All of the resistors except R1 and R2 should have equal values (R). The two capacitors should also be equal (C). The cut-off frequency for all three filter sections is determined by the following formula:

$$F = \frac{1}{2\pi RC}$$

Equation 35-18

Output A is a high-pass filter, output B is a band-pass filter, and output C exhibits a low-pass response.

The Q of this circuit is determined by R1 and R2. The formula is:

$$Q = \frac{1}{3} + \frac{R2}{3\ R1} \qquad \text{Equation 35-19}$$

In some applications R6 and R8 can be eliminated, and the non-inverting inputs of IC2 and IC3 would be shorted directly to ground. This will increase the *dc offset signal* for these outputs. That is, the ac output signal will be riding on a dc voltage, rather than being symmetrical around zero volts.

Filters are used whenever signals must be segregated according to frequency or when complex waveforms must be reduced to simpler signals.

Self-Test

1. What type of filter can be made up of a series resistor and a capacitor to ground?

A *Low-pass*
B *High-pass*
C *Band-pass*
D *Band-reject*
E *None of the above*

2. What is the name for a filter that rejects all but a specified continuous range of frequencies?

A *Low-pass*
B *High-pass*
C *Band-pass*
D *Band-reject*
E *None of the above*

3. What is the nominal cut-off frequency of a high-pass passive filter made up of a 680-ohm resistor and a 0.5 μF capacitor?

A *34 Hz*
B *29 Hz*
C *468 Hz*
D *4676 Hz*
E *None of the above*

4. What size capacitor should be used in a passive low-pass filter with a 470-ohm resistor for a cut-off frequency to 1,000 Hz? (Round off to the nearest standard capacitor value.)

A *0.33 μF*
B *0.022 μF*

446

C *0.033 μF*
D *0.47 μF*
E *None of the above*

    5. Refer to the active low-pass filter circuit shown in Fig. 35-4. Assume the following component values: R1 1,000 ohms, R2 2,200 ohms, R3 10,000 ohms, C1 0.5 μF, C2 0.01 μF. What is the nominal cut-off frequency?

A *1.4 MHz*
B *480 Hz*
C *4800 Hz*
D *140 kHz*
E *None of the above*

    6. How can a passive band-pass filter be created?

A *By placing an inductor in parallel with the capacitor in a high-pass filter*
B *By placing an inductor in series with the capacitor in a low-pass filter*
C *By placing a capacitor in parallel with the resistor in a low-pass filter*
D *By placing an inductor in parallel with the capacitor in a low-pass filter*
E *None of the above*

    7. Which of the following formulas defines the bandwidth of a passive band-pass filter?

A $BW = \dfrac{1}{4\,\pi^2\,F^2\,C}$

B $BW = \dfrac{159000}{RC}$

C $BW = \dfrac{159000}{LC}$

D $BW = \dfrac{159000}{\sqrt{RC}}$

E *None of the above*

    8. Which of the following formulas defines the Q of a band-pass filter?

A $Q = \dfrac{Fr}{BW}$

B $\quad Q = \dfrac{FrBW}{BW}$

C $\quad Q = \dfrac{BW}{Fr}$

D $\quad Q = BW \sqrt{Fr}$

E *None of the above*

9. What is another name for a band-reject filter?

A *Notch filter*
B *Attenuation filter*
C *State variable filter*
D *Q filter*
E *None of the above*

10. Assume a passive low-pass filter that is made up of a 390-ohm resistor and a 0.1 μF capacitor. What size inductor should be added for a center frequency of 12,000 Hz?

A *4 mH*
B *0.000002 mH*
C *1.8 mH*
D *18 mH*
E *None of the above*

36

Modulation

If one ac signal is used to control some aspect of a second ac signal, the process is called *modulation*. There are several types of modulation, each type having its own unique characteristics.

Modulation is often used in such applications as electronic music, and data storage, but, by far, its most common application is in radio transmission. This will be discussed in the next chapter.

AMPLITUDE MODULATION

Perhaps the most basic type of modulation is called *amplitude modulation*. This is often abbreviated as *AM*. As the name indicates, one ac signal modulates, or controls, the amplitude, or level, of a second ac signal. This is illustrated in Fig. 36-1.

The controlled signal is called the *carrier*, and the controlling signal is called the *program signal*. The program signal is superimposed onto the carrier signal.

The process of amplitude modulation produces additional signals that are not present in either of the input signals. These newly created frequencies are called *sidebands*.

For simplicity, we'll first consider what happens when both the carrier and the program are pure sine waves. In this case there will be two sidebands. The upper sideband's frequency is determined by the sum of the carrier and program frequencies. That is:

$$F_{us} = F_c + F_p$$

Equation 36-1

448

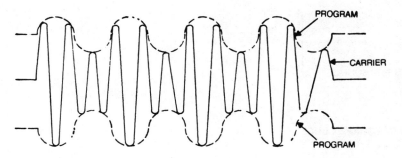

Fig. 36-1. Amplitude modulation.

where F_{us} is the frequency of the upper sideband, F_c is the carrier frequency, and F_p is the frequency of the program signal. All three quantities are in Hertz, of course.

The lower sideband is the difference between the carrier and program frequencies. That is:

$$F_{ls} = |F_c - F_p| \qquad \textbf{Equation 36-2}$$

F_{ls}, of course, is the frequency of the lower sideband. The two vertical lines around the right side of the equation indicate that the result is the *absolute value* of the equation. This means the answer will always be positive, even if F_p is greater than F_c. Negative frequencies, of course, are impossible.

If one, or both, of the input signals is a complex waveform, rather than a pure sine wave, the number of sidebands increases. Each individual harmonic will produce its own set of sidebands with each and every harmonic in the other input signal. Obviously, the resultant signal can be quite complex.

Let's assume we have a 150 Hz sine wave program signal amplitude modulating a 1,000 Hz sine wave carrier signal. The output would consist of the 1,000 Hz carrier with 150 amplitude fluctuations per second, plus weaker signals at 1,150 Hz (upper sideband) and 850 Hz (lower sideband).

Now, suppose the program signal is changed to a 150 Hz square wave. A square wave, of course, contains all of the odd harmonics. The fundamental will behave in the same way as a simple sine wave. That is, the output will contain signals at 850 Hz, 1,000 Hz, and 1,150 Hz. The third harmonic (450 Hz) will also combine with the 1,000 Hz carrier to produce additional sidebands at 550 Hz, and 1,450 Hz. The fifth harmonic (750 Hz) will produce sidebands at 250 Hz, and 1,750 Hz. And so forth. If both the carrier and the program signals are complex waveforms, the number of sidebands becomes very large.

Note that the sidebands don't necessarily (and probably will not) bear any harmonic relationship to either of the input signals.

AM is achieved by using a circuit called a *Voltage Controlled Amplifier* (or *VCA*). The program signal is used as an automatic volume control, determining the instantaneous level of the carrier signal. The basic setup is shown in Fig. 36-2.

The carrier signal is usually generated by a sine wave oscillator, while the program signal can be virtually any electrical signal, such as the output from a microphone.

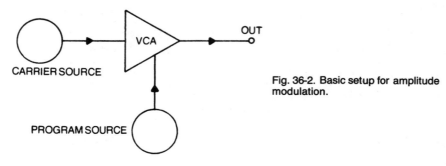

Fig. 36-2. Basic setup for amplitude modulation.

Balanced Modulation

A variation of the amplitude modulation concept is *balanced modulation*. A simple passive balanced modulator circuit is shown in Fig. 36-3. In a balanced modulator the original input signals are suppressed—only the sidebands are present at the output. For example, if the input signals are 700 Hz, and 1,300 Hz sine waves, the output would consist of just 600 Hz (lower sideband) and a 2,000 Hz (upper sideband) sine wave.

Balanced modulation is also sometimes called *ring modulation*.

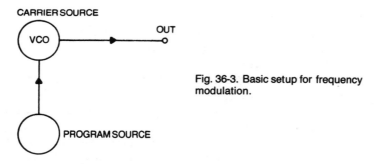

Fig. 36-3. Basic setup for frequency modulation.

Heterodyning

Closely related to amplitude modulation is the process known as *heterodyning*, or *non-linear mixing*. If two signals whose frequencies are very close to each other are superimposed onto each other, they will mix into an apparently single tone with amplitude fluctuations equal to the difference between the two frequencies. For example, a 998 Hz signal and a 1,004 Hz signal would combine with amplitude fluctuations at a 6 Hz rate.

Heterodyning is often used to fine tune two signals to the same frequency, by adjusting for the *zero beat* (no amplification fluctuations). Heterodyning is also widely used in radio receivers. This will be discussed in the next chapter.

FREQUENCY MODULATION

If one ac signal is used to control the frequency of a second oscillator (i.e., a VCO), *frequency modulation* takes place. This is often shortened to *FM*.

The basic setup for frequency modulation is illustrated in Fig. 36-3, and typical signals are shown in Fig. 36-4.

451

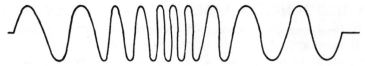

Fig. 36-4. Frequency modulation.

As with amplitude modulation, sidebands are produced by the process of frequency modulation. But there are many more sidebands produced with frequency modulation. FM sidebands appear above and below the carrier frequency at multiples of the program signal.

For example, if the carrier signal is a 1,000 Hz sine wave, and the program signal is a 60 Hz sine wave, the first set of sidebands would appear at 940 Hz, and 1,060 Hz. The second set of sidebands would be at 880 Hz, and 1,120 Hz. The third set of sidebands would be at 820 Hz, and 1,180 Hz, and so forth.

Of course, if more complex input signals are used, the pattern of the sidebands will become much more dense and complicated. FM sidebands have different amplitudes, which are at their lowest levels at frequencies farthest from the carrier frequency.

The number of sidebands produced in a given FM signal is determined by a factor called the *modulation index*. This quantity can be mathematically found with the following formula:

$$MI = \frac{\Delta F}{F_p} \qquad \textbf{Equation 36-3}$$

MI is the modulation index. ΔF is the *peak frequency deviation*. This is the difference between the modulated carrier frequency and the unmodulated carrier frequency. This is a constantly changing value. For example, a 1,000 Hz carrier signal that is modulated so that it fluctuates between 800 Hz and 1,200 Hz has a peak frequency deviation of 200 Hz. F_p is the frequency of the program signal.

The frequency deviation at any instant is determined by the instantaneous amplitude of the program signal. A high program amplitude produces a large frequency deviation. For a given amount of input power, as the modulation index increases, the output amplitude of the carrier frequency decreases.

If, for example, a 1,000 Hz carrier signal has a peak frequency deviation of 200, and the program frequency is 40 Hz, the modulation index would be equal to 200/40, or 5. This means there would be five sidebands above the carrier frequency, and five below. In other words, a modulation index of 5 indicates a total of ten sidebands. In this example, the output would consist of frequencies at 800 Hz, 840 Hz, 920 Hz, 960 Hz, 1,000 Hz (the carrier), 1,040 Hz, 1,080 Hz, 1,120 Hz, 1,160 Hz, and 1,200 Hz.

FM equipment for radio applications tends to be somewhat more expensive and complex than AM equipment, but FM is considerably less susceptible to electrical noise. This is because most electrical noise sources (such as lightning or motor brushes) act like an AM signal, and contain very little frequency modulation.

SINGLE SIDEBAND, SUPPRESSED CARRIER MODULATION

In an AM signal the upper and lower sidebands are exact mirror images of each other. Both sidebands together can carry no more information than either one separately. Similarly, the carrier frequency, since it is a constant, contains no information.

Transmitting an extra set of sidebands and the carrier uses up power that gives no greater distance or program content. This power is, in effect, wasted.

It is possible to eliminate one of the sidebands and filter out the carrier and use all the power to transmit just one set of sidebands. This more efficient system is called *single sideband—suppressed carrier*, or *SSSC*. The carrier signal is regenerated in the receiver for *demodulation* (see the next chapter).

PHASE MODULATION

Phase modulation is similar to frequency modulation except, in this case, the phase of the carrier signal (with respect to some constant standard) is varied in step with the instantaneous amplitude of the program signal. See Fig. 36-5.

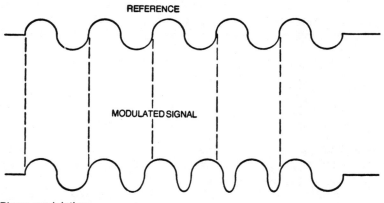

Fig. 36-5. Phase modulation.

Phase modulation is generally used in conjunction with another form of modulation when too much information for a single carrier is required. An example of the use of phase modulation is the color information signal in color television. The sound information is transmitted via frequency modulation, the basic picture by amplitude modulation, and the color by phase modulation.

PULSE CODE MODULATION

Pulse code modulation, or *PCM* converts the program signal into a string of pulses representing a binary number. Pulse code modulation is the most noise free form of transmission modulation because unless the noise coincidentally resembles one of the modulation pulses, it can easily be ignored by the receiving equipment.

Analog program signals are adapted for digital use by a circuit called an *A/D converter* (Analog to Digital Converter). See Fig. 36-6.

This is a two bit A/D converter. That is, it produces a two bit digital number that corresponds to the level of the analog input voltage. IC1, IC5, and IC9 are op amp volt-

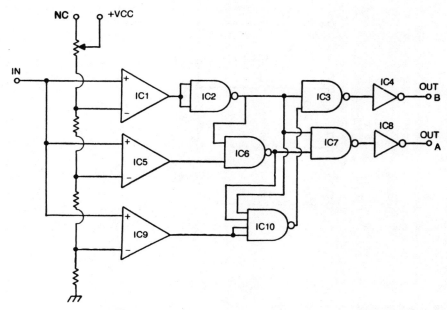

Fig. 36-6. A simple A/D converter circuit.

age comparators. If the input to one of these devices is equal or greater than a specific reference voltage, the output can be used as a digital logic 1. Otherwise, the output is a logic 0. IC2, IC3, IC6, and IC7 are 2 input NAND gates. IC10 is a 4 input NAND gate. IC4 and IC8 are simply buffers.

Table 36-1 lists some possible input/output combinations assuming IC1 has a reference voltage of 3 volts, IC5 is referenced to 2 volts, and IC9 has a reference voltage of 1 volt.

Note that since there are only two digital bits in the output, there are only four possible output conditions.

Practical A/D converters, of course, would require a much larger number of output bits to be really useful.

Table 36-1. Typical Input and Output
Signals for the Circuit Shown in Fig. 36-6.

INPUT VOLTAGE	COMPARATOR OUTPUTS			DIGITAL OUTPUTS	
	IC1	IC5	IC9	A	B
0.0	0	0	0	0	0
0.5	0	0	0	0	0
1.0	0	0	1	0	1
1.5	0	0	1	0	1
2.0	0	1	1	1	0
2.5	0	1	1	1	0
3.0	1	1	1	1	1
3.5	1	1	1	1	1

454

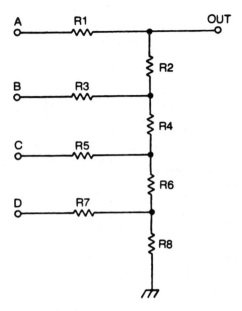

Fig. 36-7. A simple D/A converter circuit.

The digital numbers can be changed back into analog signals with a *D/A converter* (Digital to Analog converter). See Fig. 36-7.

A four bit digital number is inputted to this circuit. The output will be one of sixteen voltage levels—each corresponding to a specific digital number. A is the most significant (largest value) digit, and D is the least significant (lowest value) digit.

This type of circuit is sometimes called an "R—2R ladder network," because the effective resistance increases in a step-like manner from the most significant to the least significant digit. R2, R4 and R6 are all of equal value (R). R1, R3, R5, R7 and R8 are all twice the value of R2 (2R).

Table 36-2 shows the outputs that could be expected from such a circuit for each of the possible digital inputs. The 1 state voltage for this example is 4 volts. R is 1,000 ohms, so 2R equals 2,000 ohms.

The output from this type of circuit is usually fed through an operational amplifier for buffering and amplification.

Pulse code modulation is also useful for high quality recording. All analog recording systems produce some degree of degradation of the signal because of mechanical and electrical imperfections. If a second recording is made from the first, the distortion and noise will increase. A third copy will have an even more degraded signal. Eventually a point will be reached when the signal is no longer usable.

Pulse code modulation recordings, (or digital recordings), do not have this problem. The signal on the hundredth copy will be identical to that on the first. This is because only digital numbers are recorded. The noise and distortions are linear in nature, so they are simply ignored by the system.

**Table 36-2. Typical Input and Output
Signals for the Circuit Shown in Fig. 36-7.**

DIGITAL IN				ANALOG OUT (in volts)
A	**B**	**C**	**D**	
0	0	0	0	0.00
0	0	0	1	0.25
0	0	1	0	0.50
0	0	1	1	0.75
0	1	0	0	1.00
0	1	0	1	1.25
0	1	1	0	1.50
0	1	1	1	1.75
1	0	0	0	2.00
1	0	0	1	2.25
1	0	1	0	2.50
1	0	1	1	2.75
1	1	0	0	3.00
1	1	0	1	3.25
1	1	1	0	3.50
1	1	1	1	3.75

Self-Test

1. Which of the following is *not* a common type of signal modulation for transmission or storage of information?

A *Amplitude modulation*
B *Frequency modulation*
C *Phase modulation*
D *Ring modulation*
E *None of the above*

2. What is the frequency of the upper sideband produced when a 50,000 Hz carrier signal is amplitude modulated by a 4,500 Hz program signal?

A *54,500 Hz*
B *45,500 Hz*
C *100,623 Hz*
D *95,000 Hz*
E *None of the above*

3. How many sidebands are produced in an AM signal?

A *One*
B *Two*
C *Four*
D *It varies with the strength of the original signals*
E *None of the above.*

4. A 2,500 Hz carrier is frequency modulated by a 300 Hz program signal with an amplitude that causes the carrier signal to vary between 1,900 Hz and 3,100 Hz. How many sidebands will be produced?

A *8*
B *2*
C *4*
D *6*
E *None of the above*

5. Suppose the amplitude of the program signal in the problem described in question 4 is raised so that the carrier signal is varied between 1,000 Hz and 4,000 Hz. How many sidebands are now produced?

A *2*
B *5*
C *12*
D *10*
E *None of the above*

6. What type of modulation is used to transmit analog data in digital equipment?

A *Amplitude modulation*
B *Frequency modulation*
C *Pulse-code modulation*
D *Phase modulation*
E *None of the above*

7. How can a digital signal be converted into an analog signal?

A *With pulse-code modulation*
B *With an A/D converter*
C *With phase modulation*
D *With a D/A converter*
E *None of the above*

8. If a 15,000 Hz carrier signal is amplitude modulated by a 2,000 Hz program signal, how many sidebands are produced?

A *7*
B *15*
C *2*
D *10*
E *None of the above*

9. Two signals whose frequencies are close together are superimposed and mix into a single tone with amplitude fluctuations. What is this effect called?

A *Heterodyning*
B *D/A conversion*
C *Balanced modulation*
D *Linear mixing*
E *None of the above*

10. What determines the number of sidebands in an FM signal?

A *The difference between the carrier and program frequency*
B *The modulation index*
C *There are always two sidebands*
D *The square of the program frequency*
E *None of the above*

37

Radio

Radio is a means of communicating information from one place (*transmitter*) to another (receiver) via *electromagnetic waves*, or *electrostatic waves*.

RADIO WAVES

When ac current passes through a conductor, the conductor is surrounded by both an electromagnetic field (discussed in earlier chapters) and an *electrostatic field* (which is sometimes called just an *electric field*). These two types of fields are at right angles to each other, as shown in Fig. 37-1.

The magnetic and electrostatic lines of force are radiated outward from the conductor at the speed of light. Of course, the farther the lines of force get from the conductor, the weaker they are.

Since the strength of these fields is determined by the instantaneous strength of the current through the conductor, the lines of force will vary in step with the applied ac signal.

If the current through the conductor is high enough, and the ac frequency is within a specific range of frequencies, the lines of force can be transmitted over very long distances. These frequencies are called *radio frequencies*, or rf. The range of radio frequencies is called the *radio spectrum*.

The radio spectrum extends from approximately 20 kHz (20,000 Hz) to 30,000 MHz (30,000,000 Hz). It is divided for convenience into smaller bands, as listed in Table 37-1. These various bands have somewhat different characteristics that will be discussed in the section on *antennas*.

459

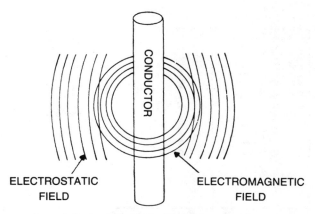

Fig. 37-1. Electromagnetic and electrostatic fields around an antenna.

Table 37-1. The Radio Spectrum.

FREQUENCY	NAME OF BAND
15 kHz to 30 kHz	Very Low Frequency- - -VLF
30 kHz to 300 kHz	Low Frequency- - -LF
300 kHz to 3 MHz	Medium Frequency- - -MF
3 MHz to 30 MHz	High Frequency- - -HF
30 MHz to 300 MHz	Very High Frequency- - -VHF
300 MHz to 3000 MHz	Ultra High Frequency- - -UHF
3000 MHz to 30,000 MHz	Super High Frequency- - -SHF

Table 37-2. Major Services Within the Radio Spectrum.

FREQUENCIES	SERVICES
535 kHz to 1605 kHz	AM BROADCAST BAND
0.2 MHz to 30 MHz	SHORT-WAVE BAND- - -INTERNATIONAL BROADCASTING AND HAM (AMATEUR) TRANSMISSION
27 MHz	CITIZEN'S BAND (CB)
54 MHz to 88 MHz	TELEVISION CHANNELS 2 THROUGH 6
88 MHz to 108 MHz	FM BROADCAST BAND
140 MHz to 170 MHz	EMERGENCY SERVICES (POLICE, FIRE, AMBULANCE, ETC.)
174 MHz to 216 MHz	TELEVISION CHANNELS 7 THROUGH 13
470 MHz to 890 MHz	UHF TELEVISION CHANNELS (14 THROUGH 83)

Portions of the radio spectrum are legally assigned for specific types of services. Some of the most important of these are listed in Table 37-2.

The earliest and simplest type of radio communication consisted simply of turning a transmitter on and off in an organized pattern and detecting the presence or absence of the carrier frequency with a receiver in another location.

A	.-	N	-.	0	-----	
B	-...	O	---	1	.----	
C	-.-.	P	.--.	2	..---	
D	-..	Q	--.-	3	...--	
E	.	R	.-.	4	....-	
F	..-.	S	...	5	.....	
G	--.	T	-	6	-....	
H	....	U	..-	7	--...	
I	..	V	...-	8	---..	
J	.---	W	.--	9	----.	
K	-.-	X	-..-			
L	.-..	Y	-.--			
M	--	Z	--..			

Table 37-3. The Morse Code.

These on/off pulses were arranged in specific patterns consisting of short and long pulses in a system called the *International Morse code*. The letters of the alphabet, numerals and a few punctuation symbols are each assigned a unique pattern of dots (short pulses) and dashes (long pulses). Each word in a message is spelled out, encoded and transmitted to a receiver where it is decoded and the original message may be read out. The Morse code is given in Table 37-3.

While the Morse code system is technically simple, and is certainly better than no long distance communication at all, it is inherently a rather cumbersome method of sending messages.

More information can be efficiently transmitted if the carrier is modulated by the signal to be transmitted. This can be speech, music or digital data. Amplitude modulation and frequency modulation are the most common types of transmission used today.

While it is definitely taking a back seat to the various modulation systems, Morse code is still used in some applications because it is simple and inexpensive, and it produces no sidebands. The transmitted bandwidth is simply the carrier frequency. Either it is transmitted, or not transmitted. Modulation always produces some sidebands. Usually these have to be limited so a reasonable number of usable frequencies can be fitted into the radio spectrum.

TRANSMITTERS

A transmitter's job is to generate the rf carrier signal, amplify it, modulate it with the program signal, and prepare it for the antenna (antennas will be discussed in the next section).

The rf carrier signal is generated by an oscillator, of course. It is always a sine wave, because harmonics would behave like additional sub-carriers at different frequencies, wasting energy, and spreading over a large bandwidth. The frequency of the oscillator must also be very tightly controlled. If the carrier frequency is allowed to drift even slightly off value, reception will be difficult, or impossible. Also, the frequency shifted signal might interfere with other signals on nearby frequencies. Radio transmitters are legally required to be on frequency within a small fraction of one per cent.

For the best possible accuracy, these rf oscillators are generally of the crystal type, and they're usually enclosed in special ovens for precise temperature control.

Phase-locked loops (or PLL's) are also frequently used. A phase-locked loop is a circuit that continuously tests the output of a voltage-controlled oscillator. If the oscillator is on frequency, nothing happens. If, however, the frequency has drifted (either increased or decreased) the phase-locked loop generates an *error voltage* that pulls the VCO back on frequency. See Fig. 37-2.

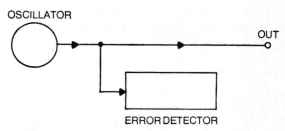

Fig. 37-2. Block diagram of a phase-locked loop (PLL).

Harmonics and stray oscillations at undesired frequencies must also be stringently avoided in transmitter circuits. Low-pass and band-pass filters are often used to block out these undesired frequencies from the final output signal.

The modulation index, and bandwidth of broadcast signals are also tightly controlled and regulated by the FCC (Federal Communication Commission). For example, all significant sidebands and radiated energy for a station within the standard AM broadcast band must be within ±5 kHz (10,000 Hz) of the nominal carrier frequency. This, of course, limits the frequency of the program signal to 15 kHz. FM and television channels are allowed wider bandwidths, but there are restrictions.

Rf amplifiers are generally similar to audio frequency (AF) amplifiers, except greater care must be taken to prevent stray capacitances. Since capacitive reactance decreases as frequency increases, an unintentional capacitance between say, two nearby circuit leads could act as a short circuit, or a feed-back path for an rf signal. This could cause parasitic oscillations, or other problems.

Figure 37-3 shows a basic block diagram for a typical AM transmitting station. The turntable and/or the microphone are the source for the program signal. They are combined

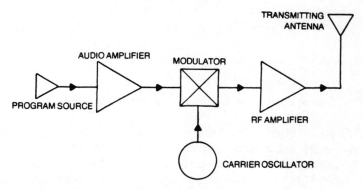

Fig. 37-3. Block diagram of a typical AM broadcasting station.

in the desired proportions in the mixer. The equalization circuit compensates for any frequency inequalities in the system. No circuit has a completely flat frequency response. Certain frequencies may be overemphasized, while others may be cut back. Equalization is a way of forcing the frequency response to be flat, by amplifying the weak frequencies, and partially filtering (i.e., attenuating) the overly strong frequencies. The program signal then passes through one or two stages of audio amplification.

Meanwhile, the crystal oscillator is generating the carrier frequency, and the phase-locked loop is helping it stay on the correct frequency. The unmodulated carrier is amplified by one or more stages of rf amplifiers.

The program is superimposed on the carrier signal in the modulator and the modulated signal is amplified again. Finally the modulated signal is fed to the transmitting antenna where it is radiated into space.

ANTENNAS

When the electromagnetic and electrostatic waves radiate out from the antenna they may follow one or more different paths to the receiving station. Each type of path has different characteristics. The paths taken depend on a number of factors—especially the carrier frequency.

The three basic types of signal paths are called *ground waves, direct waves,* and *sky waves*. See Fig. 37-4.

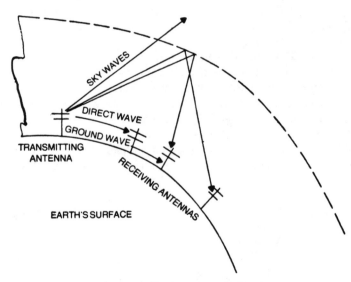

Fig. 37-4. Rf propagation—ground waves, direct wave, and sky waves.

Ground waves basically follow the curvature of the earth. Direct waves move in a direct, line-of-sight path from the transmitting antenna to the receiving antenna. Sky waves radiate upwards to the upper atmosphere. Some of the sky waves are absorbed by the atmosphere, some pass right through into space, and some bounce off the atmosphere layers and are reflected back to earth.

The atmosphere has a number of different layers. The portion we are concerned with here is the *ionosphere*. Since an *ion* is an electrically charged particle, the ionosphere is an electrically charged portion of the atmosphere.

The ionosphere is also divided into layers. There is a D layer, an E layer, and two F layers. The D and E layers exist only during daylight hours. At night they disappear.

Each frequency band in the radio spectrum is *propagated* by different combinations of these basic wave types. *Propagation* is the way a given wave travels from the transmitting antenna to the receiving antenna.

VLF (very low frequencies) and LF (low frequencies) (20,000 Hz to 300,000 Hz) are propagated mostly by the ground wave. The direct wave is generally too weak to be very useful at these frequencies, and the sky waves are virtually all absorbed by the atmosphere. The signals in this range are quite reliable, and their reception is practically independent of time of day and weather conditions. However, the range is somewhat limited.

These low frequencies tend to travel farther over sea water, because it has an extremely high amount of conductivity, so they are often used for marine applications such as ship-to-shore communications, distress signals, and weather stations.

Medium frequencies (MF) travel along a combination of ground waves and sky waves. The sky waves are reflected back to earth by the lower portions of the ionosphere, so the reception point isn't very far from the transmitting antenna. At night, when the D and E layers disappear, the sky waves are reflected from a higher angle, so the signal is propagated farther.

A transmitting station in this frequency range has three basic service areas. The primary service area is served almost entirely by the ground wave. Any sky waves in this area are extremely weak if they exist at all. The primary service area is the closest to the transmitting antenna. The secondary service area is at some distance from the transmitting antenna. This area receives relatively little signal during the day, and the station may not be receivable at all. At night, however, this area receives a strong sky wave signal, and reception is usually quite reliable.

In between the primary and secondary service areas is a region called the "fading area." Here the signal comes from both the ground wave and the sky waves. If at a given point these two signals are in phase, they will reinforce each other, if however, they are out of phase, they will partially or completely cancel each other out. Since the phase of the sky wave changes from moment to moment, as the reflection angle shifts slightly, the signal will tend to fade in and out, making reception unreliable. The standard AM broadcast band is included in this frequency range.

HF (high frequencies) (3 MHz to 30 MHz) are propagated mostly by sky waves. This frequency range is often called the short wave band. Citizens Band radios (CB) operate at approximately 27 MHz.

These frequencies are not bent at as sharp an angle by the ionosphere as MF signals, so the sky waves can reach over very large distances. This is why short wave transmissions can be used for global communications.

VHF (very high frequencies) and higher behave almost like a beam of light. They travel along a line-of-sight path. That is, the signal is propagated by the direct wave.

A few sky waves are reflected back by the upper atmosphere, but not many. Most of the sky waves pass right through the atmosphere and into space. The higher the frequency, the more easily the signal can pass through the atmosphere. Obviously signals transmitted at these extremely high frequencies can rarely be received at large distance from the transmitting antenna.

The transmitting antenna, of course, is the conductor that radiates the signal. The receiver must also have an antenna. As the signal passes the receiving antenna, a small voltage is induced into the antenna. This voltage is extremely small. Even a 50,000 watt transmitting station will rarely induce more than a microwatt (one millionth of a watt) into the receiving antenna. Often the induced signal will be only a thousandth of that.

Any antenna will be most sensitive to a specific frequency (or range of frequencies). Electrically, it behaves like a series resonant RLC circuit. The length of the antenna elements is the primary factor in determining the R, L and C values, and thus, the resonant frequency.

Remember that in a series resonant circuit, the maximum amount of current flows at the resonant frequency. Thus a signal at, or near, the antenna's resonance point will induce the strongest signal into the antenna.

Figure 37-5 shows one basic type of antenna. This is called a *Marconi antenna*. It simply consists of a conductor of the appropriate length, and a ground connection. Figure 37-6 shows the electrical equivalent circuit for a Marconi antenna.

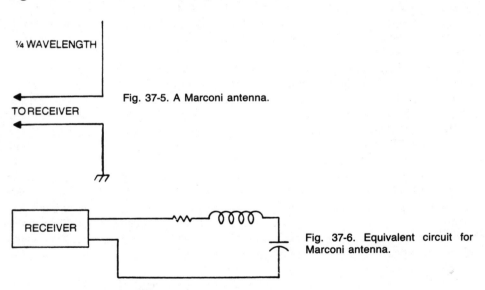

Fig. 37-5. A Marconi antenna.

Fig. 37-6. Equivalent circuit for Marconi antenna.

A Marconi antenna will be resonant at a wavelength just under four times the length of the antenna itself. For this reason, it is often called a *quarter-wave antenna*.

Another common type of receiving antenna is the *hertz antenna*, shown in Fig. 37-7. Because this type of antenna has two sections, it is also known as the *dipole antenna*. Its electrical equivalent circuit is illustrated in Fig. 37-8.

Notice that it consists essentially of two Marconi antennas. No ground connection is required. Each section is just under a quarter-wavelength long, so the entire antenna is often called a *half-wave antenna*. Typically, the actual length is about 5 percent less than one-half wave length.

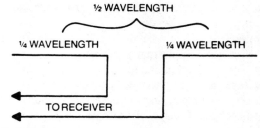

Fig. 37-7. A hertz (or dipole) antenna.

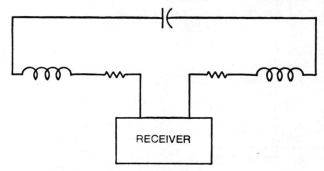

Fig. 37-8. Electrical equivalent circuit for a hertz antenna.

Of course, most practical antennas are required to receive a number of different frequencies. For example, an FM receiver needs an antenna that is functional along the entire 88 to 108 MHz (megahertz) FM broadcast band. This is a range of 20 MHz. No antenna can be truly resonant over such a wide range. Generally we have to compromise by tuning the antenna to the center of the band, so either end of the band is reasonably close to the resonance of the antenna.

A practical formula for finding the correct average length of a half-wave antenna is as follows:

$$X = \frac{468}{F}$$ **Equation 37-1**

where X is the length of the antenna in feet, and F is the frequency in megahertz (1,000,000 Hertz).

For example, let's say we need an antenna for receiving the FM broadcast band. As mentioned above, it should be resonant to the center of the band, or 98 MHz. The antenna length should therefore be equal to 468/98, or about 4.8 feet.

As a second example, let's find the length of an antenna for the standard AM broadcast band. The center of the band is 1,070 kHz (1,070,000 Hz), or 1.07 MHz. A half-wave antenna for this frequency would have length of 468/1.07, or over 437 feet. Obviously this would be extremely impractical.

For low frequencies, a coil is often added in series with the antenna, increasing the effective inductance of the circuit, and decreasing the resonant frequency. While this method is practical, a short antenna of this type will not pick up as great a signal as a true half-wave antenna would. Often portable AM broadcast band radios simply use a coil within the case of an antenna.

In a sense, an antenna acts like a band-pass filter, emphasizing only a specific set of frequencies from all of the signals that pass it.

Like a band-pass filter, an antenna has a definite Q. A high Q antenna (such as a dipole made of thin wires) has a very narrow bandwidth. Increasing the diameter of the antenna elements decreases the Q, and increases the bandwidth. The Q can also be altered by changing the physical shape of the antenna.

Antennas generally do not receive signals equally well from all directions. A half-wave dipole receives signals along its length better than signals that strike its ends. Its reception area is a figure 8 pattern, as shown in Fig. 37-9.

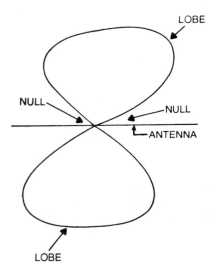

Fig. 37-9. Reception pattern for a ½ wave dipole antenna.

This pattern is true only for signals that have a wavelength that is twice the length of the antenna. At other frequencies, the reception pattern would be changed.

For example, most television antennas are one half wavelength in the middle of the band consisting of channels 2 through 6. For channels 7 through 13 the antenna is about one and a half (3/2) wavelength long. The reception pattern would resemble Fig. 37-10.

We have effectively added two additional half-wave lengths to the original dipole, and got six *lobes* (strong reception areas) and six *nulls* (weak reception areas). In general increasing the number of half-wave lengths increases the number of lobes and nulls.

Bending the antenna would make it more sensitive to signals from one side, while reducing the sensitivity on the opposite side, as shown in Fig. 37-11.

Often the *gain* of an antenna will be spoken of in technical literature. Actually an antenna has no true gain—it does not amplify the signal. Gain, in this context, is used to compare the efficiency of a specific antenna with a basic dipole antenna. An antenna with a gain of two would receive twice as much signal as a standard, simple dipole antenna.

There are basically two ways of increasing the gain of an antenna. We could connect two or more dipole sections together or we could use *reflectors*. These are simply extra rods or wires that are placed around the antenna to effectively focus the passing wave

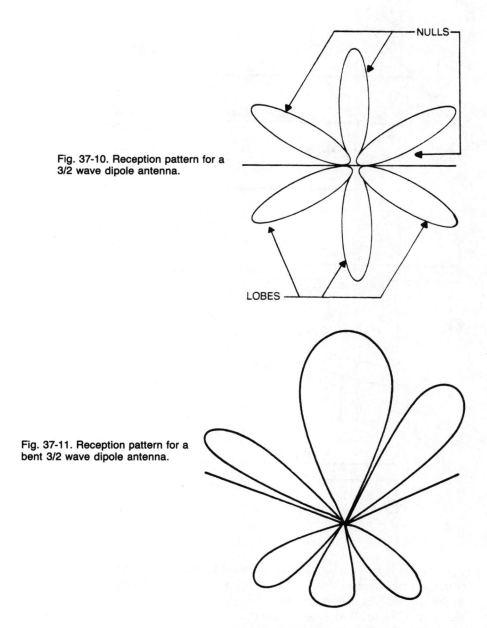

Fig. 37-10. Reception pattern for a
3/2 wave dipole antenna.

Fig. 37-11. Reception pattern for a
bent 3/2 wave dipole antenna.

onto the antenna itself. This is illustrated in Fig. 37-12. Television antennas often use
this sort of arrangement.

The reflectors are not electrically connected to the receiver. They are called *parasitic
elements*. The reflector is usually about 5% longer than the main antenna element, and
is placed about 0.15 wavelength behind the antenna itself with reference to the desired
signal source. See Fig. 37-13A. The signal that passes the antenna strikes the reflector
and bounces back to the antenna.

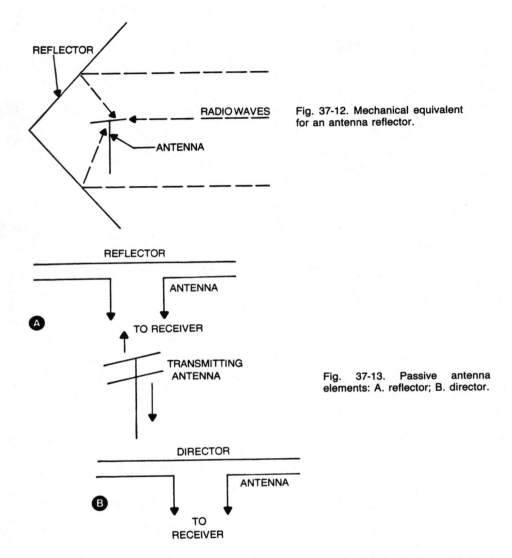

Fig. 37-12. Mechanical equivalent for an antenna reflector.

Fig. 37-13. Passive antenna elements: A. reflector; B. director.

Another type of parasitic element is called a *director*. This element is about 4% shorter than the main antenna element and is placed about a tenth of a wavelength in front of it. See Fig. 37-13B. The director focuses the passing signal onto the antenna, somewhat like an optical lens. Both types of parasitic elements can be used within a single system. This type of system is called a *Yagi antenna*.

Note that these parasitic elements increase directivity as well as gain. They make the antenna sensitive to essentially only those signals arriving from one specific side.

SWR

An antenna has its own specific impedance, that should match the impedance of the circuit it is connected to for maximum power transfer.

The cable connecting the antenna and the circuit (either receiver, or transmitter—in this case, we are more concerned with the transmitter although the following discussion will apply to receivers too, but to a lesser extent) is called the *transmission line*. It also has a specific inherent impedance. This value is constant regardless of the length of the cable. This impedance should closely match the impedances of the antenna and the transmitter. For a receiver the impedances only have to be approximately equal.

When these impedances are mismatched, *standing waves* will appear along the line. These standing waves consist of power that is reflected back from the antenna to the transmitter. Obviously, this power is serving no useful purpose, and is therefore wasted.

This effect is measured by a factor called the *standing-wave ratio*, or *SWR*. An ideal SWR would be 1:1. This would mean all of the power from the transmitter is reaching the antenna. Unfortunately, this ideal is impossible to achieve in practical circuits. Generally a 1:1.1 to a 1:1.5 SWR is the best that can be reached. Usually anything better than 1:2.5 is good enough for most applications.

In receiving only application, SWR is generally ignored beyond just roughly matching up the nominal impedances. You would not get very good results from a 75 ohm antenna connected to a 300 ohm receiver, but if the antenna's actual impedance was 250 ohms, there would probably be no noticeable problems in a 300 ohm system.

RECEIVERS

A *receiver* accepts the signal from an antenna, amplifies, and demodulates it to retrieve the original message, or other program signal.

Most modern receivers are of the *superheterodyne* type. This is often shortened to *superhet*. Figure 37-14 shows the block diagram for a superheterodyne receiver.

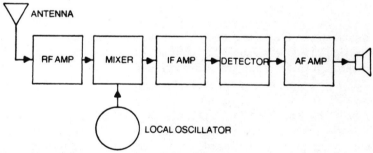

Fig. 37-14. Block diagram for a superheterodyne receiver.

Not all superheterodyne receivers include an rf amplifier stage. Inexpensive AM radios almost always leave this stage out, but it is generally included in better receivers.

A superheterodyne receiver is very sensitive. That is, it can operate on a fairly weak input signal. The rf amplifier boosts the incoming signal from the antenna, so even weaker signals can be received.

The rf amplifier stage also helps eliminate a problem called *image interference*. A latter stage in the receiver converts the rf signal to a lower, intermediate frequency (IF). Another station operating at a frequency that is twice this intermediate frequency could interfere with the desired signal.

The rf amplifier is tuned to the desired frequency and is set up to reject this potentially interfering frequency.

Strong signals can be applied directly to the mixer without an rf amplifier stage, but a weak signal could be drowned out by internal noise in the mixer circuit. An rf amplifier stage increases sensitivity.

Figure 37-15 shows the schematic diagram for a typical rf amplifier stage in an AM receiver. Capacitor C1 is tunable, so the C1/L1 parallel tank circuit can be made resonant at the desired frequency. L1 is wound around a ferrite rod, and acts as the antenna. No external antenna is required for this circuit. The ferrite core in the coil gives it a very high Q, so the resonant circuit is quite selective, rejecting near-by frequencies almost totally.

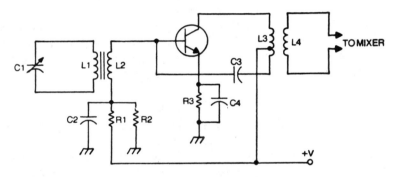

Fig. 37-15. A typical AM rf amplifier stage.

L2 is wound around the same core as L1, so the signal flowing through L1 is induced into L2. This signal is, in turn, applied directly to the base of the transistor. The signal is also sent through C2 to ground, and from ground through C4 to the emitter of the transistor. R3 helps prevent thermal runaway.

Resistors R1 and R2 form a voltage divider to forward bias the transistor.

The amplified rf signal from the collector of the transistor passes through L3 where it induces the signal into L4 which is the input of the next stage in the receiver. This may be either another stage of rf amplification (uncommon) or the mixer.

C3 is vitally important in an rf amplifier. Internal capacitances within the transistor itself could provide a feed-back path for the rf signal from the collector back to the base. This could cause the stage to break into oscillation. C3 feeds back a small out of phase signal to compensate for this. This component is called a neutralizing capacitor.

Next the rf signal is fed to a mixer stage where it is combined with the signal from a local (contained in the receiver) oscillator. The local oscillator actually modulates the received rf signal again, producing two sidebands—the received frequency plus the oscillator frequency, and the received frequency minus the oscillator frequency. Everything except this lower sideband is filtered out.

If the received signal is 1,000 kHz, and the local oscillator frequency is 545 kHz, the lower sideband would be 455 kHz. This is the intermediate frequency (IF). It is constant for all stations received by the radio. The oscillator is tuned so that it is always equal to:

$$F_o = rf - i\text{-}f$$ **Equation 37-2**

where F_o is the frequency of the local oscillator, rf is the radio frequency of the desired signal, and i-f is the intermediate frequency. The i-f used in most AM radios is 455 kHz. FM radios usually have an i-f of 10.7 MHz.

There are two major advantages of converting the rf signal to a lower i-f signal. One is that it is easier to amplify lower frequencies without the circuit breaking into oscillation. The components in an i-f amplifier can usually be less expensive types than are typically found in comparable rf amplifiers.

The other advantage is that each stage in the radio does not have to be returned each time we want to receive a different frequency. All of the stages past the mixer can be tuned to a single frequency and left there. The process of tuning the i-f stage is called *alignment*. Only the rf amplifier and the local oscillator have to be retuned for each received frequency. This is usually done via a single, ganged two section capacitor.

Figure 37-16 shows a typical mixer/oscillator circuit. Q1 is the mixer and Q2 is the oscillator.

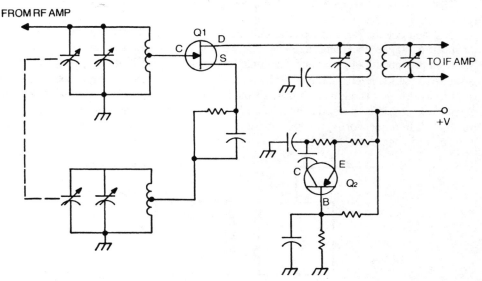

Fig. 37-16. A typical mixer/oscillator circuit.

Figure 37-17 is a similar circuit that uses a single transistor for both the mixer and the oscillator functions. L3, C3, and C4 are the oscillator's frequency determining tank circuit. L2 provides feedback compensation. R3 is fairly small and R4 is rather large. The large resistance of R4 prevents energy fed through C5 from feeding back into the power supply. L4 and L5 couple the signal into the first i-f amplifier.

I-f amplifiers are usually quite similar to rf amplifiers, but neutralizing capacitors generally aren't needed, thanks to the lower frequencies involved.

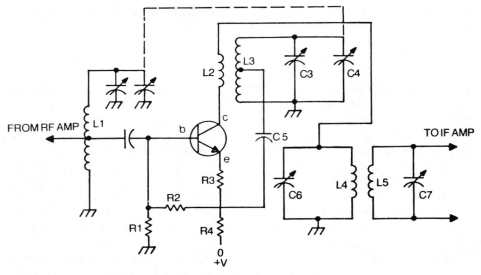

Fig. 37-17. A single transistor mixer/oscillator circuit.

Usually two or more stages of i-f amplification are included, at least in better quality receivers.

A tuned coupling circuit like the one illustrated in Fig. 37-18 is included between each pair of stages (including the mixer/first i-f amplifier connection, and the connection between the final i-f amplifier and the demodulator stage). Each of these coupling tanks is tuned to the i-f. Each additional stage increases the Q of the filtering action, reducing the bandwidth that is allowed to pass through the circuit. By making the bandwidth as narrow as possible, potential interference problems from adjacent frequencies are minimized. In other words, the selectivity of the receiver is increased by increasing the number of i-f stages.

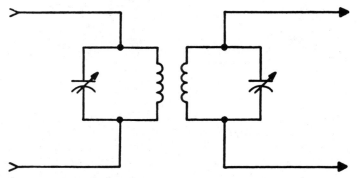

Fig. 37-18. A tuned i-f coupling circuit.

Because only a single frequency is allowed to pass through the i-f stages, they only have to be tuned during alignment. In operation they are left alone.

Next comes the *detector*, or *demodulator*. This stage recovers the original program information from the carrier. The mixing process converted the rf carrier to the i-f, but the signal modulation remained unchanged. Semi-conductor diodes are usually used for signal demodulation.

Figure 37-19 shows a simple AM detector circuit. The incoming signal resembles Fig. 37-20A. The diode will conduct only when it is forward biased by a signal with the correct polarity. This means the output consists of series of pulses whose amplitude varies in step with the original program signal. See Fig. 37-20B.

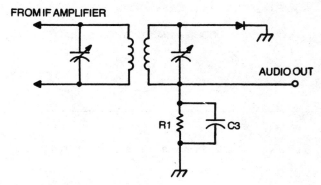

FROM IF AMPLIFIER

AUDIO OUT

R1   C3

Fig. 37-19. A simple AM detector circuit.

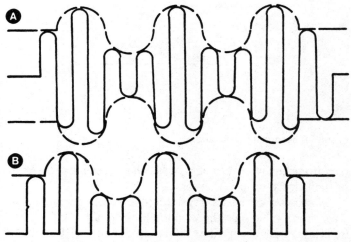

Ⓐ

Ⓑ

Fig. 37-20. Input and output signals for the circuit of Fig. 37-19: A. incoming signal; B. output signal.

These variable amplitude current pulses produce similar voltage pulses across R1. These pulses charge C3. In between pulses this capacitor discharges slightly. This produces a smoothing action on the output voltage, as shown in Fig. 37-21. This capacitor is relatively large, so it will discharge only a small amount between pulses. The output voltage follows the amplitude of the carrier. It is an ac signal superimposed on a dc signal. See Fig. 37-22. The dc voltage is proportional to the strength of the received signal. The dc voltage is often tapped off to adjust the gain of previous amplifier stages to boost

474

Fig. 37-21. AM detector output when smoothed by a large filter capacitor.

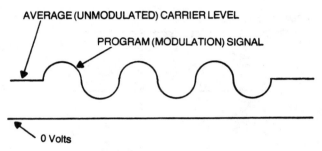

Fig. 37-22. The output of an AM detector is an ac signal superimposed on a dc voltage.

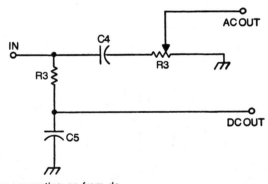

Fig. 37-23. A circuit for separating ac from dc.

weak signals, or attenuate overly strong signals. The ac portion of the output, of course, is a replica of the original program signal.

Figure 37-23 shows a circuit for separating the ac and dc signals. Dc cannot pass through the capacitor C4, so only the ac portion of the signal can be taken off of R2, R3 and C5 from a frequency sensitive voltage divider. At dc the value of R3 is much lower than the reactance of C5, so most of the voltage drop is across the capacitor, and can be taken off at the junction between these two components. At audio frequencies, however, the situation reverses. R3's resistance is considerably larger than C5's reactance at audio frequencies, so most of the ac signal is dropped across R3, and never reaches the R3/C5 junction. There will be some ac at the dc output, but it will be negligible value.

FM is somewhat more difficult to demodulate. This is due to both the type of modulation, and the allowable bandwidth (150 kHz for standard FM broadcasting). We'll look at three typical FM demodulator circuits.

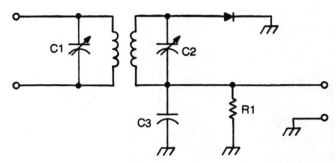

Fig. 37-24. A slope detector.

The simplest type of FM detector is shown in Fig. 37-24. This is called a *slope detector*. Notice that it is quite similar to the basic AM detector.

The resonant circuits in the slope detector are tuned slightly away from the i-f. To see why this is necessary, let's first assume it is tuned for resonance at the i-f. We'll arbitrarily assume the i-f is 500 kHz (usually FM receivers use an i-f of 10.7 MHz). The signal passing through the resonant circuits will be at its maximum level when it is at the i-f. Increasing the frequency to 510 kHz would cause the output level to drop off somewhat. Similarly, decreasing the frequency to 490 kHz, would also decrease the output level. The frequency response graph is shown in Fig. 37-25. This circuit would be of no use in recovering the program signal because it can only detect if the frequency has changed—it has no way of determining whether the frequency has increased or decreased.

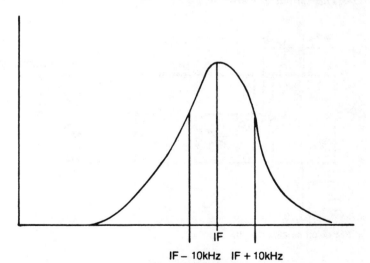

Fig. 37-25. Frequency response graph for a tuned slope detector.

But, if the circuit was detuned slightly, so that it was resonant at 510 kHz, the frequency response graph would look like Fig. 37-26. Now the unmodulated carrier would produce a specific output voltage. Increasing the frequency would increase the output. Decreasing the frequency would decrease the output level. The circuit can now determine

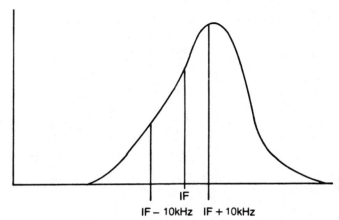

Fig. 37-26. Frequency response graph for a slightly detuned slope detector.

the direction of frequency change, and the program signal can be recovered. Because the unmodulated carrier frequency falls midway down the slope, this is called a *slope detector*.

While this circuit is fairly simple, it is rarely used because its fidelity is not very good. It is certainly unsuitable for music. Sometimes, however, it is used for narrow-band communications systems, where fidelity isn't particularly important.

Another type of FM detector is the *FM discriminator*, or the *Foster-Seeley discriminator*. This circuit is shown in Fig. 37-27.

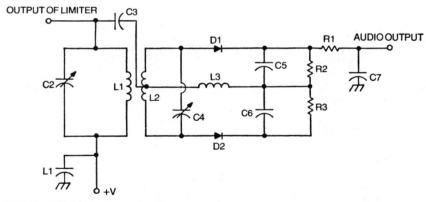

Fig. 37-27. An FM discriminator circuit.

Actually this circuit detects changes in phase rather than changes in frequency, but it really amounts to the same thing since when the frequency changes, its phase will also have to shift.

The FM discriminator is quite sensitive to amplitude variations in the signal. Every effort is made at the transmitting station to hold the carrier at a constant amplitude level, but as the signal is propagated, it may pick up some fluctuations. For this reason this type of detector is always preceded by one or more *limiter* stages. A limiter is a circuit that removes amplitude variations from a signal. It limits the amplitude to a single level.

The FM signal is demodulated by the discriminator by creating an output voltage that is proportional to the difference between the phase at the output of the limiter, and the phase across coil L2. The amount of conduction through the diodes is controlled by these phase relationships.

The voltage drop across R2 is proportional to the conduction level of D1. Similarly, R3's voltage is determined by the conduction of D2. As the input signal deviates from one nominal i-f, one diode will start to conduct more than the other, which causes the voltage drop across the resistors to change. The voltage drop across R2 and R3 will vary in step with the original program signal.

The most commonly used type of FM detector is the *ratio detector*, shown in Fig. 37-28. This circuit does not require a preceding limiter stage.

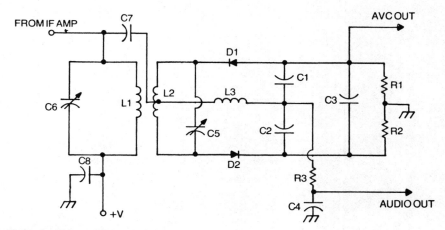

Fig. 37-28. An FM autodetector circuit.

Notice that this circuit is extremely similar to the FM discriminator, except the polarity of one of the diodes is reversed, and a large electrolytic capacitor (C3) is bridged across the output resistors. This is a *stabilizing capacitor*. It keeps the total voltage dropped across the resistors constant, even though the voltage dropped across them individually varies. This provides a self-limiting action. Any amplitude variations in the input signal are blocked from the output.

Dc voltages can flow through this circuit in only one direction. See Fig. 37-29. The ac signal paths are illustrated in Fig. 37-30. At the i-f (unmodulated carrier) the voltage

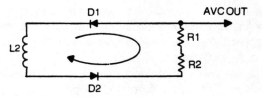

Fig. 37-29. Dc path through a ratio detector.

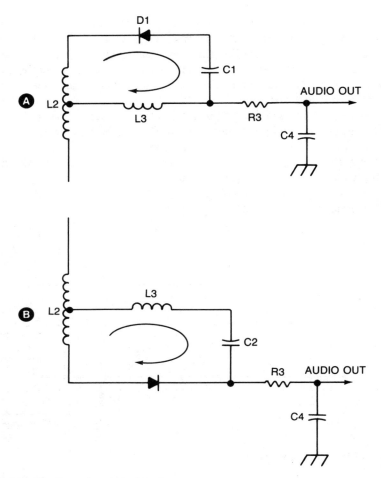

Fig. 37-30. Ac paths through a ratio detector.

across C1 equals the voltage across C2. If the voltage across C1 is 10 volts, the voltage across C1 equals the voltage across C2 which equals 5 volts. R1 and R2 each also drop 5 volts.

If the input signal drops in frequency, the impedance of the circuit becomes primarily capacitive, so the current through the secondary of the transformer leads the voltage. This means the voltage applied to D1 is greater than the voltage applied to D2. The voltage across C1 is therefore greater than the voltage across C2 (the voltages across C3 and its resistors remain unchanged). The audio output signal will be a negative voltage that is equal to one half the difference between C1's voltage and C2's voltage.

The opposite happens if the input signal increases in frequency. The voltage across C2 is greater than the voltage across C1, and the output signal is positive.

The output voltages of the ratio detector tend to be somewhat lower than those of the FM discriminator. This means more audio amplification is required. R1 and R2 vary in response to the average level of the carrier signal, producing the AGC signal.

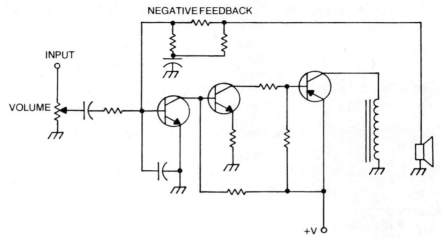

Fig. 37-31. A typical audio amplifier circuit.

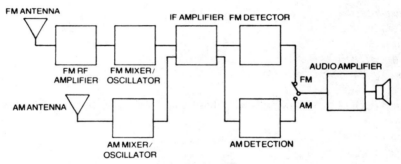

Fig. 37-32. Block diagram for a typical combined AM/FM receiver.

Regardless of the method of demodulation, we now have an audio signal that is the same as the original program signal. We merely need to boost the signal level with an audio amplifier, like the one shown in Fig. 37-31, and apply the output signal to a speaker, and we can hear the original program.

Often a single radio is designed for AM and FM reception. While these two types of signals require entirely different circuitry for certain stages, some stages can be shared, as shown in the block diagram of an AM/FM receiver in Fig. 37-32.

Self-Test

1. What type of waves are used for radio transmission?

A *Sine waves*
B *Electromagnetic waves*
C *Asymmetrical waves*
D *Symmetrical waves*
E *None of the above*

2. What is the name for the portion of the radio spectrum that includes 500 kHz?

A *LF*
B *MF*
C *HF*
D *VHF*
D *None of the above*

3. Which of the following is used in the simplest form of radio transmission?

A *AM*
B *FM*
C *PCM*
D *Morse code*
E *None of the above*

4. What is used to hold a VCO on frequency?

A *Phase-locked loop*
B *Heterodyning*
C *Modulator*
D *Bridge rectifier*
E *None of the above*

5. Which of the following is *not* a basic type of signal path for radio transmission?

A *Ground waves*
B *Direct waves*
C *Reflected waves*
D *Sky waves*
E *None of the above*

6. What should be the length of a half-wave antenna to receive signals at 65,000,000 Hz?

A *15.4 feet*
B *7.2 feet*
C *8.6 feet*
D *0.0086 inches*
E *None of the above*

7. How many lobes does a half-wave dipole antenna exhibit?

A *One*
B *Two*
C *Four*

D *Six*
E *None of the above*

      8. What should be the length of a reflector for a half-wave antenna for 100 MHz signals?

A *4.45 feet*
B *4.68 feet*
C *4.91 feet*
D *9.36 feet*
E *None of the above*

      9. What causes SWR?

A *Heterodyning with nearby frequencies*
B *Reflected duplicate signals*
C *Mismatched impedances*
D *Incorrect positioning of the antenna*
E *None of the above*

      10. What is used for demodulating AM signals?

A *I-f coupling circuit*
B *Slope detector*
C *Foster-Seeley discriminator*
D *Diode*
E *None of the above*

38

Stereo

If an audio amplifier system puts out an audio signal to just a single speaker (or the identical signal to a number of speakers) it is a *monaural*, or *mono* system. All of the sounds will come from a single source.

Real world sounds, however, arrive at our ears from various directions. Listening to a monaural recording of an orchestra or chorus has been compared to listening through a hole in the wall, as shown in Fig. 38-1. While the sound may be extremely good, with little or no distortion, it will still lack something in realism.

A *stereophonic* (or just *stereo*) system, on the other hand, provides two sound sources (a right speaker and a left speaker) for a more realistic effect. A stereo system can apparently place the instruments or voices in a number of different positions. See Fig. 38-2.

At first glance, it might seem that stereo would be like two holes in the wall, with nothing in between. But that is only the case if there is no duplication between the signals fed to the two speakers. This rarely occurs in practice.

Let's suppose we are making a recording of a trio consisting of a violin, a cello, and a viola. We'll assume each of the instruments plays a brief solo passage, each at the same volume, so that if this was a monaural recording, the speaker would be fed a constant 3 watts.

For our stereo recording we will record the violin so that its full 3 watts is applied to the right speaker. Similarly, for the cello's solo, the left speaker will receive 3 watts from the amplifier, and right will receive no signal. So far, we have just the two holes in the wall effect.

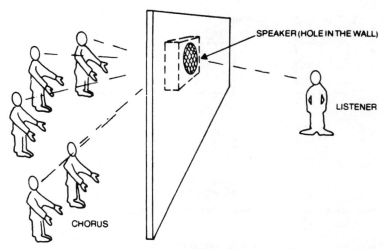

Fig. 38-1. Effective acoustic image of a monaural sound system—the hole in the wall effect.

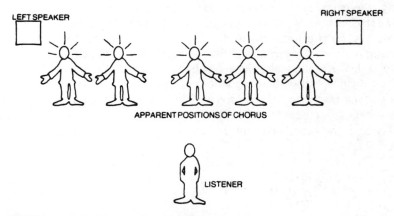

Fig. 38-2. Effective acoustical image of a stereo sound system.

The viola solo, however, is recorded so that each of the speakers receives 1.5 watt of signal (for a total of 3 watts). Since the sound coming from the two sources is identical, our ears will interpret this sound as coming from a non-existent single source between the two speakers. The acoustic image would appear to resemble the drawing in Fig. 38-3.

In this way, various sounds can be apparently placed anywhere between the two speakers simply by controlling the amount of signal fed to each speaker. Of course, this is purely an illusion, but it is quite effective in simulating realism. Virtually all modern high fidelity (realistic reproduction) sound systems are stereophonic.

Until recently, there were three basic signal sources for home stereo systems. These were magnetic tape recordings, phonograph records, and stereo FM radio broadcasts. In the last few years, a fourth stereo signal source has appeared and gained enormous popularity. This is the CD, or *Compact Disc*. Each of these signal sources will be discussed in turn through the remainder of this chapter.

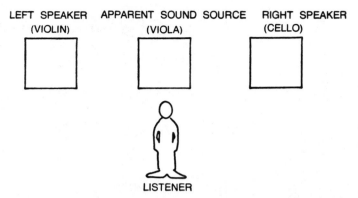

LEFT SPEAKER   APPARENT SOUND SOURCE   RIGHT SPEAKER
   (VIOLIN)   (VIOLA)   (CELLO)

LISTENER

Fig. 38-3. How a stereo system can create more than two apparent sound sources.

TAPE RECORDING

Magnetic tape consists of a thin strip of plastic that is coated with magnetic particles. Ordinarily, the particles are randomly placed, as shown in Fig. 38-4.

If this strip of tape is passed through a strong magnetic field, the particles will be aligned into a coherent pattern, as shown in Fig. 38-5. The degree of alignment will depend on the strength of the magnetizing field. By varying the magnetic field at an audio frequency (40 Hz to 16,000 Hz), an audio signal can be recorded on the tape. The particles will retain their alignment after the magnetic field is removed.

The varying magnetic field is produced by a device called a *recording head*. An electrical voltage resembling the audio signal to be recorded is fed to a coil in the recording head. Of course, the varying current through the coil will cause it to generate a varying magnetic field.

Fig. 38-4. Unrecorded magnetic tape particles are randomly scattered.

Fig. 38-5. Magnetic tape with a recorded signal.

Now, if the recorded tape is passed over a second type of head, called a *playback head*, the varying magnetic fields of the passing magnetic particles will induce a voltage into a coil within the playback head. This voltage will be a replica of the original signal. Thus, a signal can be recorded onto a strip of magnetic tape and played back. Often a single head is used for both the recording and playback functions, but most high quality *tape decks* use separate heads.

The highest frequency that can be recorded depends on the *gap* in the head's protective shield (essentially a small window that allows it to look at only a small portion of tape at any given instant) and the speed at which the tape is moving past the head. The highest possible frequency can be increased by either narrowing the head's gap, or by increasing the speed of the tape.

The three most common tape speeds are 1⅞ inches per second (*ips*), 3¾ ips, and 7½ ips. Some professional recording machines operate at 15 ips or 30 ips. Obviously, the faster the speed, the more tape that will be used up within a given time, so a trade-off is involved between frequency response and tape economy.

A recorded tape can be reused, by rescrambling the magnetic particles and recording the new signal. This rescrambling is done with a third kind of head called an *erase head*. The signal provided to the head is called *bias* and is produced by an internal *bias oscillator*. The bias is a very high level, high frequency signal. The high level of the magnetic field level forces the magnetic particles out of alignment, but the frequency is too high to give them time to realign themselves, so they end up randomly placed. The erase head is always separate from the record and playback heads.

Figure 38-6 shows the arrangement of these three heads in a tape recorder. They are always in this order. A two-headed machine (record and playback heads combined into one) is illustrated in Fig. 38-7.

The simplest way of putting a signal on a strip of tape is to use the entire width of the tape, as shown in Fig. 38-8. This is called *one-track mono recording*. Typically,

486

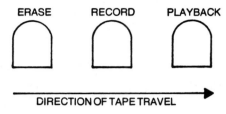

Fig. 38-6. Positions for erase, record, and playback heads.

Fig. 38-7. Positions for erase and record/playback heads.

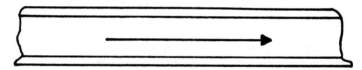

Fig. 38-8. One track monaural recording—the tape is usually ¼ inch wide.

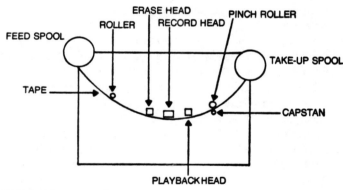

Fig. 38-9. A reel-to-reel tape recorder.

most tapes are ¼ inch wide. The tape is wound onto a spool or reel. The free end is connected to a second reel (this is called a *reel-to-reel recorder*). As the tape is pulled off the first reel by the *capstan* and *pincher wheel*, it is moved past the heads, and wound onto the second reel. See Fig. 38-9. The capstan turns at a fixed speed and controls the speed of the tape past the heads.

Once the tape is wound entirely onto the second spool, it is rewound back onto the first wheel and fed through the system once again for playback. Of course, this takes extra time, and is a bit inconvenient. Besides, the full width of the tape usually isn't needed to record the signal, so tape is wasted.

Some recorders are designed so that when the tape is wound completely onto the second spool, the two spools are reversed, and the tape is sent past the heads again in the opposite direction, recording a second program. See Fig. 38-10.

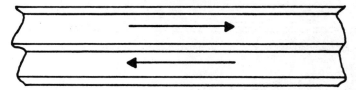

Fig. 38-10. Two track monaural recording.

Alternatively, we can record both tracks simultaneously with the two halves of a stereo program. See Fig. 38-11. On playback, the signal from the upper track is fed to the right speaker, and the lower track's signal is sent to the left speaker. These two systems are called *two-track mono*, and *two-track stereo*. Of course the stereo form must be rewound between recording and each playback as with one-track mono.

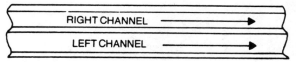

Fig. 38-11. Two track stereo recording.

Two direction stereo recording can be achieved by recording on four tracks. There are two ways of doing this. In Fig. 38-12 the tracks of the two programs are interweaved. This helps cut down on crosstalk (part of one channel bleeding over into the other channel, blurring stereo imaging). This is the system used in most four-track stereo reel-to-reel recorders.

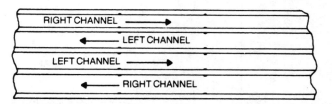

Fig. 38-12. Four track stereo recording, as used in reel-to-reel systems.

The other method shown in Fig. 38-13, is used in *cassettes*. A cassette, as illustrated in Fig. 38-14, is basically a miniature reel-to-reel system with both reels enclosed in a single plastic housing. The tape is only ⅛ inch wide, and it only moves at 1⅞ ips, so fidelity isn't quite as good as with reel-to-reel systems, but modern circuitry can compensate for these disadvantages. Modern cassette systems can often produce sound that is virtually as good as a comparable reel-to-reel system.

Ordinarily, cassettes use the two track mono method of recording shown before in Fig. 38-10. This is what they were originally designed for. The four track stereo

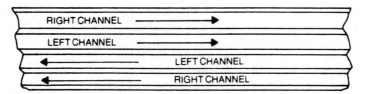

Fig. 38-13. Four track stereo recording, as used in cassette systems.

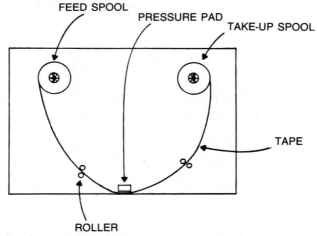

Fig. 38-14. A tape cassette.

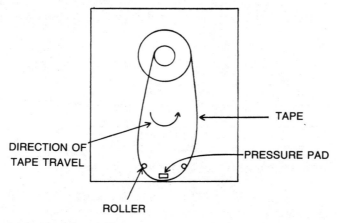

Fig. 38-15. An eight-track tape cartridge.

arrangement of Fig. 38-13 allows cassettes recorded on stereo machines to be directly interchangeable with mono machines.

A third type of system is the *eight-track stereo cartridge*. See Fig. 38-15. The tape is wound into an endless loop, with the ends spliced together with a piece of aluminum.

When this aluminum band passes a special sensor, the head is mechanically moved to play the next track, so the four programs are played sequentially.

At the end of the fourth program, the head is returned to the first program position, and starts over. This will continue for as long as the cartridge is plugged into the machine. Another sensor shuts the machine off when the tape is removed.

It is impossible to rewind an eight-track stereo cartridge. Usually a switch is provided to manually select programs, and sometimes a fast forward control is included. The arrangement of tracks is illustrated in Fig. 38-16. Notice that all of the tracks run in the same direction. The motion of the tape can never reverse direction.

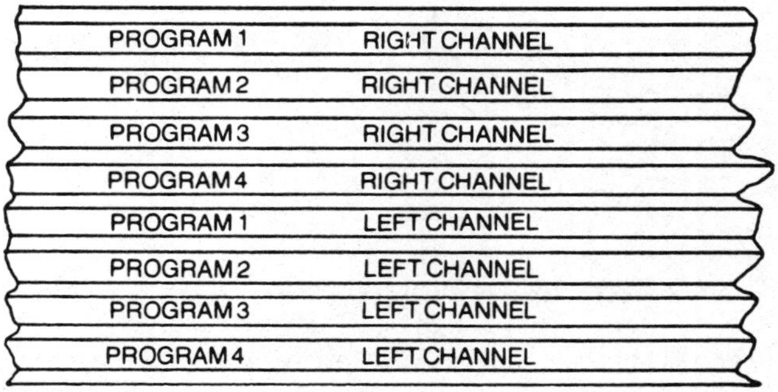

PROGRAM 1	RIGHT CHANNEL
PROGRAM 2	RIGHT CHANNEL
PROGRAM 3	RIGHT CHANNEL
PROGRAM 4	RIGHT CHANNEL
PROGRAM 1	LEFT CHANNEL
PROGRAM 2	LEFT CHANNEL
PROGRAM 3	LEFT CHANNEL
PROGRAM 4	LEFT CHANNEL

Fig. 38-16. Eight track stereo recording as used in cartridge systems.

Eight-track cartridges are popular in car stereo systems because the driver can just plug in a tape and let it play over and over while he concentrates on driving. Some automobile cassette players will automatically reverse direction at the end of a tape for the same kind of advantage, others have to be manually turned over at the end of a side.

PHONOGRAPH RECORDS

A phonograph record is a vinyl disc with tiny grooves cut into its surface. The disc is placed on a platter (called a *turntable* which rotates at a fixed speed. A sapphire or diamond needle, or *stylus* rides in the grooves. The grooves force the stylus to move back and forth, and these variations from a straight path are used to generate an electrical signal.

Figure 38-17 shows the way these undulating grooves are cut into a disc, or record. The grooves in this drawing are greatly exaggerated for clarity. The further a groove varies from its nominal, no-signal path, the higher the amplitude of the recorded signal.

The device that converts the stylus's mechanical motion into an electrical signal is called the *cartridge*. There are two basic types of cartridges used in record playing systems. These are *ceramic cartridges*, and *magnetic cartridges*.

Ceramic Cartridges

A ceramic cartridge works by the piezoelectric effect (see Chapter 11). The motion of the stylus places a mechanical stress along one axis of a piece of ceramic material. This causes a similar electrical stress to be generated across another axis in a different plane.

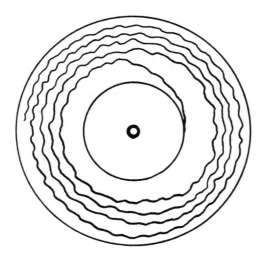

Fig. 38-17. Grooves cut in a record—greatly exaggerated.

Ceramic cartridges have a number of advantages. They are inexpensive and sturdy, and they produce a strong output signal. These advantages are offset by certain disadvantages. The disadvantages include a relatively poor frequency response and a high *tracking force*. The tracking force of a cartridge is the amount of pressure (or, the effective weight) exerted by the stylus on the record's grooves. Obviously, a high tracking force will wear down the grooves (thus rendering the record useless) faster than a light tracking force would. Tracking force is measured in grams. Most ceramic cartridges exert a tracking force of about five grams.

Ceramic cartridges are found in most inexpensive record players. Better systems generally use magnetic cartridges.

Magnetic Cartridges

In a magnetic cartridge the stylus is attached to a piece of magnetic material, which is moved between two small coils. Of course, this moving magnetic field induces an ac voltage into the coils. See Fig. 38-18. This ac voltage is a reconstruction of the original signal encoded into the grooves of the record.

Magnetic cartridges typically offer a much better high frequency response, and a lower tracking force (¾ to about 3 grams is the typical range) than ceramic cartridges. However, they are generally much more expensive, rather delicate, and they have a much lower output signal level. This type of cartridge requires an extra stage of *preamplification* before the signal is strong enough to be fed to the regular amplifier.

The grooves in a record can be cut at virtually any angle with respect to the surface of the disc. For example, the stylus could be forced by the undulations to move laterally (from side to side), or vertically (up and down). In either case, the output from the cartridge would be exactly the same. In practice, vertical groove records are fairly difficult to manufacture, so lateral motion is used for monaural recordings.

Now, suppose we had a groove with both lateral and vertical modulations. Since the two motions are in different planes (at 90° angles to each other) two separate signals could be detected by a single stylus, even if both motions take place simultaneously.

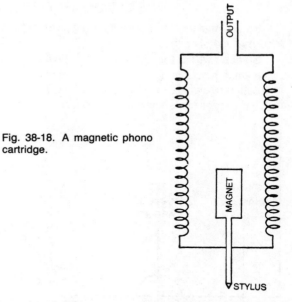

Fig. 38-18. A magnetic phono cartridge.

Because of the difficulty of cutting variable depth grooves, and to ensure compatibility with existing monaural equipment, both sets of groove undulations are rotated with respect to the disc, so that they are both at 45° angles to the surface of the disc. See Fig. 38-19. In this way, stereo signals can be recorded onto vinyl discs. A monaural cartridge would combine both signals into a single output.

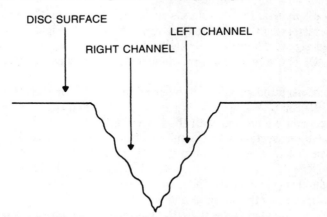

Fig. 38-19. How two sets of grooves are cut into a record disc for stereo.

A stereo cartridge, on the other hand, would translate one set of stylus motions to information for the right channel, and the other set of stylus motions to information for the left channel. In practical equipment there is some leakage, or *crosstalk* between the two channels. That is, if a disc is recorded with only right channel groove modulations, there will still be some (albeit, probably weak) signal in the left channel output too. Better cartridges usually have lower crosstalk.

STEREO BROADCASTING

Stereo signals can also be sent out over radio waves. Most stereo systems include a stereo FM receiver.

Stations within the FM broadcast band are allowed a bandwidth that is more than twice what is needed for transmitting a monaural signal. Therefore they could transmit two separate signals—one for the right channel, and one for the left channel—each with its own carrier, as shown in Fig. 38-20.

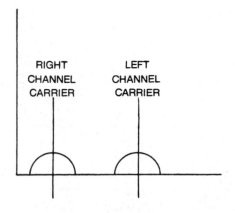

Fig. 38-20. Theoretical stereo broadcasting system.

Unfortunately, a monaural receiver would only receive the right channel information. We need some method of broadcasting a high fidelity stereo signal that is fully compatible with existing monaural FM receivers.

The answer is *matrixing*. Two signals are transmitted. One (the only one received by monaural receivers) is the full monaural signal, that is, the right channel signal, and the left channel signal. The other signal consists of the difference between the two channels (L − R). These two signals are combined in the receiver to recreate a regular stereo output.

If the two transmitted signals are combined in phase with each other we have (L + R) + (L − R), or L + L + R − R, or simply 2L. The right channel information is cancelled out, and we are left with the left channel.

Similarly, another section of the circuit combines the two transmitted signals 180° out of phase with each other. The output in this case equals (L + R) − (L − R) = L + R − L + R = L − L + R + R = 2R. The left channel information is canceled out and the right channel remains. Thus, we have a way of transmitting a stereo signal that is fully compatible with monaural equipment.

At the transmitting station the (L − R) signal is produced by a balanced modulator (see Chapter 36). This circuit produces no output unless a modulating program signal is applied. In other words, the unmodulated carrier for the (L − R) signal (which is 38 kHz higher than the (L + R) carrier) is never actually transmitted. This type of transmission is called double sideband, suppressed carrier (DSSC) transmission. Because the (L − R) carrier is not transmitted, it must be recreated at the receiver for demodulation to take place.

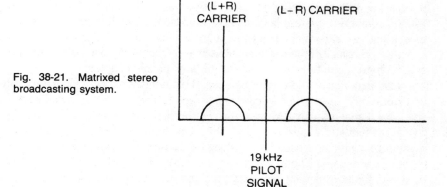

Fig. 38-21. Matrixed stereo broadcasting system.

To let the receiver know when it is receiving a stereo transmission, a 19 kHz *pilot* signal is also transmitted. This is not modulated. It is simply either present (stereo transmission) or not present (monaural transmission). Note that this 19 kHz signal is exactly one half of the required suppressed carrier frequency (38 kHz). The received pilot signal can be passed through a frequency doubler and reinserted into the (L − R) signal. Figure 38-21 shows the complete stereo FM signal as transmitted.

AM STEREOS

For the past few decades, stereo has been available in FM only. AM radios were designed for only monaural broadcasts. Stereo AM transmission was not allowed by the FCC.

In recent years, portable FM radios, and high quality FM receivers have put a severe dent in the audience for AM radio. To boost their audiences (and their profits) AM broadcasters have been pushing for the FCC to legalize stereo broadcasting for AM, as well as FM. This is not a new idea. AM stereo was proposed to the FCC in 1958. The FCC rejected the proposal in 1961, primarily because of technical problems.

Several manufacturers were not entirely willing to give up on the idea and they continued development of AM stereo transmission systems. The AM stereo proposal was again placed before the FCC in 1977. No fewer than five systems were proposed. They were developed by the following corporations: Belar, Harris, Kahn, Magnavox, and Motorola. The FCC was to consider the various proposed systems, and select one as a standard.

Recently, the FCC voted to allow AM stereo transmission, but they decided not to declare a standard. The FCC ruling stated that the best of the various systems should be selected in the marketplace, rather than prematurely limiting the industry to a single standard.

There is certainly something to be said for this idea. Television transmission standards have recently been found to leave quite a bit to be desired, but they can't be easily changed without making millions of existing sets obsolete.

On the other hand, the lack of a consistent standard, may end up killing off AM stereo altogether. The public may be reluctant to invest in a receiver designed to receive only one of the noncompatible systems, when their choice might be the loser in the marketplace war. They'd be left with an obsolete and useless piece of equipment when a competing system becomes the standard. If enough prospective buyers decide to wait until a clear winner emerges, there may be no winner at all. Sales might be poor enough to kill off all of the systems. This is essentially what happened with the competing quadraphonic sound systems (discussed later in this chapter).

This may not be the case, however. Two noncompatible videotape systems (VHS and Beta) have managed to find a solid place in the market. We will have to wait and see what happens with AM stereo over the next few years.

Many industry experts consider the front runner to be the Magnavox system. This system was initially chosen as the standard by the FCC, before they decided to opt for no official standard. The Magnavox system combines phase modulation with the standard AM signal. Adapting broadcast transmitters would be relatively simple. Receivers will require major redesign. Of course, all of the proposed systems are completely compatible with existing monaural AM radios. Directional antennas will have to be precisely phased and positioned for correct reception of the stereo signal.

The combined monaural signal (L + R) will be transmitted as ordinary AM signals, to ensure compatibility. The phase of the carrier will be modulated with the difference (L − R) signal. A 5 Hz pilot signal is also transmitted to identify the signal as a stereophonic broadcast. The pilot signal could also be adapted in the near future to transmit low speed digital information such as station identification, time, weather, traffic information, and so forth. There are no firm plans in this area at the moment.

A broadcasting system using the Magnavox AM stereo system is shown in block diagram form in Fig. 38-22. The receiver's block diagram is illustrated in Fig. 38-23.

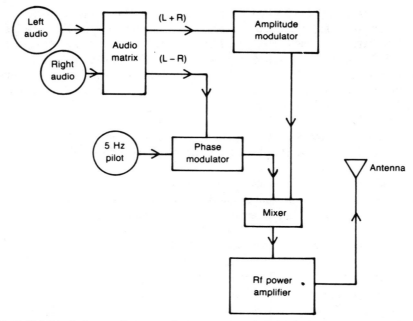

Fig. 38-22. This block diagram illustrates the basics of an AM stereo transmitter using the Magnavox system.

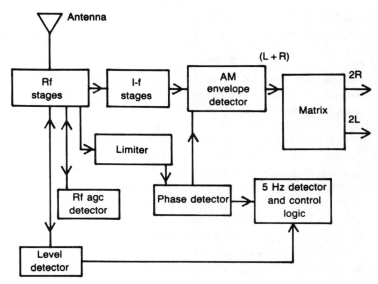

Fig. 38-23. The block diagram for an AM stereo receiver using the Magnavox system.

Motorola's proposal is called C-Quam, or Compatible Quadrature Amplitude Modulation. Two separate carriers are transmitted at the same frequency in phase-quadrature, and modulated with separate left and right signals. The transmitter outputs are tied together and fed to a common antenna. A standard monaural radio will demodulate the combined right and left signals together, for a full monaural signal. In a stereo receiver, on the other hand, the two carriers are independently demodulated by two synchronous detectors.

The Harris system uses carrier phase modulation, but in a somewhat different manner than used by Magnavox. Independent sidebands are employed in the Kahn system. Belar Electronics proposes a combination AM/FM signal for stereo broadcasting.

Which system is the best? Well, they all have their own advantages and disadvantages. Which one will win out in the end (if any)? Only time can tell.

SIMULATED STEREO

Sometimes it might be nice to be able to convert a monaural source into sound like stereo. This is often done by record companies when older, monophonic recordings are re-released.

Most systems for simulating stereo employ some form of filtering. The simplest approach would be a single simple crossover network that routes the low frequencies (bass) to one channel and the high frequencies (treble) to the other. This works, but doesn't sound very much like true stereo.

More advanced systems would use a series of bandpass filters, whose outputs are added or subtracted from each of the two channels. The filtered frequencies are increased in volume through one speaker, and decreased in volume in the other. This gives a vague, somewhat muddy stereo-like effect. Some sounds will appear to come from the right, others from the left, and still others from somewhere in-between the two speakers.

The best results are obtained with a method known as comb filtering. This is the method used professionally in recordings that are rechanneled to simulate stereo. The monaural signal is delayed slightly and mixed with itself. Some frequencies will partially cancel each other out, and others will be emphasized (increasing their apparent volume). This allows much better separation at more frequency points than using independent bandpass filters.

The frequency crossover points should not be too close together, or stereo separation will suffer. Slight variations in frequency from note to note would make each instrument seem to wander all over the apparent sound field. Generally speaking, best results are achieved if the audio range is split up into 1 kHz chunks. This can be accomplished by delaying the monaural signal by 0.5 milliseconds. Similar techniques are occasionally used to simulate quadraphonic sound (see below) from stereophonic sources.

QUADRAPHONIC SOUND

Just as the two sound sources in a stereophonic system give greater realism than a single monaural speaker, an even greater sense of reality can be achieved with a four sound source system. This is called a *quadraphonic system*, or simply *quad*. See Fig. 38-24.

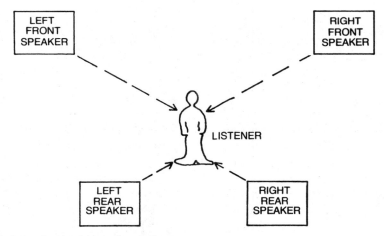

Fig. 38-24. A quadraphonic sound system.

Unfortunately, quad has virtually died out in the marketplace because of noncompatible competing systems. The two most common systems were CD-4 and SQ™ (SQ is a Trademark of Columbia Broadcasting System, Inc.).

CD-4

The CD-4 system is the simplest quadraphonic system. The four channels are simply kept separate on tape or on a record disc. Figure 38-25 shows the way a CD-4 signal would be recorded on tape. Notice that only a single direction of playback is possible. The full width of the tape is used for one program. CD-4 cannot be transmitted by radio.

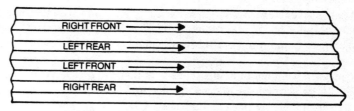

Fig. 38-25. A four track CD-4 recording tape.

SQ

SQ (stereophonic-quadraphonic) is a compatible matrixing system. An SQ program can be played on standard stereo equipment without losing any of the program signal.

The left channel contains portions of both rear channels as well as the left front channel information. These signals are combined in the following formula:

$$L_T = L_F - j0.707L_r + 0.707R_r \qquad \textbf{Equation 38-1}$$

where L_T is the total left channel signal, L_F is the left front channel, L_r is the left rear channel, and R_r is the right rear channel. The letter j indicates a 90° phase shift.

Similarly, the contents of the right channel signal (R_T are given in the following formula:

$$R_T = R_F + j0.707L_r - 0.707L_r \qquad \textbf{Equation 38-2}$$

A phase shift circuit in the amplifier decodes the rear signals from the front signals. The rear channels are amplified by only about 0.707 times the amplification of the front channel. They are effectively at half the volume of the front channel signals (the ear does not hear volume linearly).

For broadcasting purposes, the signal is further matrixed into standard stereo signals. That is, $L_T + R_T$, and $L_T - R_T$. Filling in from Equations 38-1 and 38-2, the formulas become:

$$\text{Lower Channel} = L_F + R_F - j0.707L_r - 0.707L_r + 0.707R_r + j0.707R_r$$
$$\textbf{Equation 38-3}$$
$$\text{Upper Channel} = L_f - R_F - j0.707L_r + 0.707L_r + 0.707R_r - j0.707R_f$$
$$\textbf{Equation 38-4}$$

While these formulas are rather lengthy and complex, it is easy for the circuitry to decode.

COMPACT DISCS

The compact disc, or CD has gone from an audiophile novelty to one of the most popular recorded sound sources available. The recorded sounds are stored in digital form as pits and islands on a 5 inch metal disc. The pits and islands are detected by a small laser in the CD player. Additional circuitry converts the digital data back into analog form, or sound.

The CD offers an extremely wide dynamic range, as compared to other audio signal sources. There is virtually no background noise. The clarity of sound on a well recorded CD is incredibly good, and often difficult to distinguish from live musicians right in the room with you.

Another major advantage of compact discs is that the discs themselves are nearly indestructible. Large scratches can cause problems in playback, but for the most part, you really have to try pretty hard to damage a CD.

Self-Test

1. What is the name for a system with two sound sources to allow more realistic effects?

A *Monophonic*
B *Stereophonic*
C *Quadraphonic*
D *Biphonic*
E *None of the above*

2. What determines the highest frequency that can be recorded on a magnetic tape recorder?

A *The recording head gap*
B *The tape width*
C *The bias oscillator frequency*
D *The bias level*
E *None of the above*

3. Which of the following is *not* a standard speed for tape recorders?

A *3.75 ips*
B *33.33 ips*
C *15 ips*
D *1.875 ips*
E *None of the above*

4. How many channels are used on a four-track stereo tape?

A *One*
B *Two*
C *Four*
D *Eight*
E *None of the above*

5. What two signals are transmitted by a stereo FM station?

A *L and R*
B *2L and 2R*
C *(L − R) and (L × R)*
D *(L + R) and (L − R)*
E *None of the above*

6. What is the frequency of the pilot signal in an FM stereo broadcast signal?

A *5 Hz*
B *5 kHz*
C *19 kHz*
D *38 kHz*
E *None of the above*

7. Which of the following types of cartridges requires a preamplifier?

A *Magnetic*
B *Ceramic*
C *Crystal*
D *None of the above*

8. What signal is received by a monaural FM receiver when tuned to a stereo FM station?

A *L*
B *R*
C *(L + R)*
D *(L − R)*
E *None of the above*

9. Which of the following is the matrixing system for quadraphonic sound?

A *CD-4*
B *SQ*
C *Magnavox*
D *Kahn*
E *None of the above*

10. Which of the following is the legal standard for AM stereo broadcasts?

A *Magnavox*
B *Motorola*
C *Belar*
D *Kahn*
E *None of the above*

39

Troubleshooting
and Servicing

An important part of any electronics technician's battery of skills is the ability to locate and correct any problems that may show up in a circuit. A newly designed prototype circuit may not function as the designer intended. He must determine what went wrong and correct the error in a new prototype.

An experimenter may build a presumedly debugged circuit from a magazine, but it doesn't work. He will need to determine if the error is in the published schematic or if he has made an error in construction. In either case, he needs to find and correct the problem, or his previous efforts will have been wasted.

A piece of electronic equipment that once worked properly may start malfunctioning. A technician must locate and repair the portion of the circuit that has become defective.

The process of locating the problem in a malfunctioning electronic circuit is known as *troubleshooting*. Repairing the defect is known as *servicing*. Troubleshooting a complex piece of equipment may seem like a hopeless task at first. But a little common sense can reduce the job to manageable levels. A color television set, for example, may have several hundred discrete components. Do we check each one to make sure that it is within the manufacturer's specifications? Presumedly it would be possible to locate most defects that way, but it would be extremely inefficient. It could take hours, or even days to find the bad component.

Most complex electronic devices are made up of many less complex stages. By analyzing the functions of these stages, we can narrow down the potential cause of the defect. For example, let's say the picture on a color TV is breaking up into broad diagonal stripes. We can see that objects are being displayed, although they are distorted and broken up because of the striping. Colors appear normal. The sound is fine. Where is the problem? As a first step, we look at the block diagram for the set. A typical color TV block diagram is shown in Fig. 39-1. Let's consider each stage and see if it is the problem.

Because the set is operating, we can assume the power supply circuits are working. They may be putting out incorrect voltages, but in this particular case, that doesn't seem

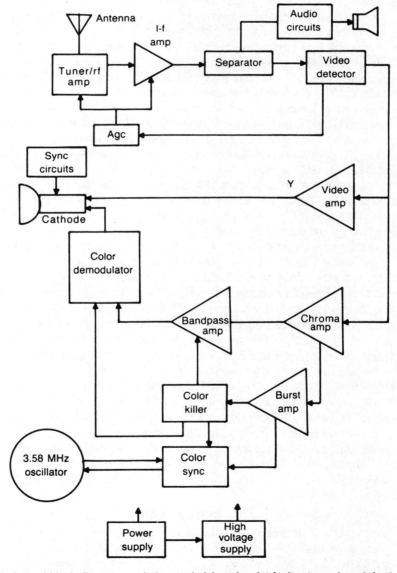

Fig. 39-1. A good block diagram can help a technician pinpoint faulty stages in a defective circuit.

a likely cause of the problem at hand. If you are unsure, it would only take a few moments to measure the supply voltages. If they are reasonably close to their specified values, we can move on. Some technicians always measure supply voltages for all problems. Too high and too low voltages can result in some very odd problems. In our example, we will assume the supply voltages are correct.

We are getting a raster (the CRT is lighting up) so we can eliminate the high voltage supply and the CRT as suspects. A signal is getting through, so the tuner, rf amplifiers, i-f amplifiers, agc, and separator circuits are probably OK. We have correct sound, so the audio circuits and speaker are not involved. We do have a picture, although it is broken up. This indicates that there isn't a problem in the video detector or video amplifier.

Since the color seems to be all right, the defect is probably not in any of the color circuits (chroma amplifier, bandpass amplifier, burst amplifier, color killer, color sync, and the 3.58 MHz oscillator). What does that leave? There is only one section of the block diagram we haven't at least tentatively ruled out. The sync circuits! An experienced technician would probably immediately suspect the sync circuits. The problem we have described is a loss of horizontal synchronization.

In some tough cases, the circuit you suspect may check out fine. There could be a problem in another portion of the circuitry that is causing a surprising symptom. For instance, the sync pulses may be distorted or attenuated in some manner in the video amplifier stage. This isn't as likely, but it could happen.

If a set has multiple symptoms, try to isolate the single stage that could be responsible for all or most of the symptoms. There may be defects in more than one stage, but the odds are that there is a common problem, or one component has failed, causing an overload to damage another component.

Let's consider another problem. The screen is completely dark, and there is no sound at all. The set just sits there and does absolutely nothing. The average person might consider this a more serious problem. After all, the set isn't working at all. But an experienced technician will be unconcerned. It will probably be a relatively easy repair.

What could cause the symptoms described? There could be several combinations of multiple causes. For example, the CRT *and* one of the audio circuits could be defective. But it's more likely that all of the symptoms are resulting from a problem in a single stage. What stage could account for all of the symptoms (dark screen and no sound)?

There is one stage common to everything in the set—the power supply. The other stages probably aren't getting any power. The culprit could be a blown fuse, or a burnt-out rectifier diode, or a shorted filter capacitor, or a broken wire, among other possibilities. (We'll assume the set is plugged in, of course. It saves time to always check out the simple and the obvious first. Dumb mistakes do happen.)

A little bit later in this chapter we will briefly cover some of the more common types of defects. But first, let's examine some of the important weapons in the technician's troubleshooting arsenal.

TEST EQUIPMENT

To determine the source of a problem in an electronic circuit, the technician employs one or more pieces of test equipment. There is no single all-purpose tester that can solve all problems. Each piece of test equipment is designed for specific purposes.

The most basic pieces of test equipment are the technician's eyes, ears, and brain. I am not being facetious here. The technician observes the symptoms with his eyes and ears and analyzes them with his brain. His brain tells him what stage(s) is a likely culprit.

Moreover, his eyes can be used to locate many defects. Does a component look burnt? Is a wire broken? The technician's nose can also be helpful. Does something smell burnt? The technician may also use the tip of his finger to determine if a component is running hot. Of course, the technician's brain must also analyze the data obtained from the various tests he performs. What could cause any unusual values found in the circuit? Generally, when speaking of test equipment, however, we are referring to external electronic circuits designed to measure various electrical characteristics and display their results in some manner.

The VOM

The most basic, and versatile piece of equipment available to the electronics technician is the VOM, or *Volt-Ohm-Milliammeter*. This device was introduced back in Chapter 10. A VOM measures dc and ac voltages, dc resistance, and dc current. A few VOMs also measure ac current.

Closely related to the VOM is the VTVM, or *vacuum-tube voltmeter*. Most VTVMs measure only voltage and resistance, but not current. One advantage of the VTVM over the VOM is when high frequency ac voltages must be measured. A VOM's reading is not usually reliable for frequencies above a few hundred hertz. A VTVM has a far wider frequency response. The VTVM can also usually measure a wider range of resistances.

The average VTVM is well protected against overloading the meter movement and burning out component parts from excessive test voltages. The meter in a VOM may be damaged if, for example, 120 volts was applied when the meter was set to a 10 volt range. In a VTVM the meter itself is isolated from the input by the vacuum-tube amplifier and scaling circuitry.

A VTVM usually runs off of 120 volts ac, which means it can only be used where an electrical socket is available. Most standard VOMs have a small battery for resistance measurements and require no further power source. VTVMs are also more bulky than most VOMs. In addition, the tube circuitry in the VTVM gives off heat.

Since a VOM is not plugged into the electrical socket along with the equipment being tested, there is better isolation, ensuring greater safety for the technician. This also tends to minimize interference problems from stray rf signals and ac hum.

The tubes in a VTVM need a certain amount of time to warm up before correct readings can be obtained. In addition, readings may tend to drift slightly, especially as the tubes age. This does not happen with a VOM, which is basically a passive device. Since the VOM involves simpler circuitry than the VTVM, it also tends to be more reliable.

The main important difference between meters is sensitivity, or input impedance. The higher the input impedance of the meter, the less it will load down the circuit being tested. This is because the meter is placed in parallel with part of the circuit to measure voltage.

A typical VOM has a sensitivity of approximately 20,000 ohms-per-volt. Some go as high as 50,000 ohms-per-volt. Cheap VOMs with an input sensitivity of only 1,000 ohms per volt are also available. These can be useful to an electrician, but an electronics

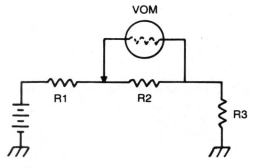

VOM

Fig. 39-2. A voltmeter can affect the circuit being tested by acting as a parallel resistance.

technician should steer clear of them—the sensitivity is far too poor for electronics circuitry.

A typical VTVM will have much greater sensitivity than a VOM. Input impedances of 11 megohms (11,000,000 ohms) are common for VTVMs.

A cross-breed type meter is the FET VOM. This is just a regular VOM with a FET amplifier input stage. This offers the advantage of isolating the meter movement from the input signal, but a power source is required for voltage and current measurements as well as for resistance measurements. This can be supplied by a small internal battery, so portability is not compromised. The sensitivity of a FET VOM is around 1 megohm (1,000,000 ohms) per-volt.

Let's examine why input sensitivity is so important. Take a look at the simple circuit shown in Fig. 39-2. Assume that the battery is putting out 9 volts, and the three resistors have the following values:

$$R1 = 22 \text{ k } (22,000 \text{ ohms})$$
$$R2 = 47 \text{ k } (47,000 \text{ ohms})$$
$$R3 = 33 \text{ k } (33,000 \text{ ohms})$$

Total circuit resistance is 102,000 ohms. According to Ohm's law, the current drawn through the circuit works out to:

$$= \frac{E}{R} = \frac{9}{102000} \approx 0.0000882 \text{ amp} = 0.0882 \text{ mA}$$

The same amount of current flows through each resistor. We can now calculate the nominal voltage drop across resistor R2:

$$E = IR = 0.0000882 \times 47000 \approx 4.15 \text{ volts}$$

Now, let's say we hook up a 1000 ohm-per-volt VOM in parallel with R2 to measure the voltage drop.

The input resistance of the VOM acts as a parallel resistance with R2. This makes the apparent value of R2 drop to:

$$\frac{1}{R_t} = \frac{1}{R2} + \frac{1}{R_m} = \frac{1}{47000} + \frac{1}{1000}$$

$$= 0.0000213 + 0.01 = 0.0010213 = \frac{1}{R_t}$$

$$R_t = \frac{1}{0.0010213} \approx 979 \text{ ohms}$$

That's quite a change.

The total circuit resistance is now:

$$R_t = 22000 + 979 + 33000 = 55,979 \text{ ohms}$$

This changes the current flowing through the circuit to:

$$I = \frac{E}{R} = \frac{9}{55979} \approx 0.00016 \text{ amp} = 0.16 \text{ mA}$$

The measured voltage drop across R2 works out to:

$$E = IR = 0.00016 \times 979 \approx 0.157 \text{ volts}$$

That is certainly an unacceptable error when we are expecting a reading of 4.15 volts!

If the input resistance of the VOM is 20,000 ohms, things will look a little better, although the error will still be significant. First, solving for the effective resistance of the R2/meter combination:

$$\frac{1}{R_t} = \frac{1}{R2} + \frac{1}{R_m} = \frac{1}{4700} = \frac{1}{20000}$$

$$= 0.0000213 + 0.00005$$

$$= 0.0000713 = \frac{0}{R_t}$$

$$R_t = \frac{1}{0.0000713} \approx 14000 \text{ ohms}$$

Total effective circuit resistance this time is about 69000 ohms, making the current flow equal to:

$$I = \frac{9}{69000} \approx 0.00013 \text{ amps} = 0.13 \text{ mA}$$

The measured voltage drop across R2 using a 20,000 ohm meter works out to:

$$E = 0.00013 \times 14000 \approx 1.8 \text{ volts}$$

Better, but still not all that good.

Now, let's assume we are using a VTVM with an input sensitivity of 11 megohms (11,000,000). The parallel combination of the meter and R2 in this case becomes:

$$\frac{1}{R_t} = \frac{1}{47000} + \frac{1}{11000000}$$

$$= 0.0000213 + 0.00000009091$$

$$\approx 0.0000214 = \frac{1}{R_t}$$

$$R_t = \frac{1}{0.0000214} \approx 46,800 \text{ ohms}$$

Notice that this is very close to the original nominal value of R2 alone (47,000 ohms). As you might suspect, there is only a small change in the current flowing through the circuit:

$$I = \frac{9}{101800} = 0.0000884 \text{ amps} = 0.0884 \text{ mA}$$

The measured voltage drop across R2 this time would work out to:

$$E = 0.0000884 \times 46800 = 4.14 \text{ volts}$$

That is very close to our calculated nominal value of 4.15 volts dropped across R2.

The larger the meter's input resistance is in comparison with that of the component being measured, the smaller the error in the reading.

The 20,000 ohm VOM didn't fare too well in that example. It would work better in a circuit with smaller resistances. For example, let's change the resistance values in Fig. 39-2 as follows:

$$R1 = 680 \text{ ohms}$$
$$R2 = 220 \text{ ohms}$$
$$R3 = 390 \text{ ohms}$$

The total circuit resistance is 1290 ohms, so the unmetered current flow through the circuit is:

$$I = \frac{E}{R} = \frac{9}{1290} = 0.00698 \text{ amps} = 6.98 \text{ mA}$$

This makes the nominal voltage drop across R2 equal to:

$$E = IR = 0.00698 \times 220 \approx 1.53 \text{ volts}$$

If a 20,000 ohm input VOM is placed in parallel with R2 to measure its voltage drop, the effective resistance changes according to the following calculations:

$$\frac{1}{R_t} = \frac{1}{R2} + \frac{1}{R_m} = \frac{1}{220} + \frac{1}{20000}$$

$$= 0.0045455 + 0.00005 = 0.0045955 = \frac{1}{R_t}$$

$$R_t = \frac{1}{0.0045955} \approx 218 \text{ ohms}$$

The meter's resistance is so large in comparison to that of R2, it only has a small effect on the value of the parallel combination. The total effective circuit resistance is 1288 ohms, instead of the nominal 1290 ohms calculated above. This two ohm difference won't change the current flowing through the circuit by much:

$$I = \frac{9}{1288} = 0.00699 \text{ amp} = 6.99$$

A difference of a mere 0.01 mA.

The measured voltage drop across R2 would therefore work out to be equal to:

$$E = IR = 0.00699 \times 218 \approx 1.52 \text{ volts}$$

Certainly that should be close enough for most purposes.

Many servicing schematics have voltages marked that were made with a 20,000 ohms-per-volt meter. In this case, a meter with higher sensitivity would give poorer results. Be sure you know what is being used as the standard.

The Digital Voltmeter

In recent years the VOM has been joined, and in some cases replaced, by the *Digital Voltmeter*, or *DVM*. Instead of the measured quantity being indicated by the position of a mechanical pointer on a dial face, the values are read out directly in numbers using seven-segment LED or LCD displays, like those used on a calculator. A DVM can measure over a much wider range than a typical VOM. The input sensitivity can be as high as that of a VTVM, but the instrument remains compact, lightweight and portable.

A DVM usually offers good protection against input overloads, and often has built-in polarity indication. If the leads are hooked up backwards (the red lead is made more negative than the black lead), a minus sign will be included on the display.

The better DVMs often include a number of special functions not found on standard VOMs. These include automatic range selection and indication, and the capability for measuring additional electrical quantities, such as conductance (the reciprocal of resistance) and capacitance. Some sort of indicator is generally included to warn the operator when the battery voltage is getting low.

Some of the newest DVMs incorporate computer circuitry to become extremely powerful pieces of test equipment. Not only are measurements displayed digitally, alphanumeric characters are also displayed to help explain what the numbers mean.

The Oscilloscope

If a VOM is figuratively a technician's right arm, his left arm would be the oscilloscope. These are the two pieces of equipment that are mandatory for anyone doing anything beyond casual puttering with electronics.

The oscilloscope was briefly introduced in Chapter 17. This is an instrument that can display the waveshape of a signal on a CRT screen. Some typical waveforms, as they appear on an oscilloscope display are shown in Fig. 39-3. The frequency of any ac waveform can be found by comparing the number of cycles displayed with the oscilloscope's vertical sweep frequency (which can be set to any of a number of values).

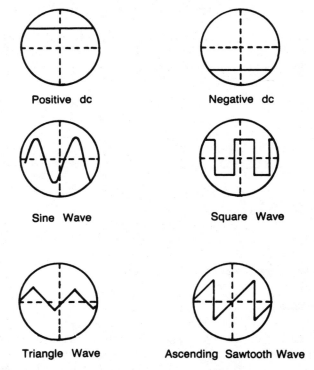

Fig. 39-3. Waveforms can be displayed directly on an oscilloscope.

The controls on an oscilloscope vary somewhat with the model, but they will usually include the following. (A few manufacturers may use different terms for some of these functions.)

☐ Vertical gain—How much deflection from the base line a given input voltage will cause.

☐ Vertical position—This allows the base-line to be moved up and down on the CRT screen.

☐ Horizontal gain—How much deflection will be caused in the right/left dimension when an external signal is used to control the sweep frequency rather than the built-in oscillator.

- ☐ Horizontal position—Moves the displayed waveform from right to left on the CRT screen.
- ☐ Coarse frequency—Sets the range for the built-in sweep oscillator.
- ☐ Fine frequency—Sets the exact frequency of the built-in sweep oscillator.
- ☐ Sync—Synchronizes the display with another signal source.
- ☐ Intensity—Determines how bright the displayed line(s) will be.
- ☐ Focus—Adjusts the clarity of the displayed line(s).

Some oscilloscopes can simultaneously display two signals. This type of machine is called a *dual-trace oscilloscope*. This is extremely useful for comparing input and output signals, or signals from different parts of the circuit being tested.

Signal Tracer

A signal tracer is basically an amplifier and output device (such as a speaker), which is used to indicate whether or not a signal is present at the stage of the circuit currently being tested.

Signal Injector

A signal injector is the reverse of a signal tracer. It allows the technician to insert a known signal into a given circuit stage to determine if it is working properly. Signal injectors are available for both audio (af) and radio (rf) frequencies.

Function Generator

A function generator is a multiple waveform oscillator or signal generator (see Chapter 34). It can produce a clean, known signal at a wide variety of frequencies. If the known signal is fed to the input of a suspected circuit, we can examine the output signal with an oscilloscope to determine what if any distortion has occurred in the circuit being tested.

Frequency Counter

A frequency counter is used to measure the frequency of an ac signal. It usually has an input stage that converts the input waveform to a clean rectangle wave that can be reliably recognized by the digital circuitry within the frequency counter.

Capacitance Meter

A capacitance meter is used to determine the value of an unknown capacitor. It can also be used to check old capacitors for leakage and other problems, or to locate stray capacitances within a circuit.

Most capacitance meters work by charging the unknown capacitor to a specific level, and then measure the time it takes for the capacitor to discharge. For more information on the time constant of a capacitor, refer back to Chapter 6.

Other Test Equipment

Other pieces of test equipment are manufactured for various special purposes. The ones discussed here are the most widely available and generally useful. A technician's workbench should also include a good variable-output power supply.

COMMON PROBLEMS

Every problem that crops up in electronic circuitry is more or less unique. Nevertheless, there are some fairly common sources of problems that you should be familiar with.

A resistor may change its value with age. This is especially true if an overvoltage or physical jolt has damaged the resistor. If you suspect a resistor problem, check for resistors that are discolored, or look burnt. Also watch out for tiny hairline cracks. These cracks may not be readily visible. Often a light tap with a screwdriver will cause a cracked resistor to break apart, but a good resistor won't be harmed.

Cracks can also cause problems in capacitors. Especially watch out for ceramic disc capacitors that often tend to crack near their leads. Electrolytic capacitors often become leaky or their electrolyte might dry up as they age. If you see a whitish powder around an electrolytic capacitor, it is probably dried electrolyte. Usually, however, this kind of fault will not be visible to the naked eye. Test equipment will have to be brought in to locate the culprit.

Of course, if the insulation between windings of a coil gets damaged or chips off, the inductor will appear to change the number of turns in the coil as adjacent turns short together. This will cause the inductance to change.

If a PC board (printed circuit) is used, a tiny hairline crack may cause an open circuit. Similarly, since the copper traces are closely spaced, a scrap of wire or solder could land between adjacent paths and cause a short.

Stray capacitances, inductances, and resistances may crop up when a new circuit is built, or they may appear in a formerly working circuit when one or more components goes bad or changes value.

Of course, active components can also go bad. Tubes often need to be replaced. Semiconductors are more reliable, but they can be damaged by hard physical jolts, runaway current, over voltages, or excessive heat. Special-purpose transistor testers are widely available. Basic semiconductor junctions (in diodes and bipolar transistors) can also easily be tested with a VOM. The resistance section of the VOM is used for this. Remember that an ohmmeter puts out a specific test voltage and measures the drop across the unknown resistance. If the polarity is applied to the semiconductor junction in one direction the ohmmeter will indicate a very low resistance. Reversing the polarity will cause a much larger resistance reading. For a bipolar transistor measure between the emitter and the base and between the base and the collector.

In most cases the semiconductor component being tested will have to be removed from the circuit because the rest of the circuit may have additional resistances in parallel with the junction, which could significantly change the readings.

ICs are treated as "black boxes." If the signals at all of the input pins are correct, but one or more of the output signals is wrong, it may be reasonable to assume that the IC is bad. Watch out though—some circuits can really fool you. If the input impedance of the next stage has changed due to some defect, the difference in the loading could cause the output being measured to be misread by the meter or oscilloscope.

Especially watch out for signs of any previous repair attempts. A lot of people who don't really know what they are doing often attempt to repair electronic equipment, and their unknowledgeable repair may have caused additional problems. A component may be physically damaged or installed with the wrong polarity. A thin lead or wire may be broken. PC board traces may be chipped or shorted.

A major headache is the cold solder joint. This is when an air bubble has formed under the solder, making an unreliable connection. If a solder joint does not look smooth or shiny, suspect a cold solder joint. Unfortunately, the visual test is not always helpful. Many cold solder joints look just fine, but the electrical connection may be poor, exhibiting an open circuit or a high resistance, or (most frustrating of all) an intermittent connection. See Fig. 39-4.

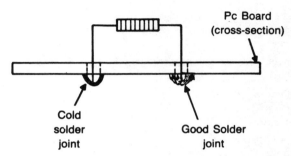

Fig. 39-4. Cold solder joints may not be visually apparent, but they can result in poor electrical connections and can cause many problems.

A cold solder joint will usually cause problems right away, so the problem most commonly comes up right after a circuit is constructed or repaired. If you suspect a cold solder joint, it may be a good idea to simply reheat all of the solder joints in the area, just to be sure.

Occasionally a cold solder joint will not cause problems until the equipment has been in use for months, or even years. They are not terribly common in commercially available equipment, but they do occur from time to time.

The bane of any technician's existence is the intermittent. This is a defect that comes and goes. In some cases you can cause the problem to appear so you can troubleshoot it by simulating specific environmental conditions (high or low ambient temperatures are frequent culprits). All too often the equipment will refuse to misbehave for the technician, and you can't very well locate a problem that isn't there when you're looking.

Often a technician will operate the equipment under extreme stress to force the intermittent fault to become permanent so that it can be easily pinpointed. This is rather risky unless you know what you are doing and are very careful. However, the component causing the intermittent problem is presumedly weaker than any of the good components, so it should fail completely before any other components are damaged.

Self-Test

1. How can a block diagram be used in troubleshooting?

A *To determine specific operating voltages*
B *By visually comparing it with the circuit under test*
C *By narrowing down the stages that could cause the fault in question*

D *By allowing the technician to systematically check out each stage of the circuit*
E *None of the above*

2. If a television set doesn't work at all (dark screen, no sound) which stage is the most likely source of the problem?

A *The power supply*
B *The video amplifier*
C *The tuner*
D *The color killer*
E *None of the above*

3. Which of the following voltmeters would give the most accurate readings?

A *20,000 ohms-per-volt VOM*
B *11 megohm VTVM*
C *1000 ohms-per-volt VOM*
D *50,000 ohms-per-volt VOM*
E *None of the above*

4. Which of the following can *not* be measured with a standard VOM?

A *Current*
B *Inductance*
C *Resistance*
D *Voltage*
E *None of the above*

5. Which of the following is an advantage of a VTVM over a VOM?

A *More portable*
B *Measures current*
C *Less expensive*
D *Operates with less heat*
E *None of the above*

6. What should be the relationship between a meter's input resistance (R_m) to the resistance the voltage drop is being measured across (R_x)?

A *R_x must be much larger than R_m*
B *They should be approximately equal*
C *R_x must be much smaller than R_m*
D *None of the above*

7. Which of the following is the second most important piece of test equipment for the average electronics technician?

A *Oscilloscope*
B *Frequency counter*
C *Signal injector*
D *Capacitance counter*
E *None of the above*

8. Which control on an oscilloscope is used to calibrate the displayed input voltage a specific distance over or below the baseline?

A *Coarse frequency*
B *Vertical gain*
C *Horizontal gain*
D *Horizontal position*
E *None of the above*

9. Which of the following is not a common defect in a resistor?

A *Changed value*
B *Burnt*
C *Cracked*
D *Leaky*
E *None of the above*

10. If an ohmmeter measures a small resistance when it is hooked up across a semiconductor diode, what would happen if the polarity of the ohmmeter leads are reversed?

A *A higher resistance will be indicated on the meter*
B *The same resistance will be measured*
C *The semiconductor junction will probably be damaged*
D *It doesn't matter—the diode must be bad*
E *None of the above.*

Appendix

Appendix

Answers to Self-Help Tests

Chapter 2

1. A
2. D
3. D
4. A
5. B

6. A
7. D
8. C
9. A
10. D

Chapter 3

1. A
2. C
3. E
4. A
5. C

6. A
7. B
8. D
9. A
10. D

Chapter 4

1. B
2. D
3. C
4. A
5. B

6. D
7. C
8. D
9. C
10. A

Chapter 5

1. B	6. B
2. A	7. E (6 volts)
3. D	8. B
4. B	9. A
5. ~~A~~ E	10. D

Chapter 6

1. C	6. C
2. D	7. B
3. A	8. C
4. E (infinity)	9. A
5. D	10. A

Chapter 7

1. C	6. B
2. A	7. A
3. B	8. C
4. C	9. D
5. A	10. B

Chapter 8

1. B	6. B
2. C	7. D
3. B	8. A
4. A	9. C
5. C	10. C

Chapter 9

1. B	6. D
2. A	7. C
3. D	8. B
4. B	9. C
5. A	10. A

Chapter 10

1. C	6. A
2. A	7. B
3. C	8. D
4. D	9. B
5. B	10. A

Chapter 11

1.	B	6.	C
2.	D	7.	B
3.	A	8.	C
4.	A	9.	C
5.	D	10.	B

Chapter 12

1.	C	6.	A
2.	A	7.	B
3.	D	8.	C
4.	C	9.	C
5.	B	10.	D

Chapter 13

1.	C	6.	C
2.	B	7.	D
3.	E	8.	C
4.	D	9.	D
5.	C	10.	B

Chapter 17

1.	C	6.	D
2.	A	7.	B
3.	B	8.	D
4.	B	9.	C
5.	A	10.	B

Chapter 18

1.	C	6.	A
2.	C	7.	D
3.	B	8.	B
4.	A	9.	D
5.	C	10.	C

Chapter 19

1.	C	6.	D
2.	B	7.	C
3.	A	8.	A
4.	C	9.	D
5.	C	10.	B

Chapter 20

1. B	6. C
2. A	7. A
3. D	8. B
4. B	9. D
5. C	10. B

Chapter 21

1. B	6. D
2. A	7. A
3. C	8. C
4. B	9. B
5. D	10. A

Chapter 22

1. B	6. C
2. A	7. C
3. D	8. D
4. D	9. A
5. B	10. B

Chapter 24

1. A	6. B
2. C	7. D
3. A	8. B
4. D	9. D
5. B	10. B

Chapter 25

1. B	6. B
2. A	7. A
3. C	8. B
4. D	9. C
5. C	10. C

Chapter 26

1. B	6. A
2. C	7. B
3. A	8. C
4. B	9. D
5. B	10. A

Chapter 27

1.	C	6.	D
2.	A	7.	A
3.	C	8.	D
4.	C	9.	D
5.	B	10.	A

Chapter 28

1.	A	6.	A
2.	C	7.	D
3.	B	8.	A
4.	D	9.	C
5.	B	10.	B

Chapter 29

1.	A	6.	C
2.	D	7.	A
3.	D	8.	B
4.	A	9.	E
5.	B	10.	D

Chapter 30

1.	A	6.	D
2.	C	7.	C
3.	B	8.	B
4.	A	9.	E
5.	A	10.	C

Chapter 32

1.	B	6.	D
2.	D	7.	C
3.	B	8.	C
4.	A	9.	C
5.	B	10.	B

Chapter 33

1.	C	6.	B
2.	C	7.	C
3.	B	8.	D
4.	A	9.	A
5.	D	10.	C

Chapter 34

1. B		6. D	
2. C		7. A	
3. A		8. B	
4. C		9. D	
5. A		10. D	

Chapter 35

1. A		6. D	
2. C		7. B	
3. C		8. A	
4. A		9. A	
5. B		10. C	

Chapter 36

1. E		6. C	
2. A		7. D	
3. B		8. C	
4. C		9. A	
5. D		10. B	

Chapter 37

1. B		6. B	
2. B		7. B	
3. D		8. A	
4. A		9. C	
5. C		10. D	

Chapter 38

1. B		6. C	
2. A		7. A	
3. B		8. C	
4. C		9. B	
5. D		10. E	

Chapter 39

1. C		6. C	
2. A		7. A	
3. B		8. B	
4. B		9. B	
5. E		10. A	

Index

Index